Applied Thermodynamics in Unit Operations

The growing demand for energy accounting in industries is the main challenge for academics and engineers working in chemical processing plants, food industries, and the energy sector. *Applied Thermodynamics in Unit Operations* addresses this demand and offers a clear contribution to the quantification of energy consumption in processes while also solving the economic aspects of energy that are vital in real-life industrial contexts.

Features

- Combines the energy and exergy routines to analyze utilities and unit operations in a wide range of engineering scopes: nozzles, turbines, compressors, evaporators, HVAC, drying technology, steam handling, and power generation.
- Offers a detailed procedure of finding economic wealth of energy in the operations.
- Discusses basic concepts of thermal engineering and industrial operational insights through practiced examples, schematic illustrations, and software code.
- The only book to include practical problems of industrial operations solved in detail and complementary EES codes for the solutions.
- Features examples selected from the authors' real-world experience in industrial projects.

The book is a handy reference for researchers and practitioners in the areas of process, chemical, and mechanical engineering; undergraduate and postgraduate students in those disciplines; engineers working in industry; and production managers. Some examples are solved in EES to help the audience apply computer coding for thermal calculations.

Applied Thermodynamics in Unit Operations

Solved Examples on Energy, Exergy, and Economic Analyses of Processes

Ali M. Nikbakht, Ahmad Piri, and Azharul Karim

CRC Press
Taylor & Francis Group
Boca Raton London New York

CRC Press is an imprint of the
Taylor & Francis Group, an **informa** business

First edition published 2024
by CRC Press
2385 Executive Center Drive, Suite 320, Boca Raton, FL 33431

and by CRC Press
4 Park Square, Milton Park, Abingdon, Oxon, OX14 4RN

CRC Press is an imprint of Taylor & Francis Group, LLC

ISBN: 978-1-032-54395-6 (hbk)
ISBN: 978-1-032-54397-0 (pbk)
ISBN: 978-1-003-42468-0 (ebk)

DOI: 10.1201/9781003424680

Typeset in Times
by Apex CoVantage, LLC

Contents

Preface

Processing is a vital element of all industries that deal with material preparation, pre-treatment, post-treatment, handling, and transport. However, the concept of processing encompasses a wide range of physical and chemical operations, which varies significantly depending on the material, nature of the industry, and final product. To investigate operations in detail, the process flow is segmented as unit operations as sub-elements to be analyzed. Energy, mass flow, and thermo-physical properties of the flow in unit operations need to be identified to achieve an engineering approach for scaling, modifications, and optimization. The real-life industrial context requires a quantitative model of the process and its unit operations to be flexible and dynamic and to satisfy the market demands. Additionally, a realistic picture is obtained when engineering calculations are integrated with economic assessments, known as a techno-economic model. Although thermodynamic analyses have a proven track in industrial applications, engineers have limited sources on the framework or modeling schemes to solve a practical problem with both thermodynamic and economic methodologies. This book addresses the vital demand of academicians and industrialists through introducing basic guidelines to solve the thermodynamics of unit operations, energy and mass balances, exergy, and exergo-economic balances of the flow with the developed code in EES. The book provides the audience with numerous solved examples, most of them already practiced by the authors in the processing industries.

The major purpose of the book is to combine the energy and exergy routines to analyze the utilities and unit operations in a wide range of engineering scope: nozzles, turbines, compressors, evaporators, HVAC, drying technology, steam handling, and power generation. Many examples on techno-economic analysis of unit operations are solved. Examples are intentionally selected from the authors' experience in industrial projects to give a more practical and real-life sense to the audience.

The three key features of the book are:

- Provides details of energy-exergy analysis of unit operations.
- Details the procedure of finding economic wealth of energy in operations.
- Step-by-step adoption of steady state equations from thermodynamics; heat transfer and fluid mechanics are presented for simplicity and fluency of problem solving.

Authors' Biographies

Ali M. Nikbakht is an experienced engineer in energy storage systems. He has lectured and conducted industrial-engaged research programs at Urmia University and Queensland University of Technology. Having honed his skills in the dynamic environment of internal combustion engine test rooms, Ali possesses a wealth of practical experience in testing various energy systems. His proficiency extends to meticulous techno-economic evaluations of thermal operations, underlining his commitment to a comprehensive approach. He has presented workshops and seminars in thermal engineering, steam generation and distribution, HVAC, and techno-economic evaluations in real-life industrial processes under the framework of 4E (energy, exergy, economy, and environment). He has commissioned a bio-energy laboratory where a solar collector simulator, fluid dynamics experimentation, and biomass conversion technologies (pyrolysis and transesterification) are practiced. As a senior consultant in launching the engine test room at Urmia University, he has presented novel and innovative approaches for testing, instrumentation, and validation.

Broad Areas of Research

- Energy storage systems
- Techno-economic analysis of energy systems
- Computational fluid dynamics in energy systems
- Exergy and energy in unit operations and processing industry

Ahmad Piri is a postdoctoral fellow at Urmia University, engaged in thermal evaluations of unit operations. He has a demonstrated track record of working in the drying, steel, food processing, and chemical industries. Piri has been successful in monitoring the process flow of sugar processing plants and simulation evaporation and crystallization of sugar using thermodynamics and heat/mass transfer principles. He is now consulting processing plants to develop process flow diagrams and dynamic models by Hysys and EES. Ahmad is a member of a young research team that has initiated a startup on novel drying technologies. His extensive insight on energy calculations was demonstrated by the stakeholder when giving consultancy to thermal engineering industries.

Broad Areas of Research

- Exergy and thermodynamics
- Thermal engineering in fruit/food processing
- Steady-state simulation of processes in Hysys

Azharul Karim is currently working as a professor in the School of Mechanical, Medical & Process Engineering (MMPE), Queensland University of Technology (QUT), and the director of the Advanced Drying and Sustainable Energy Research (ADSER) Group (https://research.qut.edu.au/adser/). His research is directed toward uncovering fundamental understanding of drying process by developing advanced multiscale and multiphase models using theoretical/computational and experimental methodologies. Through his scholarly, innovative, high-quality research, he has established his national and international standing in his field. He has authored over 230 peer-reviewed articles, including 140 high-quality journal papers, five books, and 13 book chapters. His papers have attracted more than 7000 scholar citations, with an h-index of 46. He is the editor/board member of six reputed journals, including *Drying Technology* and *Nature Scientific Reports.* He has been keynote/distinguished speaker at scores of international conferences, including the International Drying Symposium (2022), and an invited/keynote speaker in seminars in many reputed universities, including Oxford University (2018) and the University of Illinois (2022). He has been awarded 21 research grants amounting about A$4 million and won multiple international awards for his outstanding contributions in multidisciplinary fields. He is the recipient of the highly prestigious ARC Linkage (2021) and ARC Discovery grants (2022) as first chief investigator. He is also a leader in innovative application of lean manufacturing concepts in hospital emergency departments to reduce long waiting times and optimize resources. Dr Karim is the inventor of many innovative new products, including (1) an ultrasonic washing machine (patent WO02089652), (2) an ultrasonic dishwasher (patent WO0229148), (3) an advanced independent solar drying system (Patent 2019900943), and (4) an integrated heat pump, vacuum, and microwave convective drying system (Patent 2019900942). Dr Karim's high-level achievements in both fundamental and applied research in the important area of advanced drying and microwave heating have established him as one of the world's leading researchers in these fields, as demonstrated by his field-weighted citation impacts (FWCI) of 2.24 and 9.82, respectively.

Broad Areas of Research

- Multi-scale and physics-based modelling in drying
- Renewable energies and sustainable processing
- Artificial intelligence and advanced modelling in agri-industrial processes
- Nanofluid solar thermal storage
- Concentrating PV-thermal collector
- Thermal storage
- Lean manufacturing and healthcare systems

1 Introduction and Concepts

1.1 A SHORT REVIEW OF THERMODYNAMICS

Thermodynamics provides a rigorous mathematical formulation of the interactions between measurable quantities that are commonly used to describe energy and equilibria of macroscopic systems. Physical properties and chemical compositions of material systems are discussed in thermodynamics (Jacobs, 2013). The science of thermodynamics has had huge benefits for engineering. Industry is formed by unit operations and processes. There is a wide range of operation types, including transportation, filtering, distillation, evaporation, condensation, heat exchange, extraction, and gas absorption. Design, simulation, monitoring, control, optimization, and diagnostics in unit operations strongly require the systematic adoption of thermodynamic rules. Thus, a sufficient knowledge of thermodynamics is a vital prerequisite for any engineer working in process engineering and heat transfer applications. It should be further noted that the concepts of thermodynamics are highly geared to relevant topics, such as heat transfer and fluid mechanics. This book is not a thermodynamics textbook, so it aims to avoid redundancy and repetition and instead move to more practical subjects. However, a short recap of general topics is presented. In order to get the full benefit of the book, reviewing the subjects from textbooks is encouraged whenever required.

1.1.1 System

The concept of a "system" is the backbone of the thermal analyses widely used in this book. The reader may refer to textbooks to get a more detailed review, but, as a quick reminder, a system is defined as an enclosure of material separated from its surroundings (Abott & Van Ness, 1989). Based on this idea, systems can be classified into "open", "closed", and "isolated". As the name implies, an open system interacts with the surroundings by exchanging mass and energy. The open system is sometimes called the "control system", particularly in fluid dynamics. The famous analogy in nature is the leaf of a tree where the pigments absorb solar radiation for the photosynthesis reaction. This principle doesn't hold true within closed systems, such as a sealed vacuum flask holding hot coffee. Energy or mass exchange between the flask and its surroundings, or vice versa, only occurs upon opening the lid. It's worth noting that when the surroundings are not interacting, an accurate assumption can be made about the isolated system. Thermodynamic equilibrium is achieved when there is no significant macroscopic transfer of energy and mass to or from the system.

While we understand the concept of system for thermodynamic applications, we can imagine the walls of a system as "boundaries" and define any conditions at

DOI: 10.1201/9781003424680-1

this region as "boundary conditions". Further, all the thermodynamic variables we are studying in a system are defined as the "state" of that system. However, only a few properties or variables are independent and measurable and contribute to define the state of the system. This is good news since prevents massive calculations and overwhelming engineering work. Among such variables, temperature and pressure are most commonly included. That's why when an engineer looks up the thermodynamic tables, these two properties are the basis of finding density, specific heat, enthalpy, entropy, and so on.

1.1.2 Thermodynamic Laws

The laws of thermodynamics are an intrinsic part of any thermodynamic study. They were formulated in the 19th century, shedding light on the physics of energy and entropy. By the first law of thermodynamics, we learn that energy is neither created nor destroyed but only transfers from one form to another. In fact, the first law expresses the persistence of energy and expresses its quantity. "Conservation of energy" is another implication of the first law of thermodynamics. The second law of thermodynamics states that in a natural process, entropy always increases over time. In other words, no real thermodynamic process exists that is not accompanied by increased entropy over time. These laws are simple to state but can be far reaching when real-world applications are assessed.

1.1.3 Temperature

Aa explained earlier, temperature is a major property to define the state of a thermal system. The concept of temperature is highly complex to define, despite its apparent simplicity and the everyday experience of objects being colder or hotter. In statistical thermodynamics, the concept of temperature is defined according to the kinetic energy of the particles of the object. For thermodynamic calculations, the absolute scale of temperature is usually used:

$$T_K = T_C + 273.15 \tag{1.1}$$

1.1.4 Pressure

Talking about the second property of a system, pressure, can be again very simplistic and meanwhile deep in analysis. Fluids are composed of tiny particles that are placed next to each other at a small distance. The shorter the distance between these particles, the higher the so-called fluid density. Now, if these particles are packed together in a closed environment, as in Figure 1.1, the particles will exert forces on each other in different directions, and these forces perpendicular to the surface create static pressure. It can be said that the origin of pressure in a fluid is the movement of molecules, and since the movement of molecules is closely related to the temperature of the fluid, this pressure is called thermodynamic pressure, which is used in thermodynamic calculations.

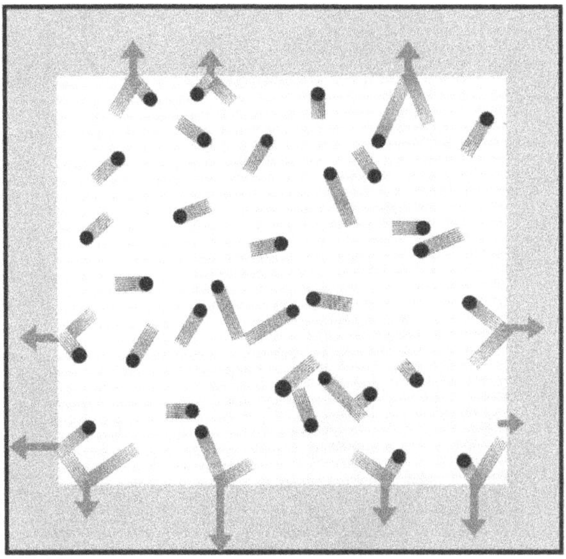

FIGURE 1.1 Concept of fluid pressure in a closed environment.

The useful representation of pressure in industrial application is relative pressure, which is correlated to the pressure of the atmosphere. As shown in Figure 1.2, relative pressure is either positive or negative depending on whether it is above or below the atmospheric pressure level. It is very clear that at the level of atmospheric pressure, the relative pressure becomes zero.

1.1.5 Intensive or Extensive?

Parameters that can be calculated in a system directly via experiments or indirectly via mathematical equations are called system properties. For a system that is divisible into sections, any property that is the sum of the property of individual sections is called "extensive". Mass, volume, enthalpy, and entropy are by this definition extensive properties. If we divide them by mass of the system, they would become "intensive". Therefore, intensive properties are independent of the mass of the system. Examples are temperature, pressure, and density. Specific volume is another intensive property that is derived by dividing volume (extensive) by mass:

$$v = \frac{V}{m} \left(m^3 / kg\right) \tag{1.2}$$

1.1.6 Energy

Energy is a core topic of this book. Transfer and storage of energy are cornerstones of energy management plans in industry. Energy cannot be observed directly, but it can be recorded and evaluated with indirect measurements. The absolute value of

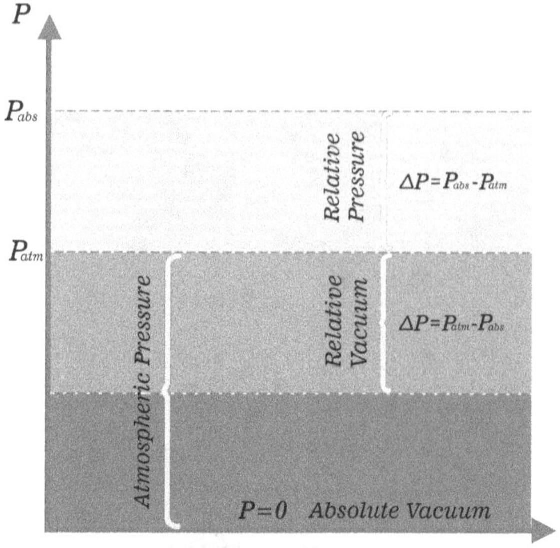

FIGURE 1.2 Pressure and its different representations.

the energy of the system is difficult to measure, while the change in the energy of the system is easily calculated. Energy can be stored in a system or transferred from one system to another; therefore, there are two categories of energy, stored energy and transferred energy.

Different forms of energy that can be stored in a system can be classified into macroscopic and microscopic groups. Macroscopic energy is evaluated with coordinates outside the object and includes kinetic energy and potential energy. Microscopic energy is related to the molecular structure of the system. Internal energy includes all microscopic forms of energy.

1.1.6.1 Kinetic Energy

The energy generated within a system due to its motion relative to a reference point is referred to as kinetic energy. When all parts of a system move at the same speed, the kinetic energy of a fixed mass is expressed as:

$$E_K = m\frac{V^2}{2}(J) \tag{1.3}$$

where m is the mass of the system in *kg*, and V is the velocity of the system in m/s relative to the external reference. Changes in kinetic energy of a fixed mass are expressed as follows:

$$\Delta E_K = \frac{1}{2}m\left(V_2^2 - V_1^2\right)(J) \tag{1.4}$$

where V_1 is the initial velocity, and V_2 is the speed of the system in the final state.

1.1.6.2 Potential Energy

The energy that a system creates as a result of height and acceleration of gravity is called potential energy and is expressed as follows:

$$E_P = mgz\,(J) \tag{1.5}$$

where g is the acceleration of gravity per unit of m/s^2, and z is the height of the center of mass of the system per unit of meters (m) relative to a selected reference surface. Changes in kinetic energy are expressed as follows:

$$\Delta E_P = mg(z_2 - z_1)\,(J) \tag{1.6}$$

where z_1 is the initial height of the system, and z_2 is the height of the system in the final state.

1.1.6.3 Internal Energy

Internal energy, denoted as (U), represents the cumulative total of all microscopic energy forms within a given system. Changes in the internal energy of a system can be caused by temperature change, state change, change in molecular arrangement (chemical reaction), change in atomic structure (nuclear fission), and the combination of smaller atoms to form larger atoms (nuclear fusion). Atoms can also have electric and magnetic energies when they are exposed to external magnetic or electric fields.

1.1.6.4 Transfer Energy

Energy can be transferred to or from a system in three ways, heat, work, and mass flow. Heat (Q) is the amount of energy that is exchanged due to the temperature difference between one object and another that is in contact with it or due to the temperature difference between the system and the environment. The driving force of heat transfer is a temperature difference, and energy transfer in the form of heat always takes place from an object with a high temperature to an object with a low temperature. The heat transfer to the system increases the energy of molecules and thus increases the internal energy of the system, and the heat transfer from the system (loss of heat) decreases the energy of molecules and consequently the internal energy. Sometimes it is desirable to use the heat transfer rate (amount of heat transferred per unit time) instead of the total heat transferred during a certain period. Q is the rate of heat transfer, whose unit is J/s or watts (W).

Work, likewise, is the exchange of energy between the system and the environment. As mentioned, energy can cross the boundary of a closed system in the form of work and heat. If the energy passing the boundary of the closed system is not the heat type, it will be the work type; therefore, we confidently say that work is the exchange of energy that is not caused by the temperature difference between the environment and the system. The rise of a piston, the rotation of an axis, and an electric wire crossing system boundaries are all related to work. The transfer of work to the system increases the energy of the system (work is done on the system), and the transfer of work from the system (work is done by the system) decreases the energy of the system. Work done is represented by *W*. The work done per unit of time is called power and is represented by *W*, whose unit is J/s or watts (W).

One of the mechanical forms of work is the work caused by the movement of the boundary of the system, which is called boundary work. To express a relation for boundary work, the expansion and compression model of a gas in a piston-cylinder apparatus is used. During the process of expansion or contraction of the gas inside the cylinder, part of the border (inner surface of the piston) moves back and forth, which is called boundary work. Consider the gas trapped in the piston–cylinder apparatus shown in Figure 1.3.

The initial gas pressure is *P*, the total volume is *V*, and the cross-sectional area of the piston is *A*. Ignoring the friction between the cylinder and the piston and assuming that the piston moves very slowly inside the cylinder, if the piston moves to a distance *ds*, the differential work done in this process is equal to:

$$\delta W_b = Fds = PAds = PdV \tag{1.7}$$

The total boundary work done during the piston movement process is obtained by summing up all the differential work done from the initial state to the final state:

$$W_b = \int_1^2 PdV \tag{1.8}$$

This integral can only be solved if the relationship between *V* and *P* during the process is determined.

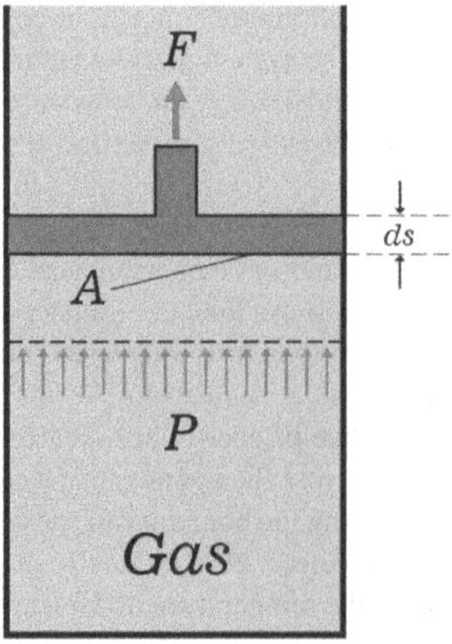

FIGURE 1.3 The gas trapped in the piston-cylinder apparatus with initial pressure *P*, total volume *V*, and cross-sectional area of the piston *A*.

Unlike closed systems, control volumes contain flow from their boundaries, and some work is required to expel or introduce fluid into the control volume. This work is known as flow work or flow energy, which is required to keep a continuous flow in a control volume constant. To obtain a suitable relation for the flow work, consider a small component (element) of fluid to volume V, as shown in Figure 1.4.

The fluid that is directly behind this component presses it to enter the control volume, and in this case, it can be considered a hypothetical piston. The fluid element is chosen to be small enough to have all the same process properties. If the fluid pressure is represented by P and the cross-sectional area of the element is represented by A (Figure 1.5), the force exerted on the fluid element by the imaginary piston is as follows:

$$F = \text{PA}$$

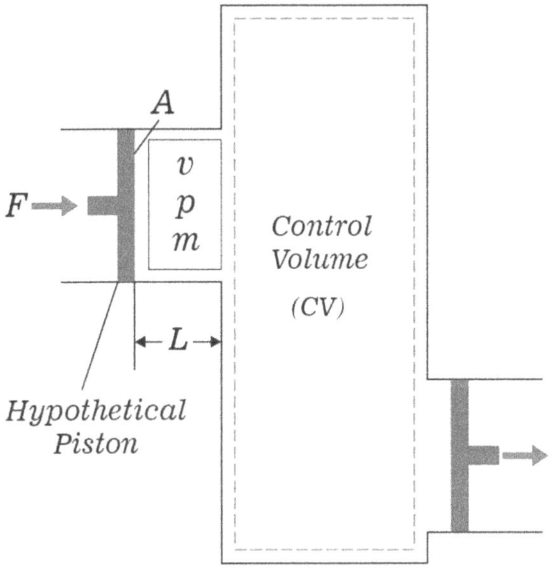

FIGURE 1.4 Schematic for flow work.

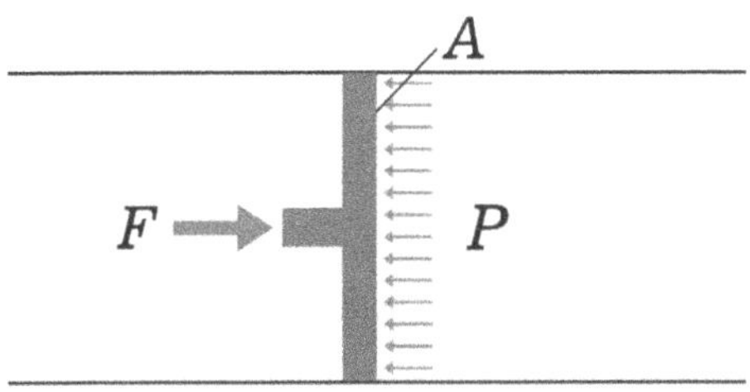

FIGURE 1.5 In the absence of acceleration, the force exerted by the piston on the fluid is equal to the force exerted on the piston by the fluid.

To drive the fluid into the control volume, this force must act along a distance L; therefore, the work done to drive the fluid element from the boundary is as follows:

$$W_{flow} = FL = PAL = PV\left(J\right) \tag{1.9}$$

The flow–work relationship is the same whether the flow is entering or leaving the control volume.

1.1.6.5 Energy of Mass Flow

When mass enters the system, the energy of the system increases because the mass carries energy with it. Likewise, when some mass is removed from the system, the energy in the system decreases because the mass removed from the system carries some energy.

The sum of the internal energy of the object and the flow energy is called enthalpy, which is written as follows:

$$H = U + PV\left(J\right) \tag{1.10}$$

Enthalpy is the function of the state and is one of the quantitative properties of the system. Enthalpy per unit mass is expressed as follows:

$$h = u + Pv \tag{1.11}$$

which is part of the intensity properties of the system. Enthalpy is considered a type of energy quantity only in certain cases. For example, the enthalpy of water inside the tank is not an energy quantity because the product of pressure and specific volume in this case is not a type of energy. The only energy related to the water tank is its internal energy. When fluid enters or exits an open system, the product of the pressure in the volume will express the energy of the flow. In this case, fluid enthalpy is obtained from the sum of internal energy and flow energy.

In the absence of electric, magnetic, surface, and other effects (simple compressible system), total energy can be defined as the sum of internal, kinetic, and potential energies as follows:

$$E = U + E_K + E_P \tag{1.12}$$

or on a per-unit mass basis:

$$e = u + e_K + e_P = u + \frac{V^2}{2} + gz \tag{1.13}$$

The fluid entering or leaving the control volume will have another type of energy: the flow energy Pv that was introduced; therefore, the total energy of a flowing fluid per mass unit will be as follows:

$$\theta = Pv + e = Pv + u + \frac{V^2}{2} + gz \tag{1.14}$$

However, the sum of $Pv + u$ was already introduced as enthalpy h; therefore, the previous equations are summarized as follows:

$$\theta = h + e_K + e_P = h + \frac{V^2}{2} + gz \tag{1.15}$$

1.1.6.6 Mechanical Energy

Mechanical energy can be defined as a form of energy that can be completely and directly converted into mechanical work by an ideal mechanical device, such as an ideal turbine. Kinetic and potential energies are familiar forms of mechanical energy, and the pressure of the flowing fluid is related to its mechanical energy. Thermal energy is not assigned to mechanical energy since it cannot be directly and entirely converted into work:

$$e_{mech} = \frac{p}{\rho} + \frac{V^2}{2} + gz \rightarrow \dot{E}_{mech} = \dot{m}\left(\frac{p}{\rho} + \frac{V^2}{2} + gz\right) \tag{1.16}$$

1.1.7 Pure Substance

A pure substance is a substance whose chemical composition is constant everywhere in a system. A pure substance can have more than one phase, provided that its chemical composition is the same in all phases. For a substance to be pure, it is not necessary to be an element or to have only one chemical compound but a mixture of elements or chemical compounds: as long as the mixture is homogeneous, the substance is pure.

Heating a pure substance leads to an increase in its temperature; at a point (at a fixed pressure), heat transfer causes a change of the substance's phase, which is called the saturation point or the point of the substance's phase change. The parameters that can be defined in the saturation range are saturation temperature, saturated liquid, compressed liquid, saturated steam, and superheated steam.

1.1.7.1 Saturation Temperature

At a certain pressure, the temperature at which a pure substance changes phase is called saturation temperature.

1.1.7.2 Saturation Liquid

If a substance is liquid at saturation temperature and pressure but is ready to evaporate, the liquid is called saturation liquid.

1.1.7.3 Subcooled or Compressed Liquid

At a temperature lower than the saturation temperature, the liquid does not show a tendency to evaporate, which is called subcooled liquid or compressed liquid.

1.1.7.4 Saturation Steam

If all the substance converts into steam at saturation temperature, it is called saturation steam. Saturation steam is about to convert to saturation liquid.

1.1.7.5 Superheated Steam

The steam that is not about to liquefy is called superheated steam.

1.1.7.6 Phase Change of a Pure Substance

Consider a system containing 1 kilogram of pure water at constant pressure. If the initial temperature of the water is 25°C, its temperature will increase due to heat transfer. As the temperature increases, the specific volume increases slightly. This process continues until the water temperature reaches 99.97°C. At this temperature, more heat transfer will lead to phase change. This phase change is the conversion of liquid to steam. During the phase change at constant pressure, the temperature inside the system is always constant, and the transferred heat is used to change the liquid to steam phase. This process continues until all the water converts into steam. After converting all the water into steam, the transferred heat is used to increase the temperature of the steam, and the water enters the superheated phase. The steps in thermal interactions can be shown in a diagram that shows changes in temperature relative to volume (Figure 1.6). In other words, the *T*-*V* diagram shows the evolution of water as a pure substance from liquid to steam.

Stages A–C: The liquid is compressed and has not yet boiled at the working pressure of the system, and only the temperature of the liquid increases.

Point C: Saturation liquid of a point where the liquid starts to boil, and the temperature of the system is constant until the liquid evaporates.

Stages C–E: The region where the fluid is boiling is called the two-phase saturation region for a saturation mixture. The pressure and temperature of the system are constant, and only the volume increases.

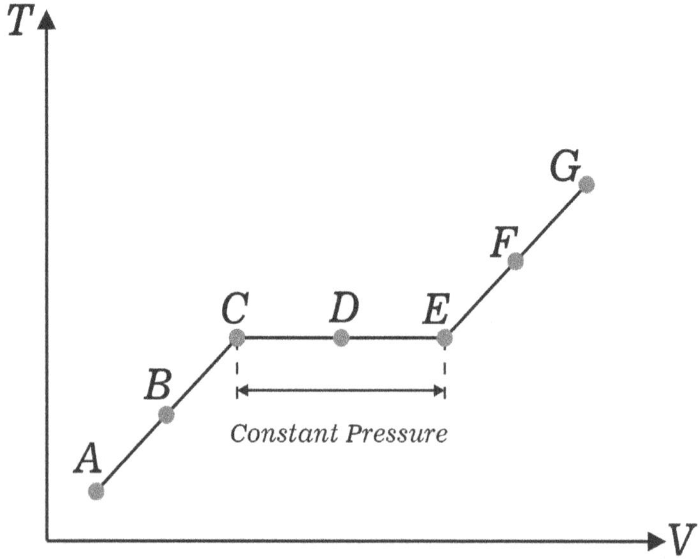

FIGURE 1.6 *T*-*V* diagram of water as a pure substance from liquid to steam state.

The heat directed towards the pure substance serves the purpose of inducing a phase transition from liquid to steam, rather than altering its temperature. In the current phase, the liquid and steam coexist simultaneously, with the liquid content continually undergoing a process of reduction and simultaneous addition to the steam phase. An essential consideration lies in the fact that the boiling point of a fluid experiences an elevation as pressure increases. Arriving at point E signifies the completion of the vaporization process for the entire pure substance. This critical juncture is referred to as the saturation steam point.

Stages E–G: The heat given to the system and the pure substance is used to increase its temperature in the steam phase. As the temperature increases, the fluid moves to the superheated steam region, and, like the compressed region, the temperature and specific volume increase.

By increasing the cylinder pressure by adding weights to the piston, these steps occur exactly above the constant pressure line of 101.325 kPa; also, if the process is carried out at a lower pressure, the steps will be repeated, with the difference that the higher the pressure, the shorter the two-phase stage, and the opposite happens when the pressure decreases. Changes in temperature and specific volume are shown in Figure 1.7.

In Figure 1.7, by connecting the points of saturation liquid and saturation steam, a dome-like shape is created, which is known as a saturation dome (Figure 1.8). As the

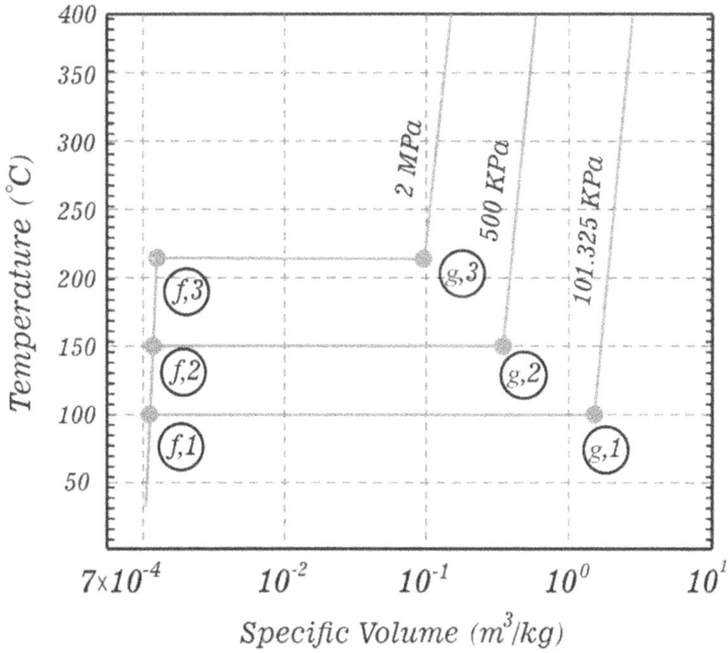

FIGURE 1.7 As the saturation pressure increases, the two-phase stage becomes shorter.

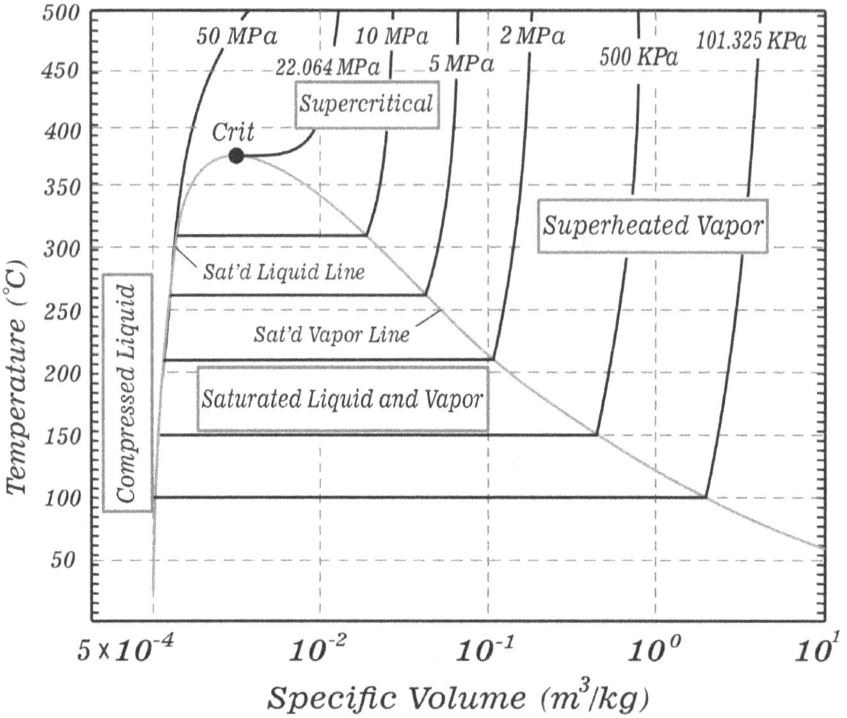

FIGURE 1.8 Saturation dome.

pressure increases, the width of the two-phase region gradually approaches zero, and this point is called the "critical point". At this point, the phase change from liquid to steam occurs directly, and the state of saturation liquid and saturation steam coincide (Figure 1.8).

At pressures above the critical pressure, there will be no process of distinct phase change; instead, the specific volume of the substance will constantly increase, and there will always be a single phase. Finally, the state of substance in these conditions will be similar to steam. Above the critical state, no border point separates the compressed liquid region from the superheated region; however, it is often customary to consider the state of a substance superheated at temperatures above the critical temperature and consider the state of a substance a compressed liquid at temperatures below the critical temperature.

All the states, including the equilibrium of the liquid and steam phase, are located under the dome-shaped curve, which is called the saturated liquid–steam mixture region or the wet region. In the saturation region, temperature and pressure are dependent on each other, which means that the state of the fluid cannot be determined with these two parameters. For example, points C and D on the curve of Figure 1.6 have the same pressure and temperature, while the state of the fluid in these two points is different. For this reason, the state of the fluid in the saturation region is not determined by temperature and pressure.

TABLE 1.1
Determining the Phase Description of the Pure Substance

Given Property in the Problem	Property	Saturation Values	State: Compressed if:	Saturated if:	Superheated if:
P,T	P	$T_{sat@P}$	$T < T_{sat@P}$	$T = T_{sat@P}$	$T > T_{sat@P}$
P,T	T	$P_{sat@T}$	$P > P_{sat@T}$	$P = P_{sat@T}$	$P < P_{sat@T}$
T,h	T	$h_{f@T}, h_{g@T}$	$h < h_f$	$h_f < h < h_g$	$h > h_g$
P,h	P	$h_{f@P}, h_{g@P}$	$h < h_f$	$h_f < h < h_g$	$h > h_g$
T,s	T	$s_{f@T}, s_{g@T}$	$s < s_f$	$s_f < s < s_g$	$s > s_g$
P,s	P	$s_{f@P}, s_{g@P}$	$s < s_f$	$s_f < s < s_g$	$s > s_g$
T,u	T	$u_{f@T}, u_{g@T}$	$u < u_f$	$u_f < u < u_g$	$u > u_g$
P,u	P	$u_{f@P}, u_{g@P}$	$u < u_f$	$u_f < u < u_g$	$u > u_g$
T,v	T	$v_{f@T}, v_{g@T}$	$v < v_f$	$v_f < v < v_g$	$v > v_g$
P,v	P	$v_{f@P}, v_{g@P}$	$v < v_f$	$v_f < v < v_g$	$v > v_g$

Thermodynamic tables are usually used to determine the exact properties of a pure substance. Typical thermodynamic tables provide values of specific volume, internal energy, enthalpy, and entropy of a pure substance at different temperature and pressure conditions.

In thermodynamic tables, index *f* is used to specify the properties of saturated liquid, and index *g* is used to specify the properties of saturated steam. Another index is *fg*, which is used to show the difference between saturated steam and saturated liquid values of a property. The first step in determining the properties of a pure substance using thermodynamic tables is to determine its phase description (whether it is compressed, saturated, or superheated). Using Table 1.1 and with two independent properties, it is possible to determine the phase description of a pure substance. With the determination of the phase description of the substance, it is possible to refer to the relevant table and extract the thermodynamic properties.

1.1.7.7 Saturated Liquid–Steam Mixture

Throughout the process of evaporation, the substance undergoes a division where one fraction retains its liquid state, while the other transforms into steam. It means that there is a mixture of saturated liquid and saturated steam. To fully investigate this mixture, the ratio of the liquid phase and steam phase in the mixture should be determined. Therefore, a new property called quality (x) is defined as the ratio of the mass of steam to the total mass of the mixture:

$$x = \frac{m_{steam}}{m_{\text{total}}} \tag{1.17}$$

such that

$$m_{\text{total}} = m_{\text{liquid}} + m_{\text{steam}} = m_f + m_g$$

Therefore, a saturated mixture can be considered a combination of two subsystems, saturated liquid and saturated steam. Since the mass of each phase is usually not known, it is often better to assume that the two phases are well mixed and form a homogeneous mixture. So, the properties of the mixture are the average properties of the saturated liquid–steam mixture. The specific volume, internal energy, enthalpy, and entropy of a saturated mixture are determined by the following equations:

$$v = v_f + xv_{fg}\left(\text{m}^3/\text{kg}\right) \tag{1.18}$$

$$u = u_f + xu_{fg}\left(\text{kJ}/\text{kg}\right) \tag{1.19}$$

$$h = h_f + xh_{fg}\left(\text{kJ}/\text{kg}\right) \tag{1.20}$$

$$s = s_f + xs_{fg}\left(\text{kJ}/\text{kg.K}\right) \tag{1.21}$$

1.1.7.8 Approximation of Compressed Liquid Properties Using Table of Saturation Properties

For a compressed liquid, suitable approximate values for u, h, v, and s can be obtained based on the saturation table. For this purpose, the following equations can be used:

1. Internal energy, entropy, and specific volume at a specific temperature change very little with pressure change; therefore:

$$v(T,P) \cong v_f(T) \tag{1.22}$$

$$u(T,P) \cong u_f(T) \tag{1.23}$$

$$s(T,P) \cong s_f(T) \tag{1.24}$$

That is, the internal energy, entropy, and specific volume for a specific temperature can be set equivalent to the saturation conditions at the same temperature.

2. Enthalpy can be determined in any of the following two ways:

$$h(T,P) \cong u_f(T) + Pv_f(T) \tag{1.25}$$

$$h(T,P) \cong h_f(T) + v_f(T)\left[P - P_{sat}(T)\right] \tag{1.26}$$

If the value of $v_f(T)\left[P - P_{sat}(T)\right]$ is small, the previous equation can be approximated as follows:

$$h(T,P) \cong h_f(T) \tag{1.27}$$

✐Example 1.1: Determine the temperature, specific volume, specific enthalpy and specific internal energy of saturated liquid water at a pressure of 200 kPa.

✍Solution

Due to the fact that the pressure is defined, we can use Table A.2 to determine the properties. In the saturation area, the temperature is equal to the saturation temperature at the given pressure:

$$T = T_{sat@P=200\text{kPa}} = 121.21°C$$

Other properties are obtained using the property values for saturated water (with *f* index):

$$v = v_{f@P=200\text{kPa}} = 0.001061 m^3/\text{kg}$$
$$h = h_{f@P=200\text{kPa}} = 504.50 \text{kJ/kg}$$
$$u = u_{f@P=200\text{kPa}} = 504.71 \text{kJ/kg}$$

✐Example 1.2: Determine the pressure, volume, enthalpy and internal energy of 2 kg of saturated water at the temperature of 250°C.

✍Solution

Given the specified temperature, we can utilize Table A.1 to ascertain the substance's properties. Within the saturation region, the pressure corresponds precisely to the saturation pressure aligned with the provided temperature:

$$P = P_{sat@T=250°C} = 3976.2 \text{ kPa}$$

Other desired properties are obtained using the values of properties for saturated steam (with index *g*):

$$V = mv = m\left(v_{g@T=250°C}\right) = (2\text{kg})\left(0.050085 \text{m}^3/\text{kg}\right) = 0.10017 \text{m}^3$$
$$H = mh = m\left(h_{g@T=250°C}\right) = (2\text{kg})(2801.0 \text{kJ}/\text{kg}) = 5602.0 \text{kJ}$$
$$U = mu = m\left(u_{g@T=250°C}\right) = (2\text{kg})(2601.8 \text{kJ}/\text{kg}) = 5203.6 \text{kJ}$$

✐Example 1.3: 200 g of saturated water at a constant pressure of 100 kPa completely evaporates and converts into saturated steam. Determine:

a) change in volume and
b) amount of received energy.

✍Solution

a) During the evaporation process, the volume change per unit mass is v_{fg}, which is defined as the difference between v_f and v_g:

$$\Delta V = V_2 - V_1 = m(v_2 - v_1) = m(v_g - v_f)$$

Using Table A.2:

$$v_{f@P=100\text{kPa}} = 0.001043\text{m}^3/\text{kg}$$

$$v_{g@P=100\text{kPa}} = 1.6941\text{m}^3/\text{kg}$$

As a result:

$$\Delta V = m(v_g - v_f) = (0.2\text{ kg})(1.6941 - 0.001043)\text{m}^3/\text{kg}$$
$$= 0.3386114\text{m}^3$$

b) The amount of energy required to evaporate a unit mass of a substance from a saturated liquid state to a saturated steam state at a certain pressure is called enthalpy of vaporization and is equal to h_{fg}. Therefore, the required energy is equal to:

$$mh_{fg@P=100\text{kPa}} = (0.2\text{kg})(2257.5\text{kJ/kg}) = 451.5\text{kJ}$$

✎Example 1.4: Calculate the energy (heat) required to convert 2 kg of water at 20°C into steam with a quality of 0.9 at a constant pressure of 5 MPa.

✍Solution

Since the process is conducted at constant pressure, the energy required for the process can be calculated from the difference in enthalpies:

$$Q = \Delta H = m\Delta h = m(h_2 - h_1)$$

According to Table A.2, the saturation temperature for 5 MPa pressure is equal to 263.94°C. The water temperature at the beginning of the process (20°C) is smaller than the saturation temperature at a pressure of 5 MPa; therefore, water would be in a compressed state. By referring to Table A.4, the enthalpy of water can be determined:

$$h_1 = 88.61\text{kJ/kg}$$

Water at the end point of the process is a mixture of saturated liquid and saturated steam; therefore, the enthalpy is calculated as follows:

$$h_2 = h_{f@P=5\text{MPa}} + xh_{fg@P=5\text{MPa}}$$

The values of enthalpy of saturated liquid and enthalpy of vaporization are determined from Table A.2:

$$h_{f@P=5MPa} = 1154.5\text{kJ/kg}, h_{fg@P=5MPa} = 1639.7\text{kJ/kg}$$

Now the enthalpy of the end point can be determined:

$$h_2 = 1154.5 + 0.9(1639.7) = 2630.23\text{kJ/kg}$$

The required energy is calculated as follows:

$$Q = m(h_2 - h_1) = 2\text{kg}(2630.23 - 88.61)\text{kJ/kg} = 2541.62\text{Kj}$$

1.1.7.9 Interpolation from Tables

Linear interpolation is employed to deduce values for temperatures that fall between the temperature values of two neighboring points within tables of thermodynamic properties. For this purpose, it is assumed that the desired properties change linearly between these two points.

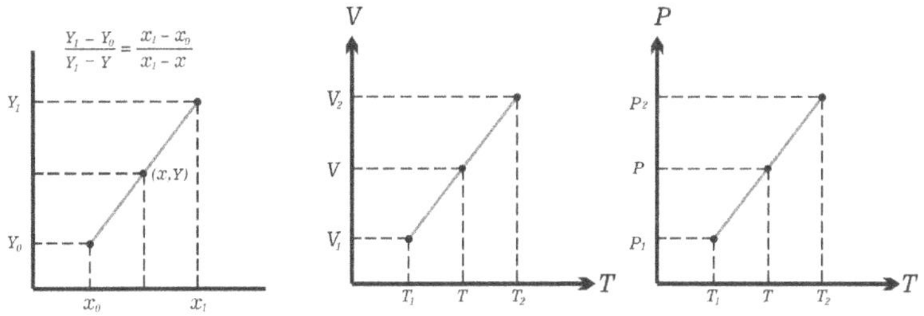

FIGURE 1.9 Linear interpolation of properties according to its characteristics between two available temperatures.

If linear changes are assumed, the slope of the graph is constant; then we can write:

$$\frac{P - P_1}{T - T_1} = \frac{P_2 - P_1}{T_2 - T_1} \rightarrow T = T_1 + \frac{P - P_1}{P_2 - P_1}(T_2 - T_1)$$

$$\frac{v - v_1}{T - T_1} = \frac{v_2 - v_1}{T_2 - T_1} \rightarrow T = T_1 + \frac{v - v_1}{v_2 - v_1}(T_2 - T_1)$$

Example 1.5: There is 0.9 kg of water in a container with a volume of 50 L at a pressure of 4 MPa. What is the temperature of the water?

Solution

The specific volume of water is calculated as follows:

$$V = \frac{V}{m} = \frac{50 \times 10^{-3}\text{m}^3}{0.9\text{kg}} = 0.05556\text{m}^3/\text{kg}$$

Using the information in Table A.2, the specific volume of saturated liquid and saturated steam at a pressure of 4 MPa is equal to:

$$v_{f@4\text{MPa}} = 0.001252\text{m}^3/\text{kg}$$

$$v_{g@4\text{MPa}} = 0.049779\text{m}^3/\text{kg}$$

Considering that the specific volume is greater than the specific volume of saturated steam at a pressure of 4 MPa, water is in a superheated state:

$$v > v_g \rightarrow \text{superheated steam}$$

From Table A.2 at a pressure of 4 MPa:

$$@4\text{MPa}\begin{cases} T_1 = 275°\text{C} \rightarrow v_1 = 0.05461\text{m}^3/\text{kg} \\ T_2 = 300°\text{C} \rightarrow v_2 = 0.05887\text{m}^3/\text{kg} \end{cases}$$

By interpolation, the temperature of the superheated steam at a specific volume of 0.05556 m³/kg is calculated:

$$\begin{aligned} T &= T_1 + \frac{v - v_1}{v_2 - v_1}(T_2 - T_1) \\ &= 275 + \frac{0.05556 - 0.05461}{0.05887 - 0.05461}(300 - 275) \\ &= 280.58°\text{C} \end{aligned}$$

✐Example 1.6: determine the temperature of saturated liquid at a pressure of 0.233 MPa.

✍Solution

In the saturated state, the temperature is equal to the saturation temperature. By referring to Table A.2, the saturation temperature at a pressure of 0.223 MPa is determined using the interpolation method:

	P (kPa)	**T (°C)**
1	200	120.21
	223	T
2	225	123.97

$$\begin{aligned} T &= T_1 + \frac{P - P_1}{P_2 - P_1}(T_2 - T_1) \\ &= 120.21 + \frac{223 - 200}{225 - 200}(123.97 - 120.21) \\ &= 123.67°\text{C} \end{aligned}$$

1.1.8 Ideal Gas

Any equation that relates the pressure, temperature, and specific volume of a substance is called an equation of state. Many equations of state can be found, among which some are simple, and some are very complex. The simplest and most famous equation is the ideal gas state. The term "steam" refers to a substance that is in the gaseous phase and is saturated or slightly warmer than the saturated state, while the same substance is referred to as gas at much higher temperatures.

The limit value of Pv/T (where v is the specific molar volume) is a constant value for a certain amount of all fluids when the density is low and the distance between the molecules is large (the pressure tends to zero or the volume tends to infinity); the value of this limit is calculated as (Shavit & Gutfinger, 2008):

$$\lim_{P\to 0}\frac{Pv}{T} = \lim_{v\to\infty}\frac{Pv}{T} = \Re = 8.314\text{kJ / kmol.K} \tag{1.28}$$

Any gas that obeys the following equation is called an ideal gas:

$$Pv = \mathcal{R}T \tag{1.29}$$

This equation of state is expressed as ideal or perfect gas law. Using the definition of specific molar volume ($v = V/n$), the previous equation can also be written as follows:

$$PV = n\mathcal{R}T \tag{1.30}$$

where n is the number of moles, and V is the gas volume in m^3. The mass of a substance is equal to the product of its molar mass by the number of its moles, that is:

$$m = Mn \to M = \frac{m}{n} \tag{1.31}$$

Also, the gas constant of each particular gas is specific to it and depends on the global constant of gases and the molar mass of that gas, as follows:

$$R = \frac{\mathcal{R}}{M} \tag{1.32}$$

Hence, the equation of ideal gas can be written as follows:

$$PV = mRT \tag{1.33}$$

The result of dividing V by m is called the specific volume and is shown by v:

$$Pv = RT \tag{1.34}$$

The specific volume is the opposite of density; the equation of ideal gas in terms of density is written as follows:

$$P = \rho RT \tag{1.35}$$

TABLE 1.2
Equations Related to the Equation of Ideal Gas State

Equation	Parameters
$PV = n\Re T$	$P\ (\text{kPa})$
	$V\ (\text{m}^3)$
	$T\ (\text{K})$
	$R = 8.134\ {}^{\text{kJ}}/_{\text{kmol°k}}$
	n number of moles: kmol
$Pv = \Re T$	$P\ (\text{kPa})$
	$T\ (\text{K})$
	$\mathcal{R} = 8.314\ {}^{\text{kJ}}/_{\text{kmol°k}}$
	v = molar specific volume (m^3/kmol)
$PV = mRT$	$P\ (\text{kPa})$
	$V\ (\text{m}^3)$
	$T\ (\text{K})$
	$m\ (\text{kg})$
	R gas constant $(\text{kJ}/\text{kg.K})$
$Pv = RT$	$P\ (\text{kPa})$
	$T\ (\text{K})$
	$R\ (\text{kJ}/\text{kg.K})$
	v specific volume (m^3/kg)
$P = \rho RT$	$P\ (\text{kPa})$
	$T\ (\text{K})$
	$R\ (\text{kJ}/\text{kg.K})$
	ρ density (kg/m^3)

Table 1.2 presents all equations related to the equation of the ideal gas state in summary.

1.1.8.1 Non-Ideal Gas

A gas whose properties do not apply to ideal gases is called an imperfect (non-ideal) gas. To check the properties of these gases, the concept of compressibility is used, which is defined as follows (Potter & Somerton, 2014):

$$z = \frac{Pv}{RT} \tag{1.36}$$

Obviously, the compressibility of ideal gases is 1.

✐Example 1.7: Using the ideal gas state equation, determine the specific volume of water at a temperature of 400°C and pressures

1. 0.01 MPa
2. 0.1 MPa
3. 20 MPa

Also, determine the specific volume using the table of thermodynamic properties of water and calculate the compressibility coefficient for each case. The molar mass of water is 18.016 g/mol.

✍Solution

The gas constant for water steam is calculated as follows:

$$R = \frac{\mathcal{R}}{M} = \frac{8.314}{18.016} = 0.4615\text{kJ/kg.K}$$

The ideal gas equation in terms of specific volume is as follows:

$$Pv = RT \rightarrow v = \frac{RT}{P}$$

Therefore

$$v_1 = \frac{RT}{P_1} = \frac{(0.4615)(400+273.15)}{10} = 31.06588\text{m}^3/\text{kg}$$

$$v_2 = \frac{RT}{P_2} = \frac{(0.4615)(400+273.15)}{100} = 3.10658\text{m}^3/\text{kg}$$

$$v_3 = \frac{RT}{P_3} = \frac{(0.4615)(400+273.15)}{20000} = 0.015533\text{m}^3/\text{kg}$$

Thermodynamic properties tables are used to determine the actual specific volume. According to Table A.2, the saturation temperature for the pressure of 10 kPa, 100 kPa, and 20,000 kPa is equal to 45.81°C, 99.61°C, and 365.75°C, respectively; therefore, water at a temperature of 400°C and at all three pressures is superheated. By referring to the superheated water table (Table A.3), the specific volume for all three cases is determined as follows:

$$v_1 = v_{@P=0.01\text{MPa},\,T=400°\text{C}} = 31.063\text{m}^3/\text{kg}$$

$$v_2 = v_{@P=0.1\text{MPa},\,T=400°\text{C}} = 3.1027\text{m}^3/\text{kg}$$

$$v_3 = v_{@P=20\text{MPa},\,T=400°\text{C}} = 0.00995\text{m}^3/\text{kg}$$

The compressibility coefficient for all three cases is calculated as follows:

$$z_1 = \frac{P_1 v_1}{RT} = \frac{(10)(31.063)}{(0.4615)(400+273.15)} = 0.99991$$

$$z_2 = \frac{P_2 v_2}{RT} = \frac{(100)(3.1027)}{(0.4615)(400 + 273.15)} = 0.99875$$

$$z_3 = \frac{P_3 v_3}{RT} = \frac{(20000)(0.00995)}{(0.4615)(400 + 273.15)} = 0.64057$$

Table 1.3 summarizes the solution of the example.

TABLE 1.3
Brief Results of Example 1.7

Case	Pressure	Calculated Specific Volume $\left(v = \frac{RT}{P}\right)$	Actual Specific Volume (from the water properties table)	Compressibility Coefficient $\left(z = \frac{Pv}{RT}\right)$
1	10 kPa	$31.06588 m^3/kg$	$31.063 m^3/kg$	0.99991
2	100 kPa	$3.10658 m^3/kg$	$3.1027 m^3/kg$	0.99875
3	20,000 kPa	$0.015533 m^3/kg$	$0.00995 m^3/kg$	0.64057

As evident, the ideal gas assumption proves highly applicable for scenarios involving lower pressures (as observed in cases 1 and 2). However, when dealing with higher pressures, such as in case 3, the behavior of steam deviates notably from the characteristics associated with ideal gas behavior.

✐Example 1.8: Repeat the previous example for the temperature of 50°C.

✍Solution

$$v_1 = \frac{RT}{P_1} = \frac{(0.4615)(50 + 273.15)}{10} = 14.91337 m^3/kg$$

$$v_2 = \frac{RT}{P_2} = \frac{(0.4615)(50 + 273.15)}{100} = 1.49134 m^3/kg$$

$$v_3 = \frac{RT}{P_3} = \frac{(0.4615)(50 + 273.15)}{20000} = 0.00746 m^3/kg$$

According to Table A.2, the saturation temperature for a pressure of 10 kPa is equal to 45.81°C; therefore, water at a temperature of 50°C and under pressure is superheated. By referring to the superheated water table (Table A.3), the specific volume of the first case is determined as follows:

$$v_1 = v_{@P=0.01MPa,\,T=50°C} = 14.867 m^3/kg$$

The saturation temperature for a pressure of 100 kPa is equal to 99.61°C; therefore, water is compressed at a temperature of 50°C and at a given pressure. Referring to the table of saturated water (Table A.1), the specific volume of the second case can be approximated as follows:

$$v_2 = v_{@P=0.1\text{MPa},\,T=50^\circ\text{C}} \cong v_{f@T=50^\circ\text{C}} = 0.001012\text{m}^3/\text{kg}$$

The saturation temperature for a pressure of 20,000 kPa is equal to 365.75°C; therefore, water is compressed at a temperature of 50°C and at a given pressure. By referring to the compressed water table, the specific volume of the third case is determined using interpolation as follows:

$$v_3 = v_{@P=20\text{MPa},\,T=50^\circ\text{C}} = 0.0010038\text{m}^3/\text{kg}$$

The compressibility coefficient for all three cases is calculated as follows:

$$z_1 = \frac{P_1 v_1}{RT} = \frac{(10)(14.867)}{(0.4615)(50+273.15)} = 0.99689$$

$$z_2 = \frac{P_2 v_2}{RT} = \frac{(100)(0.001012)}{(0.4615)(50+273.15)} = 0.00068$$

$$z_3 = \frac{P_3 v_3}{RT} = \frac{(20000)(0.0010038)}{(0.4615)(50+273.15)} = 0.13462$$

Table 1.4 summarizes the results of the example.

TABLE 1.4
Brief Summary of Results of Example 1.8

Case	Pressure	Calculated Specific Volume $\left(v = \frac{RT}{P}\right)$	Actual Specific Volume (from the water properties table)	Compressibility Coefficient $\left(z = \frac{Pv}{RT}\right)$
1	10 kPa	$14.91337\text{m}^3/\text{kg}$	$14.867\text{m}^3/\text{kg}$	0.99689
2	100 kPa	$1.49134\text{m}^3/\text{kg}$	$0.001012\text{m}^3/\text{kg}$	0.00068
3	20,000 kPa	$0.00746\text{m}^3/\text{kg}$	$0.0010038\text{m}^3/\text{kg}$	0.13462

Evidently, when the pressure is at 0.01 MPa—a level lower than other pressures—steam's behavior exhibits a closer resemblance to that of an ideal gas. Furthermore, a notable observation is that, in contrast to the previous instance, as the temperature decreases, steam's behavior increasingly diverges from ideal gas characteristics. In a broader perspective, it holds true that factors such as low density, low pressure, and high temperature collectively contribute to aligning a gas more closely with ideal gas conditions.

2 Mass Balance

2.1 MASS FLOW

The amount of mass passing through a cross-section per unit of time is called mass flow rate, and it is shown by $\dot{m}$. A fluid usually enters or exits a standard volume through pipes or conduits. The differential mass flow rate of the fluid passing through a small element of the pipe surface with area dA_c, fluid density ρ, and component of velocity perpendicular to dA_c, which is represented by V_n, is proportional and is expressed as follows:

$$\delta \dot{m} = \rho V_n dA_c \tag{2.1}$$

The mass flow through the entire cross-section of a pipe or conduit is obtained by integration:

$$\dot{m} = \int_{A_c} \delta \dot{m} = \int_{A_c} \rho V_n dA_c \quad \text{(kg/s)} \tag{2.2}$$

This equation is always valid (actually accurate), but due to its integral form, it is not always applicable for engineering analysis. It is better to express the mass flow in terms of average values on the cross-section of the pipe. In many practical applications, the density is essentially constant over the cross-sectional area of the tube and can be taken out of the integral. However, due to the non-slip condition in the walls, the velocity on the cross-section of the pipe is not uniform. Therefore, the velocity

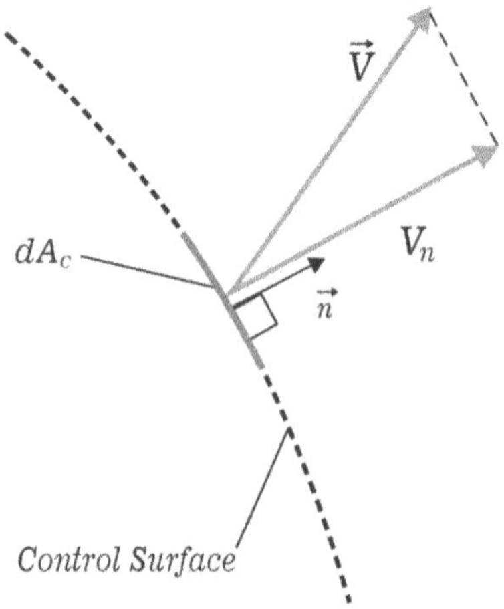

FIGURE 2.1 The vertical velocity V_n for a surface is the component of the velocity perpendicular to the surface.

DOI: 10.1201/9781003424680-2

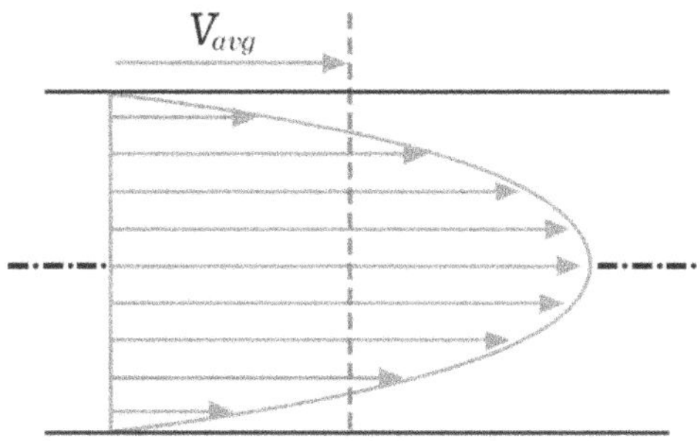

FIGURE 2.2 Average velocity in a cross-section of flow.

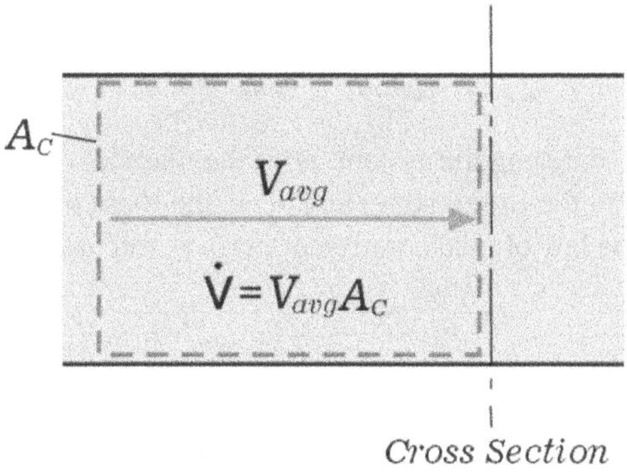

FIGURE 2.3 Volumetric flow rate is the volume of fluid flowing through a cross-section per unit of time.

changes from zero at the walls to a maximum near the center line of the tube. The average speed is defined as the average value of V_n in the cross-section of the pipe (Figure 2.2):

$$V_{avg} = \frac{1}{A_c} \int_{A_c} V_n dA_c \tag{2.3}$$

where A_c is the cross-sectional area perpendicular to the flow direction.

Now the mass flow rate can be expressed as follows:

$$\dot{m} = \rho V_{avg} A_c \ (\text{kg/s}) \tag{2.4}$$

2.2 VOLUMETRIC FLOW

The volume of fluid passing through each cross-section per unit of time is called volumetric flow (Figure 2.3) and is expressed as follows:

$$\dot{V} = \int_{A_c} V_n dA_c = V_{avg} A_c \ (\text{m}^3/\text{s}) \tag{2.5}$$

Mass flow rate and volumetric flow rate are related according to the following equation:

$$\dot{m} = \rho \dot{V} \tag{2.6}$$

2.3 MASS BALANCE

Mass is a property that remains and cannot be created or destroyed; therefore, it can be written:

System mass change = Mass leaving the system – Mass entering the system

The mathematical formulation of the law of conservation of mass is as follows:

$$\sum m_i - \sum m_o = \Delta m_{system} = m_2 - m_1 \tag{2.7}$$

where m_i is the mass entering the system, m_o is the mass leaving the system, m_1 is the mass of the system in the initial state, and m_2 is the mass of the system in the final state per kg unit. The law of conservation of mass as rate is:

$$\sum \dot{m}_i - \sum \dot{m}_o = dm_{system} / dt \tag{2.8}$$

where dm_{system}/dt is the system mass change per unit time (kg/s), m_i is the mass flow rate entering the system, and m_o is the mass flow rate leaving the system. If the flow inside the control volume is constant (steady), then the law of conservation of mass for constant flow will be as follows:

$$\sum \dot{m}_i = \sum \dot{m}_o \tag{2.9}$$

Using the definition of mass flow rate, the mass conservation law in steady state for the control volume can be written as follows:

$$\sum (\rho VA)_i = \sum (\rho VA)_o \tag{2.10}$$

In incompressible flows or flows where specific mass (density) changes in the flow can be ignored, it can be written:

$$\sum (VA)_i = \sum (VA)_o \tag{2.11}$$

As a result:

$$\sum \dot{V}_i = \sum \dot{V}_o \tag{2.12}$$

For a closed system:

$$\Delta m_{system} = m_2 - m_1 = 0 \tag{2.13}$$

Or

$$\frac{dm_{system}}{dt} = 0 \tag{2.14}$$

2.4 MATERIAL BALANCE

Equation 2.9 can be written as follows for each component (j) that enters the system, provided that there is no chemical reaction:

$$\sum X_{j,i}\dot{m}_i = \sum X_{j,o}\dot{m}_o \tag{2.15}$$

where $X_{j,i}$ is the mass fraction of component j in the input flow, and $X_{j,o}$ is its mass fraction in the output flow.

✐Example 2.1: Calculate V_2 in the stable system of the following figure. All streams are water.

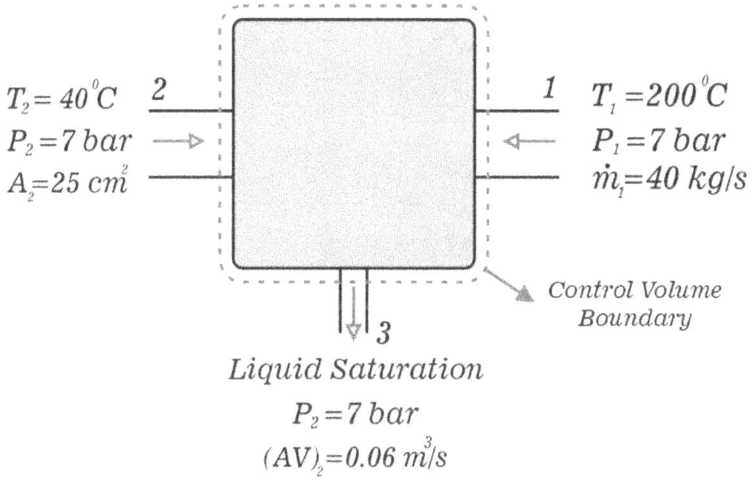

FIGURE 2.4 Diagram of Example 2.1.

✍Solution

The flow velocity at point 2 can be obtained by defining the mass flow rate as follows:

$$\dot{m} = \rho VA = \frac{VA}{v} \rightarrow V_2 = \frac{\dot{m}_2 v_2}{A_2}$$

The cross-sectional area of the pipe entering the system at point 2 is determined based on the problem definition and is equal to 0.0025 m². The specific volume of water at a temperature of 40°C and a pressure of 700 kPa is determined from the tables of thermodynamic properties of water. The saturation temperature of water at a pressure of 700 kPa is equal to 164.95°C; therefore, the water entering the system at point 2 is in a compressed state. The specific volume of compressed water can be approximated as saturated water:

$$v_2 = v_{@T=40°C,\,P=700\text{kPa}} \cong v_{f@T=40°C} = 0.001008\text{m}^3/\text{kg}$$

To calculate the mass flow in point 2, the mass balance for the system must be written. The mass balance in steady state is as follows:

$$\sum_i \dot{m} = \sum_o \dot{m} \rightarrow \dot{m}_1 + \dot{m}_2 = \dot{m}_3$$

Therefore

$$\dot{m}_2 = \dot{m}_3 - \dot{m}_1$$

The mass flow rate of stream 3 is calculated as follows:

$$\dot{m}_3 = \frac{V_3 A_3}{v_3} = \frac{(VA)_3}{v_{f@700\text{kPa}}} = \frac{0.06\text{m}^3/s}{0.001108\text{m}^3/\text{kg}} = 54.15\text{kg}/s$$

As a result:

$$\dot{m}_2 = \dot{m}_3 - \dot{m}_1 = 54.15 - 40 = 14.15\frac{\text{kg}}{\text{s}}$$

Now we can calculate flow rate 2 using the definition of volume flow rate:

$$V_2 = \frac{\dot{m}_2 v_2}{A_2} = \frac{\left(14.15\frac{\text{kg}}{\text{s}}\right)\left(0.001008\frac{\text{m}^3}{\text{kg}}\right)}{0.0025\text{m}^2} = 5.71\frac{\text{m}}{\text{s}}$$

Example 2.2: An air heater is basically a channel of constant diameter with several layers of electrical resistance in it. A small blower sucks the air, passes it over the resistor, and heats it up. Find the percentage increase in air velocity while passing through the heater.

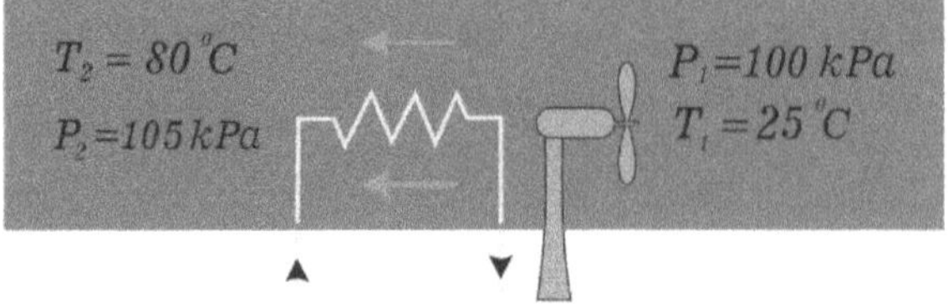

FIGURE 2.5 Diagram of Example 2.2.

✍Solution

Assumptions:

1. The process is considered steady state.
2. Air can be assumed to be an ideal gas.

The percentage of speed increase can be calculated as follows:

$$\text{diff} = \frac{V_2 - V_1}{V_1} \times 100 = \left(\frac{V_2}{V_1} - 1\right) \times 100$$

To calculate the ratio of the speed of the outlet air to the inlet air, you can use the mass balance in steady state:

$$\sum_i \dot{m} = \sum_o \dot{m} \rightarrow \sum (\rho_i V_i A_i)_{out} = \sum (\rho_i V_i A_i)_{in} \rightarrow \rho_1 A_1 V_1 = \rho_2 A_2 V_2 \rightarrow \frac{V_2}{V_1} = \frac{\rho_1}{\rho_2}$$

Considering the ideal air, the ratio of the density of the inlet air to the outlet air can be obtained as follows:

$$\frac{\rho_1}{\rho_2} = \frac{\dfrac{P_1}{RT_1}}{\dfrac{P_2}{RT_2}} = \frac{T_2 P_1}{T_1 P_2} = \frac{(80 + 273.15)(100)}{(25 + 273.15)(105)} = 1.1281$$

As a result, the percentage of speed increase can be calculated:

$$\text{diff} = \left(\frac{V_2}{V_1} - 1\right) \times 100 = (1.1281 - 1) \times 100 = 12.81\%$$

✐Example 2.3: Tomato juice flows in a tube with a flow rate of 2 kg/s and solid matter of 5% and is salted by a constant flow of salt solution of 26% of solid matter. At what rate of solution does the final product contain 2% salt? (Varzakas & Tzia, 2014)

✍Solution

Including the 2% salt added by the solution, the final product has a solid content of 7%. To calculate the mass flow rate of the saturated solution, a mass balance must be written for the system. The mass balance in steady state is as follows:

$$\sum_i \dot{m} = \sum_o \dot{m} \rightarrow \dot{m}_1 + \dot{m}_2 = \dot{m}_3$$

Therefore:

$$\dot{m}_2 = \dot{m}_3 - \dot{m}_1 = \dot{m}_3 - 2$$

Index 1 is for primary tomato juice, index 2 is for saltwater solution, and index 3 is for secondary tomato juice. This equation has two unknowns; therefore, the mass balance for the total solid must also be written:

$$TS_2\dot{m}_2 = TS_3\dot{m}_3 - TS_1\dot{m}_1 \rightarrow 0.26\dot{m}_2 = 0.07\dot{m}_3 + (0.05)(2)$$

By simultaneously solving the balance of mass and material, the mass flow rates of the solution and the final product are calculated as follows:

$$\dot{m}_2 = 0.21\frac{\text{kg}}{\text{s}}$$

$$\dot{m}_3 = 2.21\frac{\text{kg}}{\text{s}}$$

✐Example 2.4: Low-fat milk is prepared by removing some fat from whole milk. Low-fat milk contains 90.5% water, 3.5% protein, 5.1% carbohydrates, 0.1% fat, and 0.8% ash. If the whole milk contains 4.5% fat, assuming that only fat is removed to create low-fat milk and there is no decrease in the system, calculate the composition of the whole milk.

✍Solution

100 kg of low-fat milk is considered. Considering that only fat has been removed from whole milk, the mass of other components can be considered constant:

$$m_{Water} = 90.5\,\text{kg}$$

$$m_{Protein} = 3.5\,\text{kg}$$

$$m_{Carbohydrate} = 5.1\,\text{kg}$$

$$m_{Ash} = 0.8\,\text{kg}$$

Whole milk contains 4.5% fat; therefore, it can be written as follows using the definition of mass fraction:

$$X_{Fat} = \frac{m_{Fat}}{m_{Total}} = \frac{m_{Fat}}{m_{Water} + m_{Protein} + m_{Carbohydrate} + m_{Ash} + m_{Fat}}$$

$$\rightarrow 0.045 = \frac{m_{Fat}}{99.9 + m_{Fat}} \rightarrow m_{Fat} = 4.7\,\text{kg}$$

By determining the mass of fat and, as a result, the total mass of whole milk, now the mass fraction of all components can be calculated:

$$X_{Water} = \frac{m_{Water}}{m_{Total}} = \frac{90.5}{104.6} = 0.8652$$

$$X_{Protein} = \frac{m_{Protein}}{m_{Total}} = \frac{3.5}{104.6} = 0.0335$$

$$X_{Carbohydrate} = \frac{m_{Carbohydrate}}{m_{Total}} = \frac{5.1}{104.6} = 0.0488$$

$$X_{Ash} = \frac{m_{Ash}}{m_{Total}} = \frac{0.8}{104.6} = 0.0076$$

✐**Example 2.5: In the concentrate production process, sour cherry juice enters an evaporator with a flow rate of 1.45 kg/s and a solid content of 31%, and after being concentrated with a solid content of 71.51%, it comes out. Determine the flow rate of the concentrate leaving the evaporator. The mass fraction of the components of the sour cherry juice entering the evaporator is in Table 2.1; determine the mass fraction of the components of the outlet concentrate.**

TABLE 2.1
Mass Fraction of Cherry Juice Compounds in Example 2.5

X_{Water}	$X_{Protein}$	X_{Fat}	$X_{Carbohydrate}$	X_{Ash}
0.69000	0.00946	0.00026	0.28983	0.01045

✍**Solution**

By writing the material balance for the total solid material, it is possible to calculate the outlet concentration flow rate:

$$TS_1\dot{m}_1 = TS_2\dot{m}_2 \rightarrow \dot{m}_2 = \frac{TS_1\dot{m}_1}{TS_2} = \frac{(0.31)(1.45)}{0.7151} = 0.63\frac{\text{kg}}{\text{s}}$$

The mass fraction of water for concentrate is calculated as follows:

$$X_{Water,2} = 1 - TS_2 = 28.49\%$$

The mass fraction ratio of each component of sour cherry juice in the primary and secondary state is equal to the ratio of solid matter in both states:

$$\frac{X_{i,2}}{X_{i,2}} = \frac{TS_2}{TS_1} \rightarrow \begin{cases} X_{Protein,2} = X_{Protein,1} \times \dfrac{TS_2}{TS_1} = 0.00946 \times \dfrac{71.51}{31} = 0.02182 \\ X_{Fat,2} = X_{Fat,1} \times \dfrac{TS_2}{TS_1} = 0.00026 \times \dfrac{71.51}{31} = 0.00056 \\ X_{Ash,2} = X_{Ash,1} \times \dfrac{TS_2}{TS_1} = 0.01045 \times \dfrac{71.51}{31} = 0.02411 \\ X_{Carbohydrate,2} = X_{Carbohydrate,1} \times \dfrac{TS_2}{TS_1} = 0.28983 \times \dfrac{71.51}{31} = 0.66587 \end{cases}$$

3 Energy Balance

3.1 ENERGY BALANCE

The first law of thermodynamics states that energy is neither created nor destroyed but only changes from one form to another, so it can be written:

System energy change = energy leaving the system – energy entering the system

The mathematical formulation of the principle of conservation of energy is as follows:

$$E_i - E_o = E_2 - E_1 = \Delta E_{system} \tag{3.1}$$

where E_i is the energy entering the system, E_o is the energy leaving the system, E_1 is the energy of the system in the initial state, and E_2 is the energy of the system in the final state per kJ unit. Transfer energies include mass flow, work, and heat; therefore, the energy balance in the general state can be written as follows:

$$\sum(\theta m)_i + Q_i + W_i - \sum(\theta m)_o - Q_o - W_o = \Delta E_{system} \tag{3.2}$$

Or:

$$\sum\left[m\left(h+\frac{V^2}{2}+gz\right)\right]_i + Q_i + W_i - \sum\left[m\left(h+\frac{V^2}{2}+gz\right)\right]_o - Q_o - W_o = \Delta E_{system} \tag{3.3}$$

which is written as follows in the unit of time:

$$\sum\left[\dot{m}\left(h+\frac{V^2}{2}+gz\right)\right]_i + \dot{Q}_i + \dot{W}_i - \sum\left[\dot{m}\left(h+\frac{V^2}{2}+gz\right)\right]_o - \dot{Q}_o - \dot{W}_o = dE_{system}/dt \tag{3.4}$$

The energy balance for stable systems can be written as follows:

$$\sum\left[\dot{m}\left(h+\frac{V^2}{2}+gz\right)\right]_i + \dot{Q}_i + \dot{W}_i = \sum\left[\dot{m}\left(h+\frac{V^2}{2}+gz\right)\right]_o + \dot{Q}_o + \dot{W}_o \tag{3.5}$$

DOI: 10.1201/9781003424680-3

3.2 ENERGY BALANCE OF CLOSED SYSTEMS

In closed systems, there is no mass exchange between the system and the environment; energy is transferred by work and heat:

$$Q_i + W_i - Q_o - W_o = \Delta E_{system} \tag{3.6}$$

For simple compressible systems, the total energy change of a system during a process is equal to the sum of changes in internal, kinetic, and potential energies:

$$\Delta E_{system} = \Delta U + \Delta E_K + \Delta E_P \tag{3.7}$$

By combining Equations (3.6) and (3.7), the energy balance for closed systems is obtained as follows:

$$Q_i + W_i - Q_o - W_o = \Delta U + \Delta E_K + \Delta E_P \tag{3.8}$$

3.3 THERMOPHYSICAL PROPERTIES

3.3.1 Specific Heats

Specific heat refers to the quantity of energy required to raise the temperature of a unit mass of a material by 1 degree. The specific heat depends on the substance, the temperature of the substance, and how it is done. In thermodynamics, two types of specific heat are mostly used: specific heat at constant volume C_v and specific heat at constant pressure. Specific heat at constant volume is expressed as the amount of energy required to raise a unit mass of a substance by 1 degree when the volume remains constant. The amount of energy required to perform this action at constant pressure is called specific heat at constant pressure C_p. C_p is always greater than C_v because the system can be expanded at constant pressure, and the energy required for this expansion must be provided for the system. C_p and C_v are properties and depend on the system state like any other property. C_v depends on internal energy changes, and C_p depends on enthalpy changes. It is better to define C_v as the change in the specific internal energy of a substance per unit temperature change in a constant volume. Similarly, C_p is the specific enthalpy change of a substance per unit temperature change at a constant pressure. The specific heat at constant volume and constant pressure are as follows (Balmer, 2011):

$$C_v = \left(\frac{du}{dT}\right)_V \tag{3.9}$$

$$C_P = \left(\frac{dh}{dT}\right)_P \tag{3.10}$$

The unit of specific heat is $\frac{\text{kJ}}{\text{kg°C}}$ or $\frac{\text{kJ}}{\text{kg.K}}$.

3.3.2 Internal Energy, Enthalpy, and Specific Heat of Ideal Gases

For an ideal gas, the internal energy is only a function of temperature, that is:

$$u = u(T) \tag{3.11}$$

Using the definition of enthalpy and the equation of state of an ideal gas:

$$h = u + RT \tag{3.12}$$

Since R is constant and the internal energy is a function of temperature, the enthalpy of an ideal gas only must be a function of temperature:

$$h = h(T) \tag{3.13}$$

Since for an ideal gas, u and h depend only on temperature, the specific heats C_p and C_v also depend only on temperature; therefore, at a given gas temperature, an ideal gas will have constant values regardless of specific volume or pressure. According to the equations mentioned for C_p and C_v, we can write:

$$du = C_v(T)dT \tag{3.14}$$

$$dh = C_P(T)dT \tag{3.15}$$

By integrating these equations, the change in internal energy or enthalpy of the ideal gas from state 1 to state 2 during a process is calculated:

$$\Delta u = u_2 - u_1 = \int_1^2 C_v(T)dT \tag{3.16}$$

$$\Delta h = h_2 - h_1 = \int_1^2 C_P(T)dT \tag{3.17}$$

To solve these integrals, it is necessary to have equations for C_p and C_v as a function of temperature. The changes of C_p and C_v relative to temperature are curved, but for small distances, they can be approximated linearly, and specific heats can be determined with constant average values:

$$u_2 - u_1 = C_{v,ave}(T_2 - T_1) \tag{3.18}$$

$$h_2 - h_1 = C_{p,ave}(T_2 - T_1) \tag{3.19}$$

In fact, $C_{v,ave}$ and $C_{p,ave}$ are determined at the average temperature $\frac{T_2+T_1}{2}$. Another method is to determine C_p and C_v at temperatures T_1 and T_2 and use their average. In short, there are three methods to determine the internal energy and enthalpy changes of ideal gases:

1. Using tables
2. Using C_p and C_v equations as a function of temperature and integrating them
3. Using average specific heats

3.3.2.1 Equations of Specific Heat of Ideal Gases

By differentiating from the equation $h = u + RT$, a known equation between C_p and C_v for ideal gases is obtained (Borgnakke & Sonntag, 2022):

$$dh = du + RdT \tag{3.20}$$

By placing $C_p dT$ instead of dh and $C_v dT$ instead of du and dividing the resulting expression by dT:

$$C_p = C_v + R \tag{3.21}$$

Since C_v can be determined by having C_p and gas constant R, this equation is considered one of the important equations of ideal gases. Another property of ideal gas called specific heat ratio k (or atomicity coefficient γ) is defined as follows (Moran et al., 2010):

$$k = \frac{C_p}{C_v} \tag{3.22}$$

The specific heat ratio also changes with temperature, but this change is very mild. For monoatomic gases, its value will necessarily be constant at 1.667. Many diatomic gases, such as air, have a specific heat ratio of about 1.4 at room temperature.

3.3.3 Specific Heats of Liquids and Solids

A substance that has a constant specific volume (or density) is called an incompressible substance. Therefore, liquids and solids are assumed to be incompressible with a relatively suitable approximation. Assuming the volume remains constant, the energy related to volume change (such as boundary work) can be ignored compared to other forms of energy. However, this assumption will not apply to the study of thermal stresses in solids (produced by volume change with temperature) or liquid analysis in a glass thermometer. It can mathematically be shown that the specific heats at constant volume and pressure are equal for incompressible materials. Therefore, the subscripts C_p and C_v are removed for solids and liquids, and both specific heats are displayed with the symbol C, that is:

$$C_P = C_V = C \quad (3.23)$$

3.3.4 Internal Energy Changes of Solids and Liquids

The specific heats of incompressible substances, much like those of ideal gases, are solely temperature-dependent. Therefore, the ordinary differential can be used instead of partial differential in the definition equation of C_v. In this case, we can write:

$$du = C_v dT = C(T)dT \quad (3.24)$$

The internal energy change between states 1 and 2 is obtained via integration:

$$\Delta u = u_2 - u_1 = \int_1^2 C(T)dT \quad (3.25)$$

Before solving this integral, the changes of specific heat with temperature must be known, and for small temperature intervals, an average temperature can be used and considered constant:

$$\Delta u = C_{ave}(T_2 - T_1) \quad (3.26)$$

3.3.5 Enthalpy Changes of Solids and Liquids

Using the definition of enthalpy and considering that the specific volume is constant, the differential form of the enthalpy change equation of incompressible substances by derivation is defined as follows:

$$dh = du + vdP + Pdv = du + vdP \quad (3.27)$$

By integrating:

$$\Delta h = \Delta u + v\Delta P \cong C_{ave}\Delta T + v\Delta P \quad (3.28)$$

In thermal processes, the term $v\Delta P$ is not so important compared to $C_{ave}\Delta T$ and can be omitted:

$$\Delta h = \Delta u \cong C_{ave}\Delta T \quad (3.29)$$

Example 3.1: Steam flow enters a pipe with a speed of 3m/s, a pressure of 2 MPa, and a temperature of 300°C and leaves it with a pressure of 800 kPa and a temperature of 250°C. If the pipe inlet diameter is 12 cm and the steam velocity at the pipe inlet is 3 m/s, calculate the heat loss to the environment.

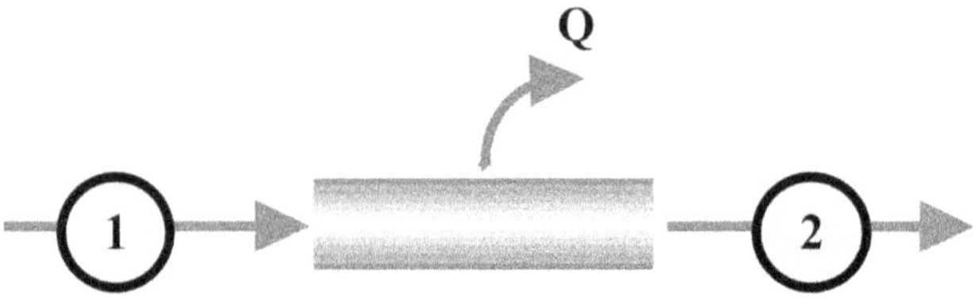

FIGURE 3.1 Diagram of Example 3.1.

Solution

Assumptions:

1. The process is considered steady state.
2. Kinetic and potential energy changes are ignored.

Determination of flow characteristics:

According to the table, the saturation temperature at 2 MPa and 800 kPa pressure is equal to 212.38°C and 170.41°C, respectively; therefore, the steam entering and exiting the pipe is in a superheated state. By referring to the superheated water table (Table A.3), you can determine the characteristics of the water:

$$\left.\begin{array}{r}2\text{MPa}\\300^\circ\text{C}\end{array}\right\}\rightarrow\begin{array}{l}v_1=0.12551\dfrac{\text{m}^3}{\text{kg}}\\h_1=3024.2\dfrac{\text{kJ}}{\text{kg}}\end{array}$$

$$\left.\begin{array}{r}800\text{kPa}\\250^\circ\text{C}\end{array}\right\}\rightarrow h_2=2950.4\frac{\text{kJ}}{\text{kg}}$$

Mass balance:

The process is considered steady state, so the mass flow rate entering the pipe is equal to the mass flow rate leaving it:

$$\sum\dot{m}_i=\sum\dot{m}_o\rightarrow\dot{m}_1=\dot{m}_2=\dot{m}$$

To determine the mass flow:

$$\dot{m} = \frac{A_1V_1}{v_1} = \frac{\left[\pi(0.06m)^2\right]\left(3\frac{m}{s}\right)}{0.12551\frac{m^3}{kg}} = 0.27\frac{kg}{s}$$

Energy balance:

Ignoring changes in kinetic and potential energy, the energy balance in the steady state for the considered tube is as follows:

$$\sum \dot{E}_i = \sum \dot{E}_o \rightarrow \dot{m}_1 h_1 = \dot{Q}_o + \dot{m}_2 h_2$$

Now, according to the mass balance, the energy balance can be expressed as follows:

$$\dot{Q}_o = \dot{m}(h_1 - h_2) = 0.27\frac{kg}{s}(3024.2 - 2950.4)\frac{kJ}{kg} = 19.93\frac{kJ}{s}$$

✐Example 3.2: Saturated liquid water with a pressure of 2 MPa and a flow rate of 4 kg/s is heated within a boiler to attain a temperature of 250°C and a pressure of 1.8 MPa. Determine the rate of heat transferred from the boiler to the water stream.

FIGURE 3.2 Diagram of Example 3.2.

✍Solution

Assumptions:

1. The process is considered steady state.
2. Kinetic and potential energy changes are ignored.
3. Heat loss from the surface of the boiler is negligible.

Determination of flow characteristics:

Water entering the boiler is in a saturated liquid state. By referring to the table of saturated water (Table A.2), the enthalpy of saturated liquid at a pressure of 2 MPa is determined:

$$h_1 = h_{f\,@\,2\mathrm{MPa}} = 908.47\frac{\mathrm{kJ}}{\mathrm{kg}}$$

According to the table, the saturation temperature at a pressure of 1.8 MPa is equal to 207.11°C; therefore, the outlet steam of the boiler is in a superheated state. By referring to the superheated water table (Table A.3), it is possible to determine the characteristics of the output water steam:

$$\left.\begin{matrix}1.8\mathrm{MPa}\\250°\mathrm{C}\end{matrix}\right\} \rightarrow h_2 = 2903.3\frac{\mathrm{kJ}}{\mathrm{kg}}$$

Mass balance:

The process is considered steady state, and the mass flow rate of inlet water to the boiler is equal to the outlet mass flow rate:

$$\sum \dot{m}_i = \sum \dot{m}_o \rightarrow \dot{m}_1 = \dot{m}_2 = \dot{m} = 4\,{}^{\mathrm{kg}}\!/_{\mathrm{s}}$$

Energy balance:

Ignoring the changes in kinetic and potential energy and heat loss from the surface of the boiler, the energy balance is as follows:

$$\sum \dot{E}_i = \sum \dot{E}_o \rightarrow \dot{m}_1 h_1 + \dot{Q}_i = \dot{m}_2 h_2$$

Now, according to the mass balance, the energy balance can be summarized as follows:

$$\dot{Q}_i = \dot{m}(h_2 - h_1) = 4\frac{\mathrm{kg}}{\mathrm{s}}(2903.3 - 908.47)\frac{\mathrm{kJ}}{\mathrm{kg}} = 7979.32\frac{\mathrm{kJ}}{\mathrm{s}}$$

✐**Example 3.3: Ambient air at a temperature of 300 K and a pressure of 100 kPa enters an electric heater with a power consumption of 1500 W and leaves it at a temperature of 80°C and a speed of 21 m/s. Determine the mass flow rate and volumetric flow rate of the air. Assume the average specific heat of air is 1.0065 kJ/kg.K.**

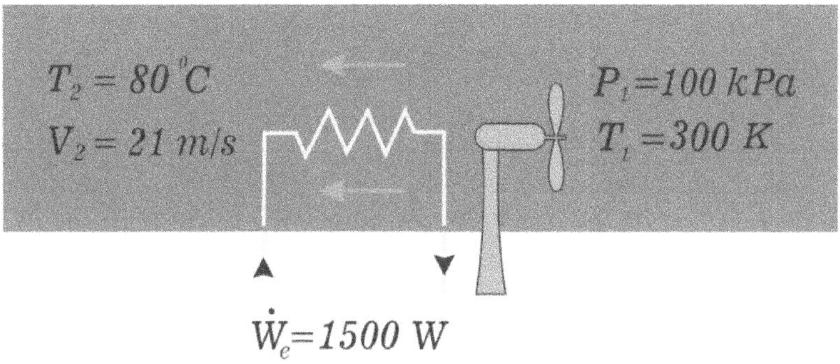

FIGURE 3.3 Diagram of Example 3.3.

✍Solution

Assumptions:

1. The process is considered steady state.
2. Potential energy changes are ignored.
3. Heat loss from the surface of the heater is negligible.
4. Air is considered an ideal gas.
5. The air velocity entering the heater is zero.

Mass balance:

The process is considered steady state, and the mass flow rate of air entering the heater is equal to the outlet mass flow rate:

$$\sum \dot{m}_i = \sum \dot{m}_o \rightarrow \dot{m}_1 = \dot{m}_2 = \dot{m}$$

Energy balance:

By ignoring potential energy changes and heat loss from the heater surface, the energy balance is as follows:

$$\sum \dot{E}_i = \sum \dot{E}_o \rightarrow \dot{m}_1\left(h_1 + \frac{V_1^2}{2}\right) + \dot{W}_i = \dot{m}_2\left(h_2 + \frac{V_2^2}{2}\right)$$

Now according to the mass balance, the energy balance can be written as follows:

$$\dot{W}_i = \dot{m}\left[h_2 - h_1 + \frac{V_2^2 - V_1^2}{2}\right] = \dot{m}\left[C_p(T_2 - T_1) + \frac{V_2^2 - V_1^2}{2}\right]$$

Now, the mass flow rate can be calculated as follows:

$$\dot{m} = \frac{\dot{W}_i}{C_p(T_2 - T_1) + \frac{V_2^2 - V_1^2}{2}}$$

$$= \frac{1.50\text{kW}}{\left(1.0065\frac{\text{kJ}}{\text{kg.K}}\right)(80 + 273.15 - 300)K + \frac{\left(21\frac{\text{m}}{\text{s}}\right)^2 + 0}{2}\left(\frac{1\frac{\text{kJ}}{\text{kg}}}{1000\frac{\text{m}^2}{\text{s}^2}}\right)} = 0.02798\frac{\text{kg}}{\text{s}}$$

Considering that air is assumed to be an ideal gas, the volumetric flow rate is calculated as follows:

$$\dot{V}_2 = \dot{m}V_2 = \dot{m}\left(\frac{RT_2}{P_2}\right) = 0.02798\left[\frac{(0.287)(80 + 273.15)}{100}\right] = 0.0284\frac{\text{m}^3}{s}$$

Example 3.4: To heat 500 kg of oil in a tank, steam flow with a temperature of 150°C and pressure of 400 kPa is used. If the initial and final temperature of the oil are 25°C and 60°C, respectively; the condensate temperature and pressure are 130°C and 350 kPa, respectively; the agitator power is 0.2 kW; and the process duration is 15 minutes, calculate the steam flow rate. The average specific heat of oil is 1.97 kJ/kg.°C.

Solution

Assumptions:

1. Kinetic and potential energy changes are ignored.
2. Heat loss from the surface of the tank is insignificant.

Determination of flow characteristics:

According to the table, the saturation temperature at the pressure of 400 kPa and 350 kPa is equal to 143.61°C and 138.86°C, respectively; therefore, the input steam to the tank is in superheated mode, and the condensate coming out of the tank is in compressed mode. By referring to the superheated water table (Table A.3), it can determine the characteristics of the steam entering the tank:

$$\left.\begin{matrix} 0.4\text{MPa} \\ 150°\text{C} \end{matrix}\right\} \rightarrow h_i = 2752.8\frac{\text{kJ}}{\text{kg}}$$

$$\left.\begin{matrix} 350\text{kPa} \\ 130°\text{C} \end{matrix}\right\} \rightarrow h_o \cong h_{f@130°C} = 546.38\frac{\text{kJ}}{\text{kg}}$$

Mass balance:

The mass of the system is constant:

$$\frac{dm_{system}}{dt} = 0 \rightarrow m_{system} = 500kg$$

The mass flow rate of condensed water leaving the system is equal to the mass flow rate of steam entering the system:

$$\dot{m}_i = \dot{m}_o = \dot{m}$$

Energy balance:

Kinetic and potential energy changes are ignored, and heat loss from the tank surface is negligible:

$$\dot{m}_i h_i + \dot{W}_i - \dot{m}_o h_o = \frac{dU_{system}}{dt}$$

According to the mass balance, the energy balance for the tank is as follows:

$$\dot{m}(h_i - h_o) + \dot{W}_i = \frac{m_{system}(u_2 - u_1)}{\Delta t} = \frac{m_{system}C_p(T_2 - T_1)}{\Delta t}$$

The steam flow rate is calculated as follows:

$$\dot{m} = \frac{\frac{m_{system}C_p(T_2 - T_1)}{\Delta t} - \dot{W}_i}{h_i - h_o} = \frac{\frac{(500\text{kg})\left(1.97\frac{\text{kJ}}{\text{kg.K}}\right)(60-25)\text{K}}{(15\times 60)\text{s}} - 0.2\text{kW}}{(2752.8 - 546.38)\frac{\text{kJ}}{\text{kg}}} = 0.017\frac{\text{kg}}{\text{s}}$$

4 Entropy

Energy is a conserved property, and within classical physics, no process contradicts the fundamental principle of the first law of thermodynamics. Processes strictly move forward and do not operate in reverse. While the first law doesn't dictate the direction of process, adhering to its conditions doesn't guarantee process occurrence. This limitation is offset by introducing another overarching principle known as the second law of thermodynamics. Breach of the second law can be readily identified through the utilization of a property termed "entropy".

4.1 REVERSIBLE AND IRREVERSIBLE PROCESS

A process is called reversible if, when it is returned to the initial state, it does not leave an impact on the environment. A reversible process is an ideal process that is used as a criterion for determining the maximum efficiency of various processes. An irreversible process is a process such that, after the process is completed, the system and its surroundings cannot return to their original state. All processes in nature are irreversible.

4.2 IRREVERSIBILITY FACTORS

All processes in nature are irreversible. Several factors cause the irreversibility of a process, the most important of which are mentioned in this section. The presence of any of these factors will indicate an irreversible process.

4.2.1 Friction

The effect of friction on irreversibility seems obvious, but for a better understanding, see Figure 4.1. Work must be done to lift a weight up an inclined plane. This work is used to increase the potential energy of the weight and overcome friction. On the way back, the heat generated due to friction is not recovered, and the friction force opposes the movement anyway.

4.2.2 Free Expansion

The iconic illustration of free expansion is exemplified by the process of a gas occupying a vacuum, as depicted in Figure 4.2. In this process, no work is done, but doing the return process requires the exchange of work and heat transfer with the environment, which means irreversibility. In conclusion, considering that the work environment has lost and gained heat, it has not returned to its original state, and therefore the process is irreversible.

 DOI: 10.1201/9781003424680-4

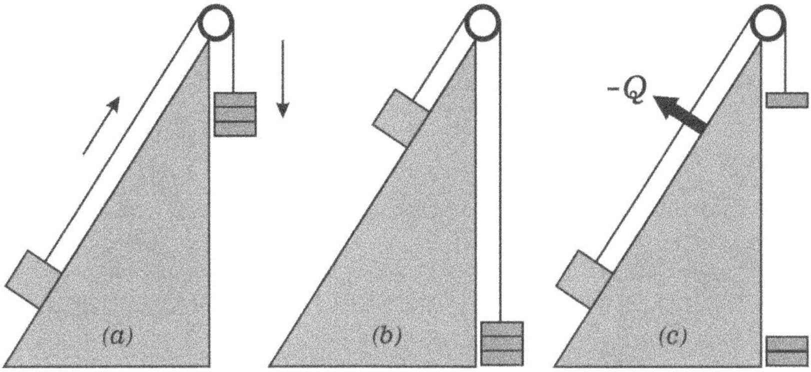

FIGURE 4.1 The effect of friction on irreversibility.

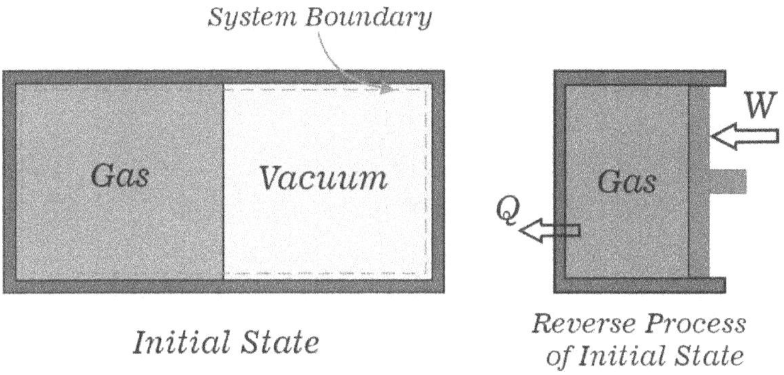

FIGURE 4.2 The effect of free expansion on irreversibility.

4.2.3 Heat Transfer Due to Specific Temperature Difference

If heat transfer occurs between two objects with different temperatures, a refrigeration cycle is required to return it, which requires receiving work; therefore, the heat transfer process is irreversible.

4.2.4 Mixing Two Different Materials

To comprehend the process of mixing two distinct materials, let's examine a scenario involving two different gases contained within a tank, separated by a membrane. When this membrane ruptures, the two gases naturally blend together. During this mixing process, no active work is performed, yet an expenditure of work becomes necessary to restore the system to its initial state.

Several additional factors contribute to irreversibility in various processes. These include residual effects, heat dissipation, the passage of electric current, and combustion. These factors lead to deviations from idealized reversible behavior, introducing inefficiencies and preventing the complete recovery of the initial state. Irreversibility emerges as a result of these underlying complexities and the tendency of natural processes to move towards greater disorder and randomness.

4.3 INTERNAL AND EXTERNAL IRREVERSIBILITY

In many cases, it is essential to consider the concept of internal and external irreversibility. Figure 4.3 shows two systems that are transferring heat. Both systems contain a pure substance in a saturated state; therefore, the temperature of the system remains constant during heat transfer. In one system, heat is transferred from a source at temperature $T + dT$ (very close to the system temperature), but in the other system, the source temperature is $T + \Delta T$ (much higher than the system temperature). The first process is a reversible process of heat transfer, and the second process is irreversible, but if we pay attention only to the system, in both, the process follows exactly the same path. In this case, it is stated that the first process is a reversible process, but the second process is internally reversible but externally irreversible because the irreversibility is created outside the system.

A process is called reversible when no irreversibility exists within the system and its environment.

4.4 ENTROPY

We are familiar with the Clausius inequality as follows:

$$\oint \frac{\delta Q}{T} \le 0 \tag{4.1}$$

This means that the cyclic integral $\delta Q/T$ is always less than or equal to 0. This inequality is true for all reversible and irreversible cycles. The integral $\delta Q/T$ can be considered the sum of all partial heat transfer values divided by the absolute temperature at the boundary. Indeed, in the context of the Clausius inequality, the principle

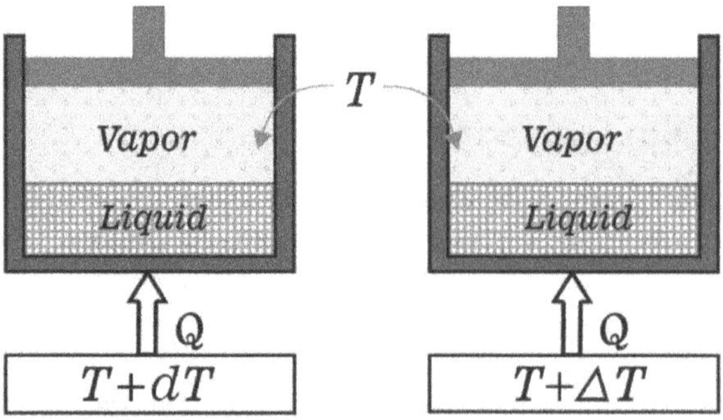

FIGURE 4.3 Internal and external reversible processes.

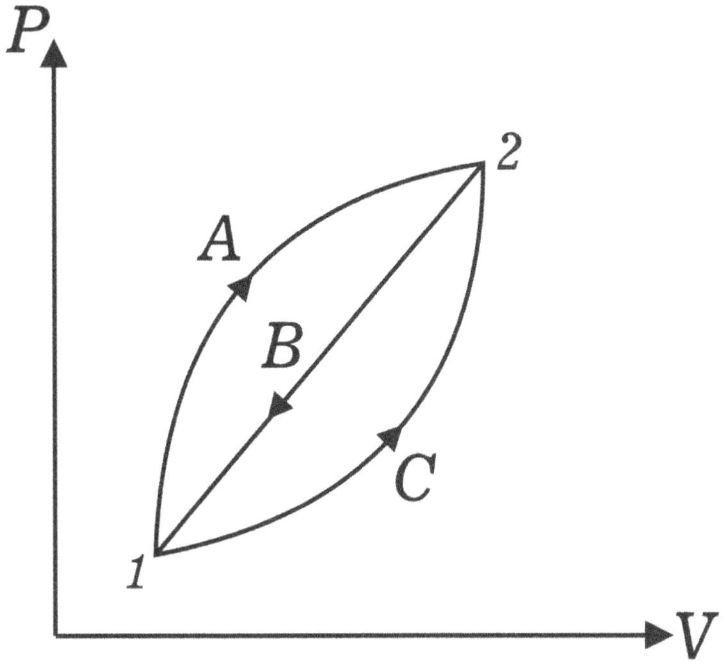

FIGURE 4.4 A–B and C–B cycles consisting of reversible processes.

holds that equality applies to internally reversible cycles, while inequality holds true for irreversible cycles. This fundamental concept highlights the distinction between these two types of thermodynamic processes based on the direction of heat transfer and the efficiency of energy conversion.

Let's imagine a scenario where a system, treated as a closed mass, undergoes a reversible process labeled as "A", transitioning from point 1 to point 2. This system completes a cycle by traversing the reversible route "B". Simultaneously, another distinct reversible cycle emerges when the system follows path "C" and subsequently retraces its steps along path "B".

According to the Clausius inequality for reversible processes, we can write:

$$\oint \frac{\delta Q}{T} = 0 = \int_1^2 \left(\frac{\delta Q}{T}\right)_A + \int_2^1 \left(\frac{\delta Q}{T}\right)_B$$

$$\oint \frac{\delta Q}{T} = 0 = \int_1^2 \left(\frac{\delta Q}{T}\right)_C + \int_2^1 \left(\frac{\delta Q}{T}\right)_B \rightarrow \int_1^2 \left(\frac{\delta Q}{T}\right)_A = \int_1^2 \left(\frac{\delta Q}{T}\right)_C$$

This equation shows that the value of the quantity inside the integral is independent of the reversible path between the initial and final points; therefore, it is considered a system property. This property is called entropy and is defined as follows:

$$dS = \left(\frac{\delta Q}{T}\right)_{\text{in rev}} \left(\text{kJ}/\text{K}\right) \tag{4.2}$$

Entropy is a broad property of the system and is sometimes referred to as total entropy. Entropy in the unit of mass with the sign s is a concentrated property and has a unit of kJ/(kg.K). The change in entropy of a system during a process can be calculated by integrating Equation 4.2 between the initial and final states:

$$\Delta S = S_2 - S_1 = \int_1^2 \left(\frac{\delta Q}{T}\right)_{\text{int rev}} \left(\frac{\text{kJ}}{K}\right) \tag{4.3}$$

In fact, the change in entropy is defined instead of entropy, similar to defining a change in energy instead of energy itself. For integration, it is necessary to have a relationship between Q and T during a process. This relation is often not available, and integration can be done for limited cases. For most cases, we have to trust entropy information tables. The entropy change between two states in a system is the same in reversible and irreversible transformations. From the previous equation, the entropy changes between two states can be obtained from a reversible path. Of course, the change in entropy between the same two states of an irreversible path is the same as the obtained value because entropy is a property and independent of the path. Entropy can be considered a measure of molecular disorder. As a system becomes more disordered, the positions of molecules become less predictable, and entropy increases.

4.4.1 Entropy Generation

Consider a cycle consisting of two processes. One of them is a reversible process, and the other is irreversible (Figure 4.5).

This cycle is an irreversible cycle because part of it is irreversible. Using the Clausius inequality:

$$\oint \frac{\delta Q}{T} \le 0 \rightarrow \int_1^2 \frac{\delta Q}{T} + \int_2^1 \left(\frac{\delta Q}{T}\right)_{\text{int rev}} \le 0$$

$$\int_2^1 \left(\frac{\delta Q}{T}\right)_{\text{int rev}} = S_2 - S_1 \rightarrow \int_1^2 \frac{\delta Q}{T} + S_1 - S_2 \le 0 \rightarrow S_2 - S_1 \ge \int_1^2 \frac{\delta Q}{T}$$

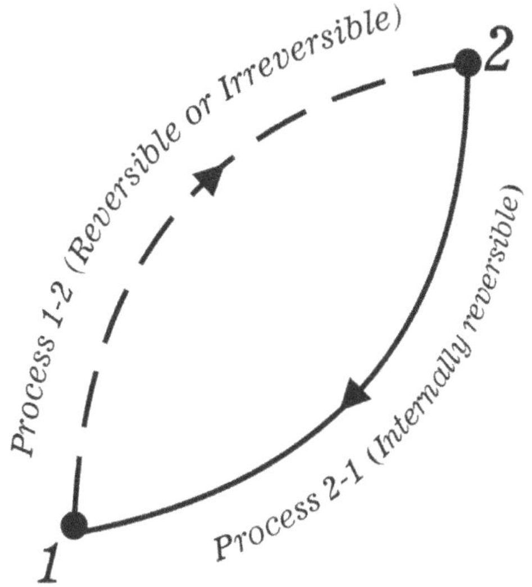

FIGURE 4.5 A cycle consisting of reversible and irreversible processes.

It can also be expressed in differential form as follows:

$$dS \geq \frac{\delta Q}{T}$$

The equality sign is for reversible processes, and the inequality sign indicates that the entropy change for a closed system during an irreversible process is always greater than the entropy transfer. This means that some entropy is generated during an irreversible process, and this product is completely dependent on the presence of irreversible factors. From the equations obtained, it can be concluded that the entropy change in an irreversible process will be greater than in the case of a reversible process with the same heat and temperature transfer. In this case, we can write:

$$\Delta S_{system} = S_2 - S_1 = \int_1^2 \frac{\delta Q}{T} + S_{gen} \tag{4.4}$$

It should be noted that always

$$S_{gen} \geq 0$$

This term is called entropy generation due to irreversibilities.

4.4.2 The Principle of Increase of Entropy

The principle of increase of entropy can be summarized as follows:

$$S_{gen}\begin{cases} >0 & \textit{Irreversible process} \\ =0 & \textit{Reversible process} \\ <0 & \textit{Impossible process} \end{cases}$$

This relationship serves as a criterion to determine whether the process is reversible, irreversible, or impossible. The principle of increase of entropy does not mean that the entropy of the system and its environment cannot decrease. While the entropy change of a system or its environment can be negative during a specific process, it's essential to note that this doesn't apply to the generation of entropy.

4.5 ISENTROPIC PROCESS

The entropy of a fixed mass will not change during an internally reversible process which is referred to as an isentropic (constant entropy) process. Consequently, a process characterized by a consistent entropy value is termed an isentropic process:

$$\Delta S = 0 \qquad \text{or} \qquad S_2 = S_1 \left(\frac{\text{kJ}}{\text{kg.K}}\right)$$

Devices that generate work, like turbines, exhibit higher work output when operating in a reversible manner. Conversely, devices that consume work, such as pumps and compressors, require less work input when they function reversibly. Any type of heat dissipation is undesirable in steady-flow devices, and ideally devices should be adiabatic. In addition, an ideal process should not include irreversible factors because these factors reduce the efficiency of the device; therefore, the desired ideal process is an isentropic process.

4.6 GIBBS EQUATIONS

The differential form of the energy conservation equation for a closed stationary system (constant mass) that includes a simple compressible substance can be expressed as follows for an internally reversible process:

$$\delta Q_{int\ rev} - \delta W_{int\ rev} = dU$$

but

$$\delta Q_{int\,rev} = T\, dS$$

$$\delta W_{int\,rev} = P\, dV$$

therefore:

$$TdS = dU + PdV\left(\text{kJ}\right) \tag{4.5}$$

or

$$Tds = du + Pdv\left(\frac{\text{kJ}}{\text{kg}}\right) \tag{4.6}$$

Note that the only type and exchange of work that may exist for a system under internally reversible process is boundary work.

$$h = u + Pv \rightarrow dh = du + Pdv + vdP$$
$$Tds = du + Pdv$$

As a result

$$Tds = dh - vdP \tag{4.7}$$

It should be noted that the obtained result is true for both reversible and irreversible processes because entropy is a property and a change in a property does not depend on the type of system process.

✐Example 4.1: Calculate the entropy change of water from saturated liquid state to saturated steam at 10°C using Gibbs equations.

✍Solution

The phase change process from saturated liquid to saturated steam takes place at constant temperature and pressure; therefore, the Gibbs equation is reduced as follows:

$$Tds = dh - vdP \rightarrow ds = \frac{dh}{T}$$

The entropy and enthalpy change from saturated liquid to saturated steam can be written as follows:

$$s_g - s_f = \frac{h_g - h_f}{T}$$

For water at a temperature of 10°C, the value of *hg* – *hf* is equal to 2477.2 kJ/kg (Table A.1), which is obtained by inserting the entropy changes in the previous equation:

$$s_g - s_f = \frac{2477.2\frac{\text{kJ}}{\text{kg}}}{(10+273.15)\text{K}} = 8.7488\frac{\text{kJ}}{\text{kg.K}}$$

This value is equal to the value read from the water properties table (Table A.1).

4.7 ENTROPY CHANGE OF LIQUIDS AND SOLIDS

Solids and liquids are incompressible systems, and their volume remains constant during a process. According to the Gibbs equation:

$$ds = \frac{du}{T} = \frac{C\,dT}{T} \tag{4.8}$$

After integrating:

$$s_2 - s_1 = \int_1^2 C(T)\frac{dT}{T} \cong C_{avg}\ \ln\frac{T_2}{T_1} \tag{4.9}$$

Therefore, the entropy of incompressible materials depends only on temperature and is independent of pressure. For isentropic processes of liquids and solids:

$$s_2 - s_1 = C_{ave}\ \ln\frac{T_2}{T_1} = 0 \rightarrow T_2 = T_1 \tag{4.10}$$

The temperature of an incompressible substance remains constant during an isentropic process; therefore, an isentropic process of an incompressible substance is constant.

✐Example 4.2: One kilogram of liquid water is heated from 20°C to 90°C. Determine the entropy change and compare the value obtained with the exact value read from the table.

✍Solution

To calculate the specific heat of water, the following equation can be used, where the temperature is in terms of degrees Celsius:

$$C_p = 4.1762 - \frac{9.0864}{10^5}T + \frac{5.4731}{10^6}T^2 \rightarrow \begin{cases} T = 20°\text{C} \rightarrow C_p = 4.1784\frac{\text{kJ}}{\text{kg.K}} \\ T = 90°\text{C} \rightarrow C_p = 4.2124\frac{\text{kJ}}{\text{kg.K}} \end{cases}$$

$$\rightarrow C_{p,\text{avg}} = \frac{C_{p@20°\text{C}} + C_{p@90°\text{C}}}{2} = \frac{4.1784 + 4.2124}{2} = 4.1954\frac{\text{kJ}}{\text{kg.K}}$$

Calculation of enthalpy change:

$$s_2 - s_1 = C_{avg} \ln\frac{T_2}{T_1} = \left(4.1954\frac{kJ}{kg.K}\right)\left[\ln\frac{(90+273.15)\,K}{(20+273.15)\,K}\right] = 0.8984\frac{kJ}{kg.K}$$

Determining the enthalpy change using the table of thermodynamic properties (Table A.1):

$$s_2 - s_1 \cong s_{f@90°C} - s_{f@20°C} = 1.1929 - 0.2965 = 0.8964\frac{kJ}{kg.K}$$

It can be seen that the obtained values are almost equal to each other; the difference between the value calculated by the formula and the exact value read from the properties table is less than 0.3%.

4.8 ENTROPY CHANGE OF IDEAL GAS

In ideal gases, the following equations are established:

$$Pv = RT,\ du = C_v(T)dT,\ dh = C_P(T)dT$$

By placing the relationships related to ideal gases in the Gibbs equations, the entropy change of the ideal gas is obtained:

$$s_2 - s_1 = \int_1^2 C_v(T)\frac{dT}{T} + R\ln\frac{v_2}{v_1} \tag{4.11}$$

$$s_2 - s_1 = \int_1^2 C_p(T)\frac{dT}{T} - R\ln\frac{P_2}{P_1} \tag{4.12}$$

To calculate these integrals, the relationship between C_v and C_p with temperature must be specified. Even when the functions $C_v(T)$ and $C_p(T)$ are available, it will be difficult to perform the integration operation. These integrations can be done with two methods. The equations for the change in entropy of an ideal gas under the assumption of constant specific heat can be easily obtained by replacing $C_v(T)$ and $C_p(T)$ with their average values, $C_{v,avg}$ and $C_{p,avg}$ and then integrating the equations. Therefore, we obtain:

$$s_2 - s_1 = C_{v,ave}\ln\frac{T_2}{T_1} + R\ln\frac{v_2}{v_1} \tag{4.13}$$

$$s_2 - s_1 = C_{p,ave}\ln\frac{T_2}{T_1} - R\ln\frac{P_2}{P_1} \tag{4.14}$$

Example 4.3: Air compresses in the initial state at 100 kPa and 17°C to the final state at 600 kPa and 57°C. It is desirable to change the entropy of air during the condensation process.

Solution

To calculate the specific heat of air, the following equation can be used, where the temperature is in terms of Kelvin:

$$C_{p,a} = \frac{28.11 + 0.1967\times10^{-2}T + 0.4802\times10^{-5}T^2 - 1.966\times10^{-9}T^3}{28.97}$$

$$\rightarrow \begin{cases} T = 17°\text{C} \rightarrow C_p = 1.002\dfrac{\text{kJ}}{\text{kg.K}} \\ T = 57°\text{C} \rightarrow C_p = 1.008\dfrac{\text{kJ}}{\text{kg.K}} \end{cases}$$

$$\rightarrow C_{p,avg} = \frac{C_{p@17°\text{C}} + C_{p@57°\text{C}}}{2} = \frac{1.002 + 1.008}{2} = 1.005\frac{\text{kJ}}{\text{kg.K}}$$

At the average temperature of 37°C, we determine the value of C_p from the table and put it in the equation:

$$s_2 - s_1 = C_{p,ave}\ln\frac{T_2}{T_1} - R\ln\frac{P_2}{P_2}\left(\frac{\text{kJ}}{\text{kg.K}}\right)$$

$$s_2 - s_1 = \left(1.005\frac{\text{kJ}}{\text{kg.K}}\right)\ln\frac{(57 + 273.15)\text{K}}{(17 + 273.15)\text{K}} - \left(0.287\frac{\text{kJ}}{\text{kg.K}}\right)\ln\frac{600\text{kPa}}{100\text{kPa}} = -0.3844\frac{\text{kJ}}{\text{kg.K}}$$

4.9 ISENTROPIC PROCESSES IN IDEAL GASES

In an isentropic process, the entropy change of the system is zero:

$$s_2 - s_1 = C_{v,avg}\ln\frac{T_2}{T_1} + R\ln\frac{v_2}{v_1} = 0 \rightarrow ln\frac{T_2}{T_1} = -\frac{R}{C_v}\ln\frac{v_2}{v_1}$$

$$\rightarrow \ln\frac{T_2}{T_1} = \ln\left(\frac{v_1}{v_2}\right)^{\frac{R}{C_v}}$$

or

$$\frac{T_2}{T_1} = \left(\frac{v_1}{v_2}\right)^{k-1} \tag{4.15}$$

$$\frac{R}{C_v} = k - 1,\ R = C_p - C_v,\ k = \frac{C_p}{C_v}$$

therefore $R = C_p - C_v$, $k = C_p/C_v$, and $R/C_v = k - 1$.

$$s_2 - s_1 = C_{p,avg} \ln \frac{T_2}{T_1} - R \ln \frac{P_2}{P_1} = 0$$

After simplifying, we can write:

$$\frac{T_2}{T_1} = \left(\frac{P_2}{P_1} \right)^{\frac{k-1}{k}} \tag{4.16}$$

By simplifying Equations 4.15 and 4.16:

$$\frac{P_2}{P_1} = \left(\frac{v_1}{v_2} \right)^{k} \tag{4.17}$$

The ratio of specific heat k usually changes with temperature, and in principle, the average value of k should be used in the given temperature range.

✐Example 4.4: Air is compressed from a temperature of 23°C and a pressure of 90 kPa to a pressure of 900 kPa during a reversible adiabatic process. Calculate the work required for this process per unit mass of air. ($C_{v,avg}$ = 0.73 kJ/(kg.K) and k = 1.393)

✍Solution

The entropy process is constant, so the final temperature can be determined from the isentropic equations of the ideal gas:

$$\frac{T_2}{T_1} = \left(\frac{P_2}{P_1} \right)^{\frac{(k-1)}{k}} \rightarrow T_2 = T_1 \left(\frac{P_2}{P_1} \right)^{\frac{(k-1)}{k}} = (295.15\text{K}) \left(\frac{900}{90} \right)^{\frac{0.393}{1.393}} = 564.9\text{K}$$

Using the energy balance, the required amount of work is calculated:

$$W_i = \Delta U \rightarrow w_i = C_v (T_2 - T_1) = 0.73 \frac{\text{kJ}}{\text{kg.K}} (564.9 - 295.15)\text{K} = 197 \frac{\text{kJ}}{\text{kg}}$$

4.10 MECHANISMS OF ENTROPY TRANSFER

Entropy can be transferred to or from the system by two mechanisms, heat transfer and mass transfer.

4.10.1 Heat Transfer

The process of heat transfer into a system leads to an elevation in its entropy, contributing to a higher degree of molecular disorder. Conversely, heat transfer out of the system leads to a reduction in entropy, corresponding to a decrease in molecular disorder. In fact, heat dissipation is the only way to reduce the entropy of a fixed mass. The ratio of heat transfer Q in a place with absolute temperature T in the same place is called entropy flow or entropy transfer and is expressed as follows:

$$S_{Heat} = \frac{Q}{T}\left(T = Constant\right) \tag{4.18}$$

The Q/T quantity indicates entropy transfer along with heat transfer, and the direction of entropy transfer is the same as the direction of heat transfer. When the temperature T is not constant, the entropy transfer during processes 1–2 can be calculated by integrating as follows:

$$S_{Heat} = \int_1^2 \frac{\delta Q}{T} \cong \sum \frac{Q_b}{T_b} \tag{4.19}$$

where Q_b is the heat transfer from the boundary at temperature T_b at location b. Work is free of entropy, and entropy is not transferred by work.

4.10.2 Mass Flow

Mass encompasses both entropy and energy within a system. The relationship among energy, entropy, and mass follows a proportional pattern, where an increase in mass corresponds to a proportional increase in the system's energy and entropy (for instance, doubling the mass doubles the energy and entropy of the system's constituents). The flow of material into or out of the system facilitates the transfer of both entropy and energy. The rate of entropy and energy transfer into or out of the system is proportional to the flow. Closed systems do not involve mass flow; therefore, no entropy transfer is carried out by the mass. However, when a mass represented by "m" enters or exits a system, it brings along an accompanying entropy denoted as "m.s", where "s" represents the specific entropy.

4.11 ENTROPY BALANCE

Entropy is a non-conservative property, and there is no such thing as the principle of conservation of entropy. The entropy balance can be expressed as follows (Dinçer & Rosen, 2015):

System entropy change = output entropy − entropy generation + input entropy

The mathematical formulation of entropy balance:

$$\sum_i (\text{ms}) + \left(\frac{Q}{T_b}\right)_i + S_{\text{gen}} - \sum_o (\text{ms}) - \left(\frac{Q}{T_b}\right)_o = \Delta S_{\text{system}} \tag{4.20}$$

Entropy balance as a rate:

$$\sum_i (\dot{m}s) + \left(\frac{\dot{Q}}{T_b}\right)_i + \dot{S}_{gen} - \sum_o (\dot{m}s) - \left(\frac{\dot{Q}}{T_b}\right)_o = \frac{dS_{system}}{dt} \tag{4.21}$$

Entropy balance for the steady state:

$$\dot{S}_{gen} = \sum_o (\dot{m}s) + \left(\frac{\dot{Q}}{T_b}\right)_o - \sum_i (\dot{m}s) - \left(\frac{\dot{Q}}{T_b}\right)_i \tag{4.22}$$

Entropy balance for the closed system:

$$S_2 - S_1 = \frac{\dot{Q}}{T_b} + S_{gen} \tag{4.23}$$

✐Example 4.5: Saturated water at a temperature of 100°C inside a rigid and insulated tank is stirred by a stirrer and turns into saturated steam. Calculate the work consumed and entropy generated per unit weight of water in this process.

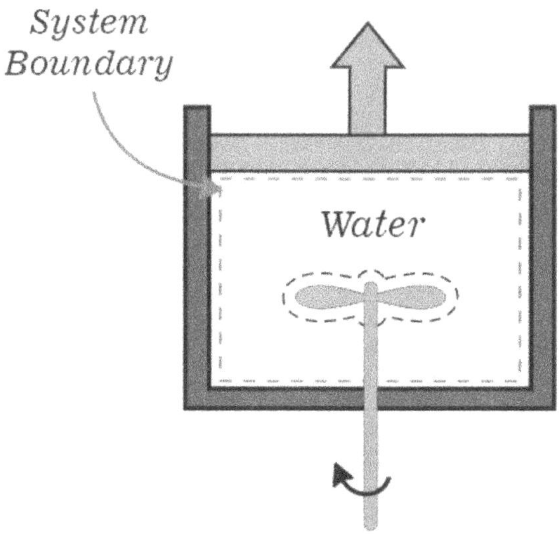

FIGURE 4.6 Diagram of Example 4.5.

✍Solution

By ignoring changes in kinetic and potential energy, the energy balance for the closed system is as follows:

$$W_i = \Delta U \rightarrow m(u_2 - u_1) = W \rightarrow \frac{W}{m} = (u_g - u_f) = u_{fg@10°C} = 2087.0\frac{kJ}{kg}$$

The entropy generated per unit mass of the closed system is also calculated as follows:

$$S_2 - S_1 = S_{gen} \rightarrow \frac{S_{gen}}{m} = s_g - s_f = s_{fg@100°C} = 6.0470\frac{kJ}{kg}.K$$

5 Exergy

It is very desirable to have a property that determines the useful work capability for a certain amount of energy in a specific state. This property is called exergy (availability). The output work is highest when the process is carried out reversibly between two specified states. Therefore, in examining the functionality, all the things that cause irreversibility are ignored. For maximum output of work, the system should be placed in a dead state at the end of the process. A system in a dead state is in thermodynamic equilibrium with its surroundings. In the dead state, the temperature and pressure of the system are equal to the environment (thermal and mechanical equilibrium), have no potential and kinetic energy relative to the surrounding environment, and do not react with the surrounding environment (chemical equilibrium). In the dead state, the system has zero exergy. The properties of a system in a dead state are written with a zero index. The dead state of temperature and pressure are $T_0 = 25°C$, and $P_0 = 1$ atm.

5.1 CYCLE

If a system returns to its initial state at the end of the process, it has gone through a cycle; this means that for a cycle, the initial and final states are the same (Figure 5.1).

5.2 THERMAL ENGINE

Converting work into different forms of energy is easily possible, but converting different types of energy into work is not easy. For example, converting heat into work

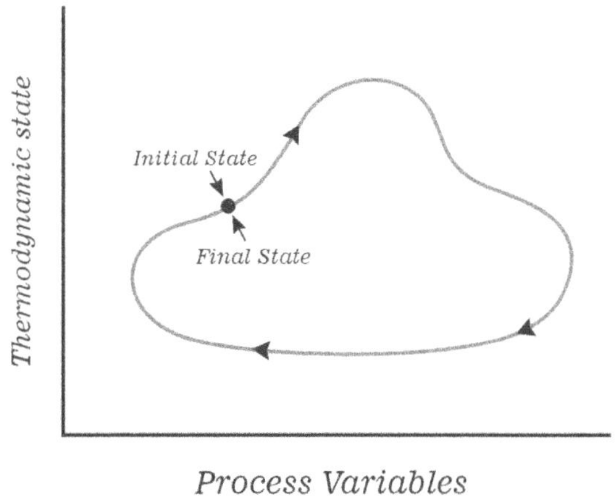

FIGURE 5.1 Thermodynamic cycle.

DOI: 10.1201/9781003424680-5

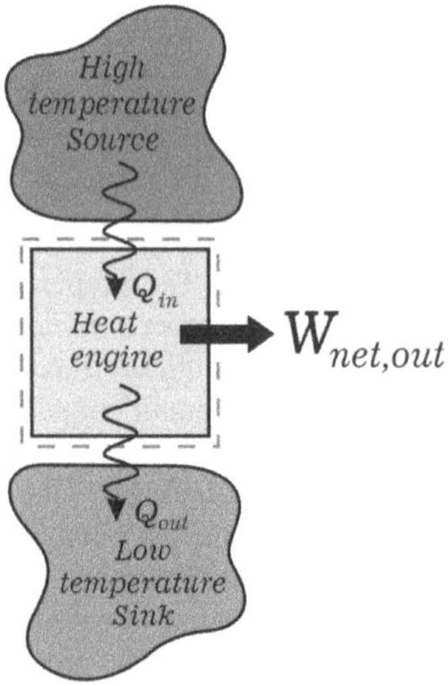

FIGURE 5.2 Schematic of a heat engine.

requires special equipment called heat engines. Heat engines are systems that, under a cyclic process, take heat from the high-temperature source in each cycle, convert some of it into work, and transfer the rest to the low-temperature sink.

5.3 CARNOT CYCLE

Consider a heat engine where every individual process is reversible. In this scenario, when all processes within the cycle are reversible, it means that the entire cycle itself is reversible. This concept leads to what is known as the "reverse Carnot cycle" or the "Carnot refrigeration cycle", where the entire cycle can be executed in the opposite direction as well. This cycle was named the Carnot cycle in honor of Sadi-Carnot. The Carnot cycle can operate in a closed system or a steady flow system. This cycle has maximum efficiency. A Carnot cycle consists of four processes:

1. A reversible isothermal process in which heat is transferred from the high-temperature source to the cycle.
2. A reversible adiabatic process in which the temperature of the cycle working fluid decreases from the temperature of the high-temperature source to the temperature of the low temperature-sink.
3. A reversible isothermal process in which heat is transferred from the cycle to the low-temperature sink.
4. A reversible adiabatic process in which the temperature of the cycle working fluid increases from the temperature of the low-temperature sink source to the temperature of the high-temperature source.

5.3.1 Carnot Heat Engine

An imaginary heat engine that operates in a Carnot cycle is called a Carnot heat engine. For reversible heat engines, the heat transfer ratio can be replaced by the ratio of the absolute temperatures of the two sources. So, the efficiency of a Carnot heat engine or any other reversible heat engine is as follows:

$$\eta_{Carnot} = 1 - \frac{T_L}{T_H} \tag{5.1}$$

This relationship is often referred to as Carnot efficiency. This is the maximum efficiency that a heat engine that operates between two heat sources with temperatures T_H and T_L can have. All irreversible (true) heat engines that operate between these two thermal ranges will be less efficient. Note that T_H and T_L are absolute temperatures. When evaluating the performance of heat engines, efficiencies should not be compared to 100% efficiency but to the efficiency of reversible heat engines operating within the same temperature range because this efficiency is the highest possible theoretically correct limit, not the 100% efficiency value.

It is clear that the efficiency of a Carnot heat engine increases with increasing temperature T_H and decreasing temperature T_L. The thermal efficiency of real heat engines can reach its maximum value by feeding the engine with the highest possible temperature (limited by the material) and removing heat from the heat engine at the lowest possible temperature (limited by the cooling medium, such as a river, sea water, or atmosphere).

✐Example 5.1: A heat engine receives the heat rate 1 MW at the temperature of 550°C and returns the energy to the environment at a temperature of 300 K. The work generation rate of the engine is 450 kW. How much heat enters the environment through the engine and what is the efficiency of the engine? Compare these values with those for a Carnot engine operating between the same two sources.

✍Solution

$$\dot{W}_{net} = \dot{Q}_{in} - \dot{Q}_{out} \rightarrow \dot{Q}_{out} = \dot{Q}_{in} - \dot{W}_{net} = 1000\,\text{kW} - 450\,\text{kW} = 550\,\text{kW}$$

$$\eta_{thermal} = 1 - \frac{\dot{Q}_{out}}{\dot{Q}_{in}} = 1 - \frac{550\ \text{kW}}{1000\ \text{kW}} = 0.45 = 45\%$$

$$\eta_{carnot} = 1 - \frac{T_L}{T_H} = 1 - \frac{300\text{K}}{(550 + 273.15)\text{K}} = 0.635 = 63.5\%$$

✐**Example 5.2: An inventor claims to have developed a power cycle whose net work output is 410 kJ for a heat transfer of 1000 kJ. This cycle receives heat from hot gases at 500 K and returns the output heat to the environment at 300 K. Evaluate this claim.**

✍**Solution**

The efficiency of this power cycle is calculated as follows:

$$\eta_{thermal} = \frac{\dot{W}_{net}}{\dot{Q}_{in}} = \frac{410\text{kJ}}{1000\text{kJ}} = 0.41 = 41\%$$

The maximum efficiency of each cycle between the temperatures 500 K and 300 K is calculated as follows:

$$\eta_{carnot} = 1 - \frac{T_L}{T_H} = 1 - \frac{300\text{K}}{500\text{K}} = 0.4 = 40\%$$

Because the efficiency of the cycle is greater than the maximum efficiency, this claim is not correct.

5.4 EXERGY OF HEAT

The workability of the energy transferred from a heat source at temperature T is the maximum work that can be obtained from the source located in an environment at temperature T_0, and this is equivalent to the work produced by a Carnot heat engine that works between the source and the work:

$$Ex_Q = \left(1 - \frac{T_0}{T_b}\right) Q \tag{5.2}$$

In this regard, Ex_Q is the heat exergy (kJ), T_b is the system boundary temperature, and T_0 is the dead state temperature in degrees Kelvin.

5.5 EXERGY OF WORK

Exergy transfer by work can be expressed as follows:

$$Ex_{work} = W \tag{5.3}$$

5.6 EXERGY OF MASS

In general, mass exergy can be divided into four separate parts: physical exergy (ex_{ph}), chemical exergy (ex_{ch}), kinetic exergy (ex_{kin}), and potential exergy (ex_{pot}). Exergy with different chemical compounds and chemical reactions is not investigated in this book.

5.6.1 Exergy of Kinetic and Potential Energy

Kinetic energy and potential are mechanical energy and therefore have the ability to be fully converted into useful work:

$$ex_{kin} = \frac{V^2}{2}\left(\frac{\text{kJ}}{\text{kg}}\right) \tag{5.4}$$

$$ex_{pot} = gz\left(\frac{\text{kJ}}{\text{kg}}\right) \tag{5.5}$$

5.6.2 Physical Exergy of Constant Mass

Consider a closed system in a specified state that reaches the ambient state through a reversible process (the final temperature and pressure of the system should be T_0 and P_0, respectively). The useful work released during this process is equal to the exergy of the system in the initial state (Figure 5.3).

Consider a cylinder-piston device containing a fluid of mass m, temperature T, and pressure P. The system (mass inside the cylinder) has volume V, internal energy U, and entropy S. Whenever volume is transferred by a differential amount dV and

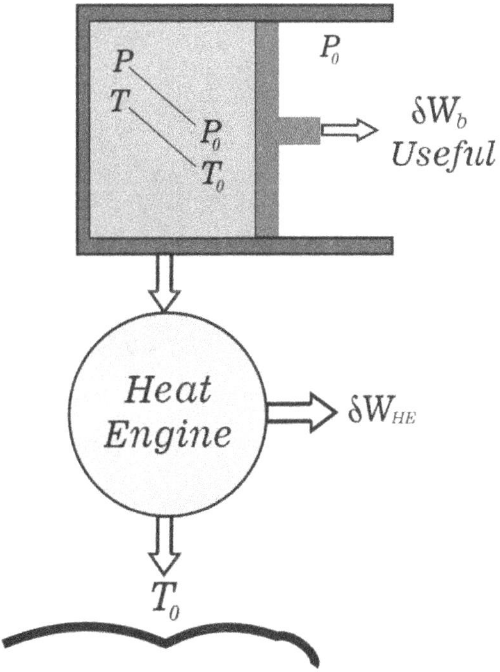

FIGURE 5.3 The exergy of a specific mass in a specific state is equal to the useful work that can be produced from the mass under a reversible process in the ambient state.

heat is transferred by a differential amount δQ, the system is allowed to undergo differential changes of state. Consider the direction of heat and work transfer from the side of the system. According to the energy balance:

$$\delta E_{in} - \delta E_{out} = \delta E_{system}$$

$$-\delta Q - \delta W = dU$$

When the only form of system energy consists of internal energy, the only form of energy transfer of a fixed mass can consist of heat and work. Also, the only work mode of a simple compressible system during a reversible process includes boundary work, which is expressed as $\delta W = P\,dV$. The pressure P in the expression $P\,dV$ is the absolute pressure measured from zero. Any useful work released by a cylinder-piston device is due to a pressure above atmospheric; therefore:

$$\delta W = P\,dV = (P - P_0)dV + P_0\,dV = \delta W_{b,useful} + P_0\,dV$$

A reversible process cannot involve any heat transfer, and therefore any heat transfer between the system at temperature T and the environment at temperature T_0 should occur in a reversible heat engine. Note that for a reversible process, $dS = \delta Q/T$, and the thermal efficiency of a reversible heat engine operating between temperatures T and T_0 is equal to $\eta_{thermal} = 1 - T_0/T$. The work differential produced by the engine due to heat transfer is:

$$\delta W_{HE} = \left(1 - \frac{T_0}{T}\right)\delta Q = \delta Q - \frac{T_0}{T}\delta Q = \delta Q - (-T_0 dS)$$

$$\delta Q = \delta W_{HE} - T_0 dS$$

$$\delta W_{total,useful} = \delta W_{HE} + \delta W_{b,useful} = -dU - P_0 dV + T_0 dS$$

By integrating the given state to the dead state, we will have:

$$W_{total,useful} = Ex_{Ph} = (U - U_0) + P_0(V - V_0) - T_0(S - S_0) \quad (5.6)$$

where U_0, V_0, and S_0 are the internal energy, volume, and enthalpy of the mass in the dead state, respectively. Physical exergy for fixed mass is expressed as follows:

$$ex_{Ph,nonflowing\ fluid} = (u - u_0) + P_0(v - v_0) - T_0(s - s_0) \quad (5.7)$$

5.6.3 Physical Exergy of Mass Flow

The flow work is equal to PV, and the work done against the atmosphere is equal to $P_0\,V$. The exergy related to the flow energy can be expressed as follows:

$$Ex_{flow} = PV - P_0 V = (P - P_0)V \quad (5.8)$$

Based on the mass unit:

$$ex_{flow} = (P - P_0)v \tag{5.9}$$

The exergy of the material flow is obtained by adding the previous flow exergy equation to the exergy equation for a non-flowing mass:

$$\begin{aligned} ex_{flowing\ fluid} &= ex_{nonflowing\ fluid} + ex_{flow} \\ &= (u - u_0) + P_0(v - v_0) - T_0(s - s_0) + (P - P_0)v \\ &= (u + Pv) - (u_0 + P_0 v_0) - T_0(s - s_0) \end{aligned}$$

Therefore, the physical exergy of the mass flow is equal to:

$$ex_{flowing\ fluid} = (h - h_0) - T_0(s - s_0) \tag{5.10}$$

where h_0 is the specific enthalpy of the mass in the dead state.

✐Example 5.3: Calculate the specific exergy of saturated water at a temperature of 120°C, a speed of 30 m/s, and a height of 6 m.

✍Solution

Determine the properties:

Properties of saturated water at 120°C (Table A.1)

$$u = u_{f@120°C} = 503.60\frac{\text{kJ}}{\text{kg}}$$

$$v = v_{f@120°C} = 0.001060\frac{\text{m}^3}{\text{kg}}$$

$$s = s_{f@120°C} = 1.5279\frac{\text{kJ}}{\text{kg}}.\text{K}$$

Properties of water in dead state

Water with a temperature of 25°C and a pressure of 100 kPa is in compressed state; the properties of compressed water can be approximated by the properties of saturated water at the same temperature (Table A.1):

$$u_0 \cong u_{f@25°C} = 140.83\frac{\text{kJ}}{\text{kg}}$$

$$v_0 \cong v_{f@25°C} = 0.001003\frac{\text{m}^3}{\text{kg}}$$

$$s_0 \cong s_{f@25°C} = 0.3672\frac{\text{kJ}}{\text{kg}}.\text{K}$$

The specific exergy of a fixed mass can be calculated as follows:

$$ex = (u - u_0) + P_0(v - v_0) - T_0(s - s_0) + \frac{V^2}{2} + gz$$

$$= \left[(503.60 - 140.83)\,{}^{\text{kJ}}\!/_{\text{kg}}\right] + \left[(100\text{kPa})(0.001060 - 0.001003)\,{}^{\text{m}^3}\!/_{\text{kg}}\right]$$

$$-\left[(25 + 273.15)\text{K}(1.5279 - 0.3672)\,{}^{\text{kJ}}\!/_{\text{kg}}.\text{K}\right] + \left[\frac{\left(30\,{}^{\text{m}}\!/_{\text{s}}\right)^2}{2} + \left(9.81\,{}^{\text{m}}\!/_{\text{s}^2}\right)(6\text{m})\right]\left|\frac{1\,{}^{\text{kJ}}\!/_{\text{kg}}}{1000\,{}^{\text{m}^2}\!/_{\text{s}^2}}\right|$$

$$17.22 = {}^{\text{kJ}}\!/_{\text{kg}}$$

✐**Example 5.4: In a power plant condenser, a stream of steam with a quality of 0.85 and a pressure of 5 kPa is converted into saturated water at the same pressure. Calculate its specific physical exergy change.**

✍**Solution**

The physical exergy change of a flow is as follows:

$$\left.\begin{aligned} ex_1 &= (h_1 - h_0) - T_0(s_1 - s_0) \\ ex_2 &= (h_2 - h_0) - T_0(s_2 - s_0) \end{aligned}\right\} \rightarrow ex_2 - ex_1$$
$$= (h_2 - h_1) - T_0(s_2 - s_1)$$

Determining enthalpy and entropy of flows (from Table A.2):

$$s_1 = s_{f@5\text{kPa}} + xs_{fg@5\text{kPa}} = 0.4762 + 0.85(7.9176) = 7.20616\frac{\text{kJ}}{\text{kg.K}}$$

$$h_1 = h_{f@5\text{kPa}} + xh_{fg@5\text{kPa}} = 137.75 + 0.85(2423.00) = 2197.3\frac{\text{kJ}}{\text{kg}}$$

$$s_2 = s_{f@5\text{kPa}} = 0.4762\frac{\text{kJ}}{\text{kg.K}},\ h_2 = h_{f@5\text{kPa}} = 137.75\frac{\text{kJ}}{\text{kg}}$$

Changing flow exergy:

$$ex_2 - ex_1 = (137.75 - 2197.3)\frac{\text{kJ}}{\text{kg}}$$
$$-298.15K(0.4762 - 7.20616)\frac{\text{kJ}}{\text{kg.K}} = -53.0124\frac{\text{kJ}}{\text{kg}}$$

5.7 EXERGY OF IDEAL GAS

The specific physical exergy of an ideal gas can be calculated as follows:

$$ex = C_{p,ave}(T - T_0) - T_0\left(C_{p,ave}\ln\frac{T}{T_0} - R\ln\frac{P}{P_0}\right) \tag{5.11}$$

5.8 THERMAL AND MECHANICAL EXERGY OF MASS FLOW

Physical exergy of mass flow can be divided into thermal $\left(ex_{ph}{}^{T}\right)$ and mechanical $\left(ex_{ph}{}^{P}\right)$ components:

$$ex_{ph} = ex_{ph}{}^{P} + ex_{ph}{}^{T} \tag{5.12}$$

$$ex_{ph}{}^{P} = \left[h(T,P) - h(T_0,P)\right] - T_0\left[s(T,P) - s(T_0,P)\right] \tag{5.13}$$

$$ex_{ph}{}^{T} = \left[h(T_0,P) - h(T_0,P_0)\right] - T_0\left[s(T_0,P) - s(T_0,P_0)\right] \tag{5.14}$$

5.9 COLD AND HOT EXERGY

The thermal exergy of a mass flow with a temperature lower than the dead state is called "cold exergy", and a flow with a temperature higher than the dead state is called "hot exergy".

5.10 DESTRUCTION OF EXERGY

Irreversible processes, characterized by factors like friction, mixing, chemical reactions, heat transfer across limited temperature gradients, sudden expansions, and non-equilibrium compressions and expansions, invariably result in the generation of entropy. Notably, any phenomenon that leads to entropy generation also corresponds to the destruction of exergy. As is known, the destroyed exergy is proportional to the entropy generation and is expressed as follows:

$$Ex_{des} = T_0 S_{gen} \tag{5.15}$$

5.11 EXERGY BALANCE

Exergy balance is expressed as follows:

Accumulation of exergy = exergy destroyed – output exergy – input exergy

The mathematical formulation of exergy balance is as follows:

$$\sum_{in} \mathrm{m}(\mathrm{ex}) + \left(\mathrm{Ex_Q}\right)_{in} + \mathrm{W_i} - \sum_{out} \mathrm{m}(\mathrm{ex}) - \left(\mathrm{Ex_Q}\right)_{out} - \mathrm{W_{out}} - \mathrm{Ex_{des}} = \Delta\mathrm{Ex_{system}} \tag{5.16}$$

This equation is stated in the rate form as follows:

$$\sum_{in}\dot{m}(ex)+\left(\dot{E}x_Q\right)_{in}+\dot{W}_{in}-\sum_{out}\dot{m}(ex)-\left(\dot{E}x_Q\right)_{out}-\dot{W}_{out}-\dot{E}x_{des}={dEx_{system}}/{dt} \quad (5.17)$$

For a steady flow process:

$$\sum_{in}\dot{m}(ex)+\left(\dot{E}x_Q\right)_{in}+\dot{W}_{in}=\sum_{out}\dot{m}(ex)+\left(\dot{E}x_Q\right)_{out}+\dot{W}_{out}+\dot{E}x_{des} \quad (5.18)$$

Exergy balance for a closed system:

$$\left(Ex_Q\right)_{in}+W_{in}-\left(Ex_Q\right)_{out}-W_{out}-Ex_{des}=Ex_2-Ex_1 \quad (5.19)$$

Example 5.5: Consider a cylinder-piston device as shown in the figure, which initially contains air at a pressure of 400 kPa. By a shaft, 50 kJ of work enters the system for every kilogram of air. Heat enters the device from a source with a temperature of 100°C, and the temperature of the gas inside the cylinder-piston remains constant at 300 K during the process. The air volume triples during this process. Calculate:

a. boundary work
b. heat entering the system
c. entropy generated
d. exergy destruction

(Dinçer, 2018).

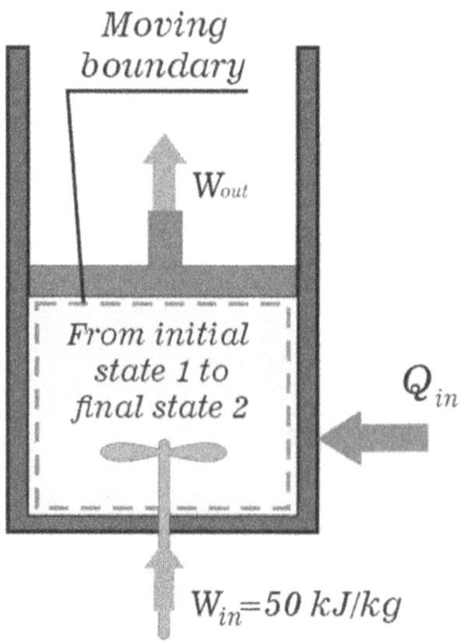

FIGURE 5.4 Diagram of Example 5.5.

✍Solution

Assumptions:

1. Air can be assumed to be an ideal gas.
2. Kinetic and potential energy changes are ignored.
3. The temperature of the dead state is considered to be 25°C.

During the process the volume triples, so we can write:

$$\frac{v_2}{v_1} = 3$$

In this process, the temperature is constant; as a result, $PV = mRT = Const$. Considering that the mass is constant during the process, the boundary work can be calculated as follows:

$$W_{12} = \int_1^2 PdV = \int_1^2 PV\frac{dV}{V} = \int_1^2 Const\frac{dV}{V} = mRT\ln\frac{V_2}{V_1}$$

The work per mass unit is as follows:

$$w_b = RT\ln\left(\frac{v_2}{v_1}\right) = 0.287\frac{kJ}{kg.K}(300K)\ln(3) = 91.62\frac{kJ}{kg}$$

By writing the energy balance, the heat input to the system is calculated:

$$q_{in} + w_{in} - w_b = \Delta u$$

$$\rightarrow q_{in} + (w_{in} - w_b) = u_2 - u_1 \rightarrow q_{in} = u_2 - u_1 + w_b - w_{in}$$

The internal energy of an ideal gas is only a function of its temperature, so given that the process takes place at constant temperature, the internal energy changes are zero:

$$q_{in} = w_b - w_{in} = 91.62 - 50 = 41.62\frac{kJ}{kg}$$

Entropy generation:

$$s_1 + \frac{q_{in}}{T_b} + s_{gen} = s_2 \rightarrow s_{gen} = (s_2 - s_1) - \frac{q_{in}}{T_b}$$

$$= \left(C_v\ln\frac{T_2}{T_1} + R\ln\frac{v_2}{v_1}\right) - \frac{q_{in}}{T_b}$$

$$= [0.287\ln(3)] - \left(\frac{41.62}{373.15}\right) = 0.2037\frac{kJ}{kg.K}$$

Exergy destruction:

$$ex_{des} = T_0 s_{gen} = (298.15K)\left(0.2037\frac{kJ}{kg.K}\right) = 60.733\frac{kJ}{kg}$$

6 Thermodynamic Analyses in Unit Operations

6.1 THERMODYNAMIC ANALYSES

Thermodynamics provides the possibility of investigating the performance of systems to convert energy from one form to another. One of the most powerful aspects of thermodynamics is its black-box approach to analyzing a system. In this approach, having information about the interactions within the system is not required, but knowing what and how much crosses the boundaries of the system is enough to analyze the system.

6.2 ENERGY ANALYSIS

A conventional thermodynamic analysis is based on the first law of thermodynamics, which states the principle of conservation of energy. The energy analysis of a system includes the simultaneous solution of the balance of substance, mass, and energy for a system and provides a report of the energy input to the system and output from the system. Output energies can be divided into two categories of products and losses. Often, energy analysis is used to determine and reduce heat losses and sometimes to increase heat loss recovery (Shukla & Kumar, 2017). However, energy analysis does not provide any information about the degradation of energy quality during the process and does not quantify the quality of energy flows; therefore, it cannot check whether the system performance is close to the ideal state. In addition, the thermodynamic inefficiencies that occur inside the system (factors that cause the performance to deviate from the ideal state) are not identified and evaluated by energy analysis (Kanoğlu et al., 2012). Therefore, the results of energy analysis are misleading and can show performance of a system different from its actual state.

6.3 EXERGY ANALYSIS

Thermodynamic inefficiencies increase the energy required by the processes in a system, followed by increased environmental effects and operating costs of the process. Exergy analysis is a thermodynamic analysis method based on the combination of the first and second laws of thermodynamics, which provide an alternative and clear tool to evaluate and compare processes and systems reasonably and logically. Exergy analysis provides results on a true scale that indicate the closeness of a real process to the ideal state and more clearly specify the causes and locations of thermodynamic losses than energy analysis. Therefore, exergy analysis overcomes the

DOI: 10.1201/9781003424680-6

shortcomings of energy analysis and can help in improving and optimizing energy systems.

6.3.1 Definition of Fuel and Product for the System

In the context of a system, the term "product" encompasses the combined value of output exergy and the exergy increment between the input and output of an exergy flow. Furthermore, "fuel" is characterized by the sum of input exergy and the reduction in exergy between the input and output of an exergy flow (Lazzaretto & Tsatsaronis, 2006). By defining the fuel and product for the system, the exergy balance for the steady flow process can be written as follows:

$$\dot{E}x_F = \dot{E}x_P + \dot{E}x_{\text{des}} + \dot{E}x_L \tag{6.1}$$

Ex_F is the exergy rate of the product, Ex_P is the exergy rate of the fuel, and Ex_L is the exergy rate of waste in kW.

6.3.2 Efficiency

In any energy conversion process, analysis and evaluation of its performance via correct efficiency is crucial. Thermodynamically, any energy conversion system is usually described by a suitable definition of system efficiency. This may be done via energy efficiency or exergy. Efficiency is often defined as the ratio of energy values and is used to evaluate different systems. However, energy efficiency is misleading because it does not show the closeness of the system performance to the ideal state. Exergy efficiency is defined based on the second law of thermodynamics, which is also called second-law efficiency or exergetic efficiency:

$$\eta_{\text{ex}} = \frac{\textit{output exergy}}{\textit{input exergy}} = \frac{\dot{E}x_{\textit{output}}}{\dot{E}x_{\textit{inlet}}} = 1 - \frac{\dot{E}x_{\text{des}}}{\dot{E}x_{\text{inlet}}} \tag{6.2}$$

By defining the fuel and product for a system, the exergy efficiency of the system can be defined as the ratio of product exergy to fuel exergy:

$$\eta_{\text{ex}} = \frac{\dot{E}x_P}{\dot{E}x_F} \tag{6.3}$$

6.4 SYSTEM LOSSES

In a subsystem, loss is heat loss from surfaces, but for a general system, loss is any energy flow (mass flow or heat) that leaves the system without any use. Therefore, in the analysis of the processing lines, losses of the subsystems will be different from the losses of the overall system.

6.5 APPLICATION OF THERMODYNAMICS IN THE FOOD INDUSTRY

Thermodynamic principles can be used to design, analyze, evaluate, and improve the performance of systems. One of the most important obstacles to the development and increase of productivity of conversion industries is the high intensity of energy consumption and environmental effects of these industries. The food industry is one of the largest industrial sectors in the world and therefore one of the largest consumers of energy. In developing the management model of agricultural product processing systems, it is necessary to consider various factors, such as optimal allocation of used resources, preservation of non-renewable resources, and reduction of environmental effects. The energy crisis, reduction of fossil fuel reserves, and price fluctuations are all considered motivations for research in the field of energy optimization in the world. For these reasons, any reduction in energy consumption in industrial lines can reduce the energy burden of production and make production more economically efficient by reducing the share of energy in the final product.

Reducing energy consumption is vital not only from an economic point of view but also from the point of view of protecting the environment. Optimum use of energy and the use of the most efficient methods imply the recognition of the factors that waste energy, as well as the negative effects on life and the environment caused by improper use of energy. The correct use of energy will lead to the continuation of life and sustainable development of society, as well as leading to energy storage for everyone and future generations and will be an obstacle to the production and spread of environmental pollution caused by incorrect energy consumption.

The goal of thermodynamic studies in the food industry is to have a scientific and accurate image of thermodynamic inefficiencies. Energy analysis alone is not able to acquire such an image. Despite specifying the energy balance and energy loss centers, energy analysis is unable to increase real and logical productivity and thus reduce energy consumption and environmental effects. Therefore, it is imperative to use new and efficient methods to monitor the situation, optimize energy consumption, reduce production costs, and reduce pollutants. By studying exergy along with energy, a detailed image of thermodynamic inefficiencies and the real efficiency of systems and subsystems in the food industry is extracted. With exergy analysis, it is possible to reveal improvement potentials and calculate the actual efficiency of production line components, identify the places of occurrence of the most losses, and find solutions to reduce them (Aghbashlo et al., 2016).

For these reasons, the technique of exergy analysis from both engineering and management angles is the best method for issues related to energy consumption optimization, and the implementation and institutionalization of its protocols is an effective step in energy consumption optimization along with economic and environmental considerations.

6.6 THERMOPHYSICAL PROPERTIES OF MATERIALS

Thermal properties play a crucial role in process engineering analysis and design. They are essential for tasks like energy balance calculations and determining the duration of thermal processes such as refrigeration, freezing, heating, and drying. Without a comprehensive understanding of thermal properties, these critical aspects of analysis and design would be impractical to achieve.

6.6.1 Specific Heat

Since the thermal properties of food are highly dependent on the chemical composition and temperature, the most suitable option to predict the thermal properties is to use relationships that include the effect of compounds and temperature. The main components of food are water, protein, fat, carbohydrates, and ash. The specific heat of agricultural products can be calculated as a function of temperature in degrees Celsius as follows (Yildirim & Genc, 2017):

$$\left(C_P\right)_{product} = \sum_i X_i C_{p.i} \tag{6.4}$$

X_i is the mass fraction of component i, and $C_{p.i}$ is the specific heat of component i (kJ/kg.K).

6.6.2 Enthalpy

The enthalpy of agricultural and food products can be obtained from the following equation:

$$h - h_0 = C_{P,\text{ave}}\left(T - T_0\right) + v\left(P - P_0\right) \tag{6.5}$$

In this equation, T_0 and P_0 are the reference temperature and pressure, respectively, and h_0 is the enthalpy in the reference state. At the basic temperature of 0°C and the basic pressure of 101.33 kPa, the enthalpy can be considered zero and the previous equation can be written as follows:

$$h = C_{P,\text{ave}}T + v\left(P - P_0\right) \tag{6.6}$$

6.6.3 Specific Volume

The specific volume of agricultural products and food can be calculated as follows (Yildirim & Genc, 2017):

$$v = \sum_i \frac{X_i}{\rho_i} \tag{6.7}$$

X_i is the mass fraction of component i, and ρ_i is the density of component i (kg/m^3). The specific heat and volume of the components of agricultural products and food are presented in Table 6.1 as a function of temperature.

TABLE 6.1
Specific Heat and Density of Components of Agricultural Products and Foods (Yildirim & Genc, 2017)

Component	Specific Heat Equation	Density Equation
Protein	$C_p = 2.0082 + \frac{1.2089}{10^3}T - \frac{1.3129}{10^6}T^2$	$\rho = 1.3299 \times 10^3 - \frac{5.184}{10}T$
Fat	$C_p = 1.9842 + \frac{1.4733}{10^3}T - \frac{4.8008}{10^6}T^2$	$\rho = 9.2559 \times 10^2 - \frac{4.1757}{10}T$
carbohydrate	$C_p = 1.5488 + \frac{1.9625}{10^3}T - \frac{5.9399}{10^6}T^2$	$\rho = 1.5991 \times 10^3 - \frac{3.1046}{10}T$
Ash	$C_p = 1.0926 + \frac{1.8896}{10^3}T - \frac{3.6817}{10^6}T^2$	$\rho = 2.4238 \times 10^3 - \frac{2.8063}{10}T$
Water	$C_p = 4.1762 - \frac{9.0864}{10^3}T + \frac{5.4731}{10^6}T^2$	$\rho = 9.9718 \times 10^2 + \frac{3.1439}{10^3}T - \frac{3.7574}{10^3}T^2$

6.6.4 Physical Exergy

Physical exergy specific for mass flow and fixed mass for food and agricultural products can be calculated as follows:

$$ex = C_{p,avg}\left(T - T_0 - T_0 \ln\left(\frac{T}{T_0}\right)\right) + v\left(P - P_0\right) \tag{6.8}$$

$$ex = C_{p,avg}\left(T - T_0 - T_0 \ln\left(\frac{T}{T_0}\right)\right) + P_0\left(v - v_0\right) \tag{6.9}$$

6.7 ENERGY ANALYSIS IN UNIT OPERATION

In this book, exergy analysis of processes involving chemical reactions is not investigated. It should be noted that for mixing and separation processes, such as evaporation and drying, the changes in chemical exergy are very small compared to physical exergy and can be ignored (Bühler et al., 2018).

6.7.1 Heat Exchangers

Heat exchangers are equipment designed for the efficient transfer of heat between two fluids of varying temperatures, all while ensuring there's no mixing between them. Importantly, heat exchangers do not facilitate the exchange of work, and any effects of kinetic and potential energies are generally negligible for each fluid flow. Typically, heat exchangers find applications in cooling a hot fluid, heating a fluid with lower temperature, or performing both functions simultaneously. These versatile devices are widely employed in various processes, including blanching, pasteurization, sterilization, as well as evaporation, drying, cooling, and freezing operations. For a visual representation of the types of heat exchangers, refer to Figure 6.1, which illustrates their classification.

6.7.1.1 Double-Tube Heat Exchanger

The simplest type of tube exchanger is the double-tube heat exchanger, which consists of two concentric tubes. The food material in the inner tube and the heating or cooling fluid in the outer pipe flow asymmetrically. The problem with double-tube heat exchangers is the limited heat transfer surface.

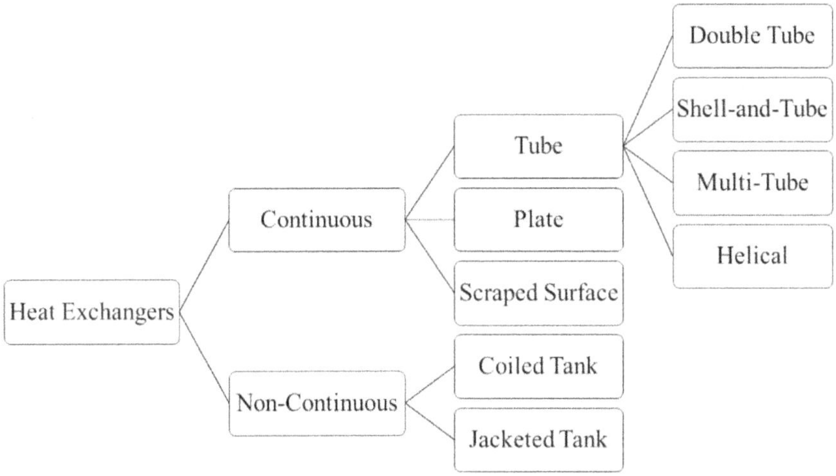

FIGURE 6.1 Types of heat exchangers.

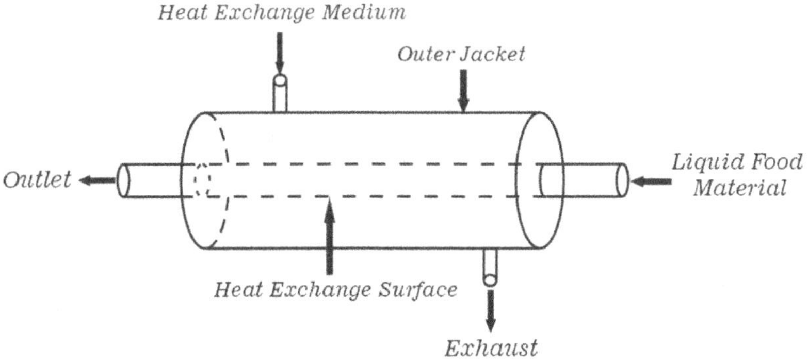

FIGURE 6.2 Double-tube heat exchanger.

6.7.1.2 Multi-Tube Heat Exchanger

This exchanger includes concentric tubes with different diameters, where the food and the heating or cooling medium are placed in these pipes alternately, and they flow unequally with each other. The existence of spiral congresses (spiral protrusions) on the inner and outer surfaces of the pipes creates turbulent conditions at lower speeds and increases the heat transfer rate.

6.7.1.3 Shell-and-Tube Heat Exchanger

Shell-and-tube heat exchangers consist of a series of parallel tubes inside a shell. The input feed to this exchanger is distributed inside these tubes. In heating applications, the cold flow is inside the tubes, and the heating medium is outside the tubes (inside the shell). These exchangers provide a larger heat transfer surface by placing a large number of tubes with a small diameter inside the shell. Barriers (baffles) are built in the shell, which cause the fluid to flow on the tubes and create turbulent conditions. To increase the circulation time of the fluid inside the heat exchanger, the desired fluid can be passed through the tubes twice or more. Shell-and-tube heat exchangers are cheaper than other types of food heat exchangers (especially when a large heat load is to be transferred). These exchangers can be used at higher pressures and temperatures (such as steam with a pressure of 6 bar), but plate heat exchangers do not have this possibility. However, the frequent passage of the fluid through the shell-tube heat exchanger causes a high pressure drop and also makes it difficult to clean and disinfect the exchanger. For this reason, these exchangers are not widely used in the food industry.

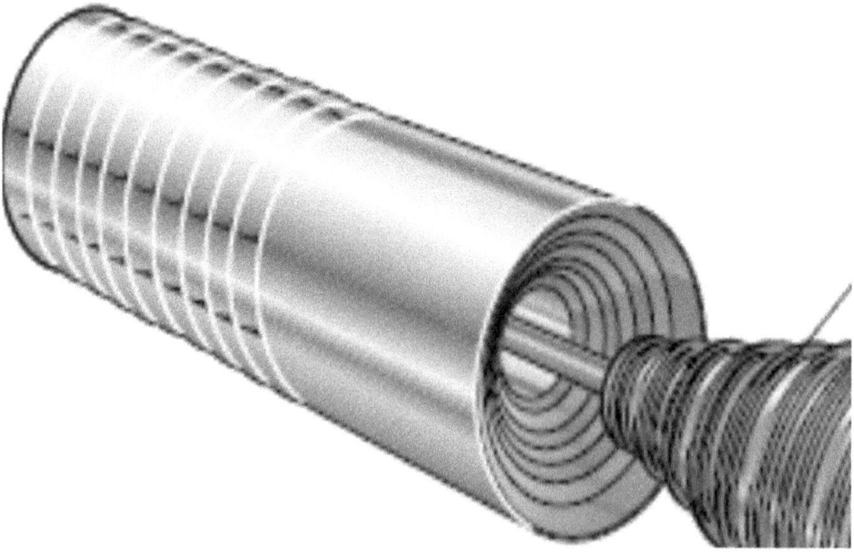

FIGURE 6.3 Multi-tube heat exchanger.

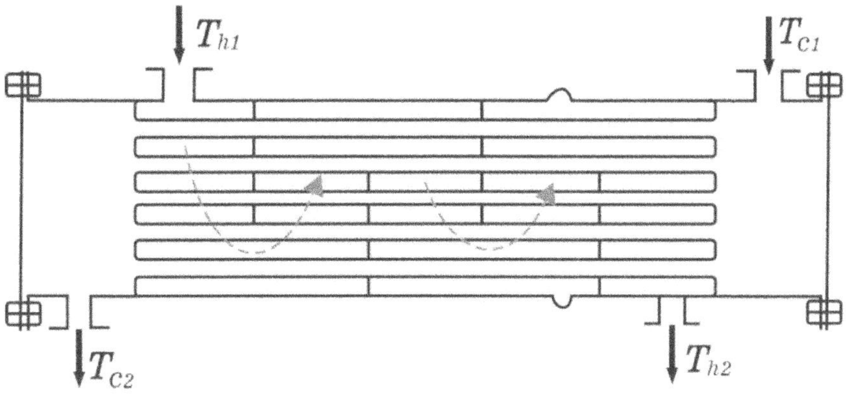

FIGURE 6.4 Shell-and-tube heat exchanger.

6.7.1.4 Helical Heat Exchanger

In this exchanger, the food material flows inside the coil (small-diameter spiral tube), and the heating medium flows in its outer jacket. Due to the high turbulence of the product, a high heat transfer coefficient is obtained in helical heat exchangers; as a result, the intensity of heat transfer in this exchanger is higher than in other exchangers, and these exchangers are used for rapid heating of liquid food.

6.7.1.5 Plate Heat Exchanger

The plate heat exchanger consists of thin vertical plates at short distances in a metal frame, and the food and heating or cooling material are pumped alternately and asymmetrically through these plates. There are holes in these plates, and when the plates are placed adjacent to each other, four ducts are formed by placing the holes together. Through one duct, the cold flow enters the space between the plates, and through another duct, it collects and exits from the space between the plates. The hot flow enters and exits through the other two ducts. In the places where the plates are connected at their edges to each other and around the ducts made of holes, washers prevent the leakage of liquids and the mixing of hot and cold flow. To improve heat transfer by creating more turbulence, the surface of the plates is grooved. Plate heat exchangers are widely used in the food industry because they have high thermal efficiency, and, in addition to being small and compact, they have a suitable design from the point of view of hygiene and washing. This type of exchanger is suitable for liquids with low viscosity; because the distance between the plates is small, it will not be suitable for viscous liquids or granular materials; such materials cause a high pressure drop and also increase the possibility of the flow path being blocked.

6.7.1.6 Scraped-Surface Heat Exchanger

A scraped-surface heat exchanger comprises two concentric cylinders, housing a rotating axis (rotor) with shaving blades inside the inner cylinder. These blades ensure even heat exchange and prevent deposits by continuously shaving the inner

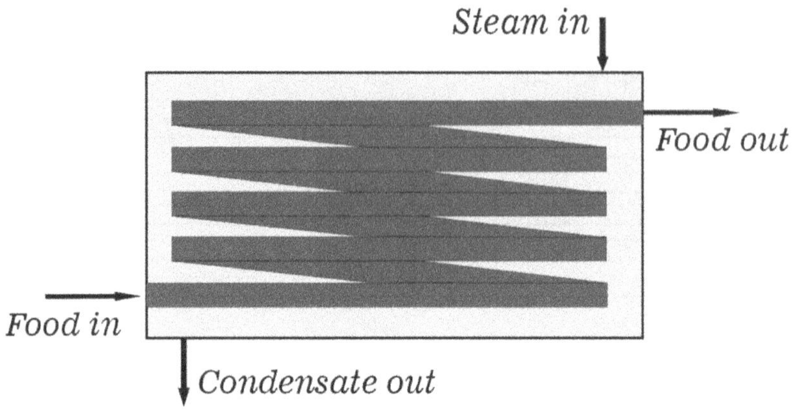

FIGURE 6.5 Helical heat exchanger.

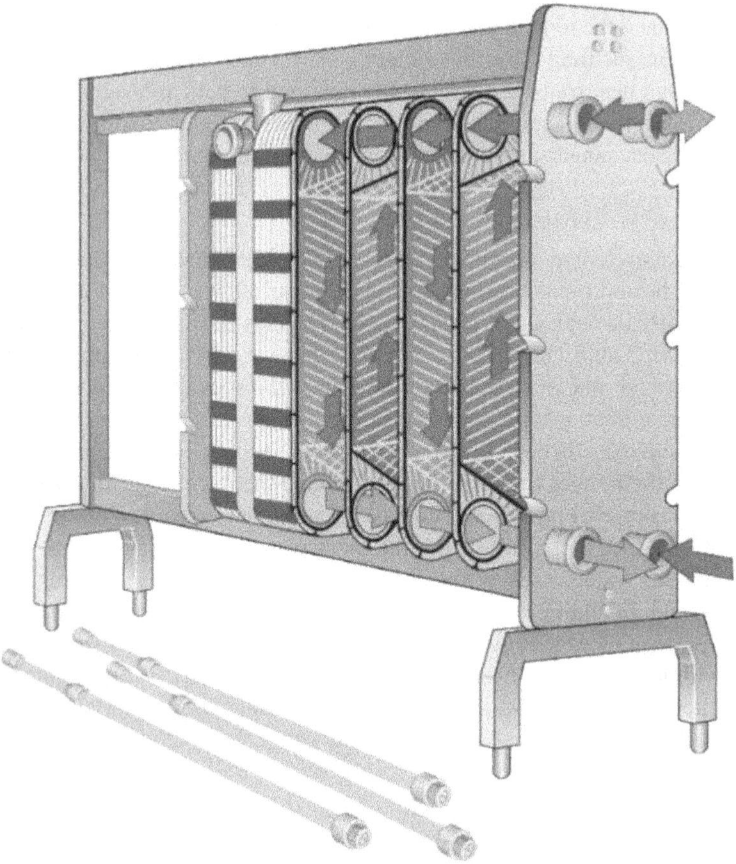

FIGURE 6.6 Plate heat exchanger.

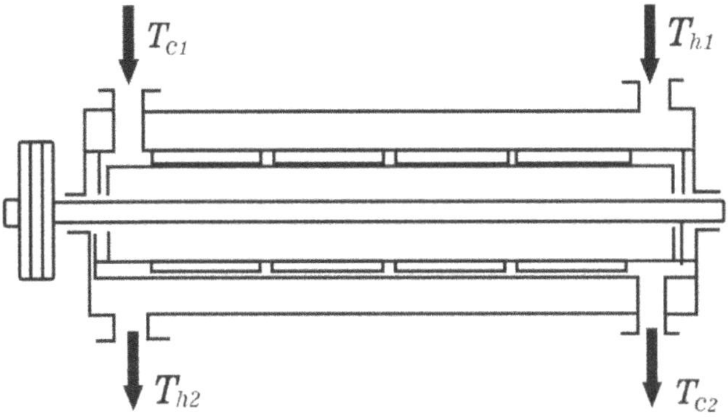

FIGURE 6.7 Scraped-surface heat exchanger.

cylinder's surface, mixing it into the product mass. This design suits heating or cooling high-viscosity and crystallizing liquid foods like ice cream and margarine. The sharpening mechanism doubles as a pump for product transfer. Due to limited heat transfer, this exchanger is used in parallel (to enhance material heating capacity) or in series (to extend residence time, elevate final temperature, and improve heat penetration into suspended particles).

6.7.1.7 Batch Heat Exchangers

In this type of coiled tank, the food inside the tank is heated or cooled by coils (small-diameter spiral pipes) in which the heating or cooling medium flows. A jacketed tank heat exchanger consists of a tank containing food that is heated or cooled by the heating and cooling medium flowing in the outer wall of the tank (jacket); an agitator is used inside these tanks for uniformity and acceleration of heat transfer in the jacketed tank. The ratio of the heat transfer surface to the food volume in the coiled tank exchanger is higher than in the jacketed tank. These heat exchangers are used for heating (or cooling) liquid foods with high viscosity.

6.7.1.8 Applications of Heat Exchangers in the Process Industry

Exchangers used for food sterilization include shell-tube exchangers, helical tube exchangers, concentric tube exchangers, and plate exchangers. Helical exchangers are a more favorable choice for preserving food quality through swift heating. In contrast, shell-tube and concentric tube exchangers take longer to heat, making them less suitable. Additionally, plate exchangers pose concerns due to their washers being sensitive to high temperatures, potentially jeopardizing their aseptic condition.

For the pasteurization of foods with low viscosity that do not have suspended particles (such as milk, cream, and clear fruit juice), the plate exchanger is used; for the pasteurization of foods with low viscosity that also have fine fibrous suspended particles (such as fruit juices with pulp and tomato paste), the shell-tube exchanger is used; and for the pasteurization of foods with high viscosity (such as fruit purees), the scraped surface exchanger is used. Blanching is applied in

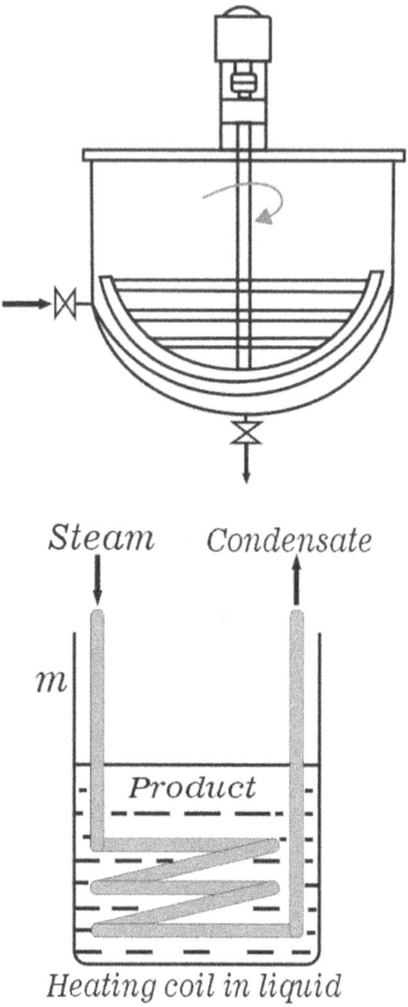

FIGURE 6.8 Batch heat exchangers. A) Jacketed tank. B) Coiled tank.

the early stages of the production line to inactivate enzymes in food, especially vegetable juice (such as tomato pulp) and some fruit juices. Since all of them are pulped in the early stages, a tubular shell heat exchanger is used. For food processing, their temperature should be increased to the appropriate level and then kept at that appropriate temperature during the application of the desired process to increase the intensity of mass transfer or the efficiency of component separation. For example, the temperature of milk before entering a creamer should be brought to 62°C–64°C by the plate exchanger to reduce energy consumption and increase the efficiency of the operation. Another example is that before each oil refining and modification operation (degumming, neutralization, decolorization, and hydrogenation), the oil temperature is increased by the shell-and-tube exchanger to the appropriate temperature to increase the intensity of the operation and reduce the operation time. In the sugar syrup purification processes, the use of shell-and-tube exchangers for this purpose is often observed. Jacketed or coiled tanks are used to keep the temperature at the right level at the same time as the implementation of operation and during the operation (such as sugar syrup and oil refining processes). Cooling of the pasteurized and sterilized product is done to prevent the spoilage of the product due to the growth of the remaining microbes and also to prevent the loss of the nutritional and sensory value of the food. For this purpose, depending on the physical characteristics of the product (viscosity and the presence or absence of suspended particles), plate or tube (shell-and-tube or concentric) exchangers or scraped surfaces are used. For the crystallization of fat or water in food, cooling must be done, and for this purpose, a scraped-surface exchanger is used, such as for the crystallization of fat in the production of margarine or the crystallization of water in the production of ice cream.

6.7.1.9 Balance Equations

Mass balance:

$$\begin{aligned} \dot{m}_1 &= \dot{m}_2 \\ \dot{m}_3 &= \dot{m}_4 \end{aligned} \tag{6.10}$$

Energy balance:

$$\dot{m}_1 h_1 + \dot{m}_3 h_3 = \dot{m}_2 h_2 + \dot{m}_4 h_4 + \dot{Q} \tag{6.11}$$

Entropy balance:

$$\dot{m}_1 s_1 + \dot{m}_3 s_3 + \dot{S}_{gen} = \dot{m}_2 s_2 + \dot{m}_4 s_4 + \frac{\dot{Q}}{T_b} \tag{6.12}$$

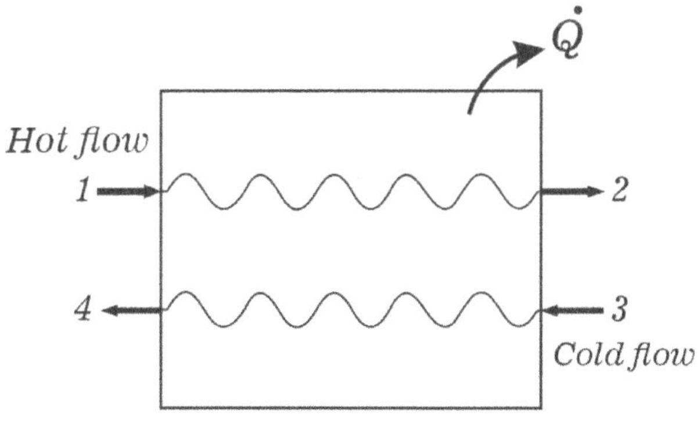

FIGURE 6.9 Schematic of a heat exchanger.

Exergy balance:

$$\dot{E}x_F = \dot{E}x_1 - \dot{E}x_2 \tag{6.13}$$

$$\dot{E}x_P = \dot{E}x_4 - \dot{E}x_3 \tag{6.14}$$

$$\dot{E}x_L = \left(1 - \frac{T_0}{T_b}\right)\dot{Q} \tag{6.15}$$

$$\dot{E}x_{des} = \left(\dot{E}x_1 - \dot{E}x_2\right) - \left(\dot{E}x_4 - \dot{E}x_3\right) - \left(1 - \frac{T_0}{T_b}\right)\dot{Q} \tag{6.16}$$

Exergy efficiency:

$$\eta_{ex} = \frac{\dot{E}x_4 - \dot{E}x_3}{\dot{E}x_1 - \dot{E}x_2} \tag{6.17}$$

6.7.1.10 Pasteurizers and Continuous Sterilizers

Industrially, pasteurization means heating food at a temperature below the boiling point and is the most common thermal process of milk and its products. During the pasteurization process, pathogenic bacteria and most non-pathogenic microorganisms are removed. Pasteurization of unpackaged food is done in two continuous and discontinuous ways.

Continuous pasteurizers are used to increase the capacity and also maintain the nutritional value of food. The most important and widely used type of continuous pasteurizer is a plate pasteurizer. The plate pasteurizer consists of three main parts: the heat recovery part, pasteurization part, and cooling part. In this system, food liquid first enters the balance tank so that the amount of liquid entering the pasteurizer is adjusted according to the capacity of the tank. Then it enters the heat recovery part and flows into the adjacent channel with the pasteurized liquid inside the pasteurizer. In this part, the heat of the pasteurized liquid is transferred to the input liquid. In fact, the input liquid is preheated and the temperature of the pasteurized liquid is greatly reduced. Then the preheated liquid enters the pasteurization part and reaches the pasteurization temperature. For the pasteurization time to pass, the food must be kept at this temperature for a certain period, which is done in the holder. Along the path of this tube, a flow return valve equipped with temperature control is installed; if the pasteurization conditions are not met, the food liquid is returned to the balance tank. After the heat recovery stage, the pasteurized liquid goes to the cooler and chiller and its temperature is reduced to the desired level by city water or cold water.

The mechanism of operation of plate sterilizers is the same as that of plate pasteurizers, with the difference that they are used at a higher temperature. To create a temperature higher than 100°C inside the system and on the other hand to prevent the boiling of the food inside the heat exchanger, the whole system must work at

FIGURE 6.10 Plate pasteurizer.

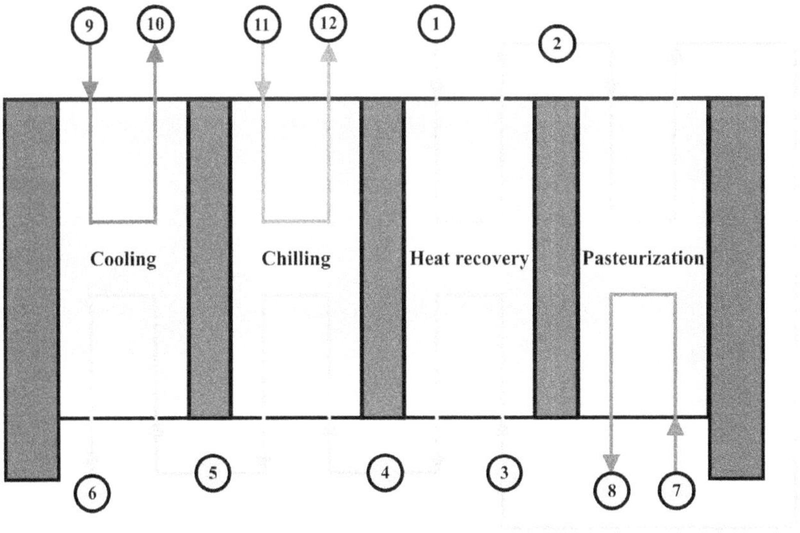

FIGURE 6.11 Schematic of a plate pasteurizer.

high pressure. This method can only be used for diluting liquids and liquids with low viscosity and is mostly used in sterilizing milk and juice.

Mass balance:

$$\dot{m}_1 = \dot{m}_2 = \dot{m}_3 = \dot{m}_4 = \dot{m}_5 = \dot{m}_6 \tag{6.18}$$

$$\dot{m}_7 = \dot{m}_8$$

$$\dot{m}_9 = \dot{m}_{10}$$

$$\dot{m}_{11} = \dot{m}_{12}$$

Energy balance:

$$\dot{m}_1 h_1 + \dot{m}_7 h_7 + \dot{m}_9 h_9 + \dot{m}_{11} h_{11} = \dot{m}_6 h_6 + \dot{m}_8 h_8 + \dot{m}_{10} h_{10} + \dot{m}_{12} h_{12} + \dot{Q} \tag{6.19}$$

Entropy balance:

$$\dot{m}_1 s_1 + \dot{m}_7 s_7 + \dot{m}_9 s_9 + \dot{m}_{11} s_{11} + \dot{S}_{\text{gen}} = \dot{m}_6 s_6 + \dot{m}_8 s_8 + \dot{m}_{10} s_{10} + \dot{m}_{12} s_{12} + \frac{\dot{Q}}{T_b} \tag{6.20}$$

Exergy balance:

$$\dot{E}x_F = \left(\dot{E}x_7 - \dot{E}x_8\right) + \left(\dot{E}x_4 - \dot{E}x_3\right) + \left(\dot{E}x_5 - \dot{E}x_4\right) + \left(\dot{E}x_6 - \dot{E}x_5\right) \tag{6.21}$$

$$\dot{E}x_P = \left(\dot{E}x_2 - \dot{E}x_1\right) + \left(\dot{E}x_3 - \dot{E}x_2\right) + \left(\dot{E}x_{10} - \dot{E}x_9\right) + \left(\dot{E}x_{12} - \dot{E}x_{11}\right) \tag{6.22}$$

$$\dot{E}x_L = \left(1 - \frac{T_0}{T_b}\right)\dot{Q} \tag{6.23}$$

$$\dot{E}x_{\text{des}} = \left[\left(\dot{E}x_7 - \dot{E}x_8\right) + \left(\dot{E}x_4 - \dot{E}x_3\right) + \left(\dot{E}x_5 - \dot{E}x_4\right) + \left(\dot{E}x_6 - \dot{E}x_5\right)\right] \tag{6.24}$$

$$-\left[\left(\dot{E}x_2 - \dot{E}x_1\right) + \left(\dot{E}x_3 - \dot{E}x_2\right) + \left(\dot{E}x_{10} - \dot{E}x_9\right) + \left(\dot{E}x_{12} - \dot{E}x_{11}\right)\right] - \left(1 - \frac{T_0}{T_b}\right)\dot{Q}$$

Exergy efficiency balance:

$$\eta_{\text{ex}} = \frac{\left(\dot{E}x_2 - \dot{E}x_1\right) + \left(\dot{E}x_3 - \dot{E}x_2\right) + \left(\dot{E}x_{10} - \dot{E}x_9\right) + \left(\dot{E}x_{12} - \dot{E}x_{11}\right)}{\left(\dot{E}x_7 - \dot{E}x_8\right) + \left(\dot{E}x_4 - \dot{E}x_3\right) + \left(\dot{E}x_5 - \dot{E}x_4\right) + \left(\dot{E}x_6 - \dot{E}x_5\right)} \tag{6.25}$$

✐Example 6.1: In a concentrate production line, condensed water is used to preheat juice with a flow rate of 3.5 kg/s. Calculate the rate of exergy destruction in this process. Information about flows is presented in Table 6.2.

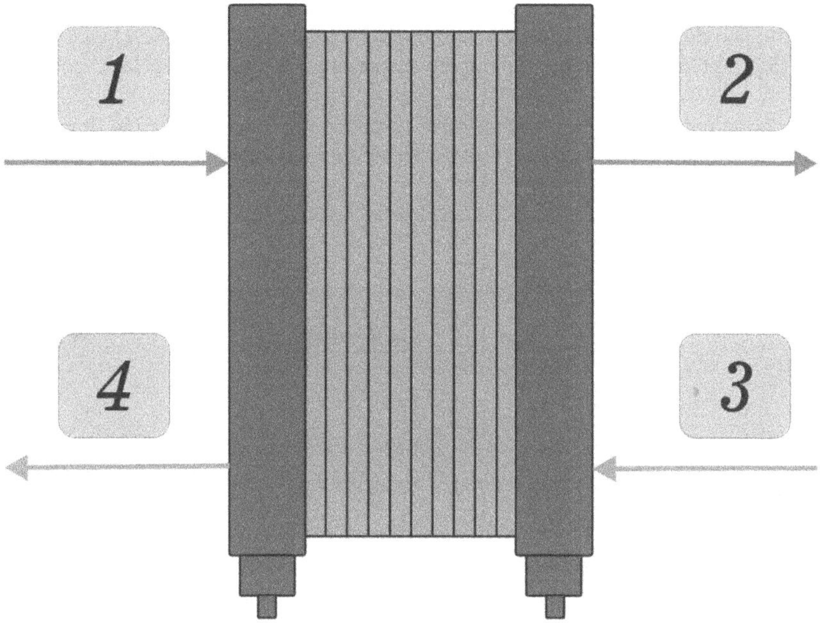

FIGURE 6.12 Schematic of the heat exchanger related to Example 6.1.

TABLE 6.2
Details of Example 6.1

State No.	Fluid Type	*T* (°C)	*P* (kPa)
1	Juice	50	551.325
2	Juice	70	451.325
3	Water	105	481.325
4	Water	85	331.325

The mass fraction of water is 0.8949, protein 0.0029, carbohydrate 0.09892, fat 0.00008, and ash 0.0032. Calculate the rate of exergy destruction and exergy efficiency in this process.

✍Solution:

Assumptions:

1. The process is considered steady state.
2. The heat loss from the surface of the heat exchanger is insignificant.
3. Kinetic and potential energy changes are ignored.
4. The dead state temperature and pressure are 25°C and 101.325 kPa.
5. The reference temperature and pressure are 0°C and 101.325 kPa.

The equations in Table 6.1 were used to calculate the specific heat and density of the juice compounds, and the results are presented in Table 6.3.

TABLE 6.3
Specific Heat and Density of Components of Juice in Example 6.1

	Protein		Fat		Carbohydrate		Ash		Water	
T (°C)	C_P	ρ	C_P	ρ	C_P	ρ	C_P	ρ	C_P	ρ
$T_{ref} = 0°C$	2.01	—	1.98	—	1.55	—	1.09	—	4.18	—
$T_0 = 25°C$	2.04	—	2.02	—	1.59	—	1.14	—	4.18	—
$T_1 = 50°C$	2.07	1303.98	2.05	904.71	1.63	1583.58	1.18	2409.77	4.19	987.94
$T_2 = 70°C$	2.09	1293.61	2.06	896.36	1.66	1577.37	1.21	2404.16	4.20	978.99

The specific heat and volume of fruit juice were determined using Equations 6.4 and 6.7, respectively, and the results are presented in Table 6.4.

TABLE 6.4
Specific Heat and Density of Streams in Example 6.1

State No.		$C_P\left(\frac{kJ}{kg.K}\right)$	$v\left(\frac{m^3}{kg}\right)$
1	$T_{ref} = 0°C$	3.90	—
	$T_0 = 25°C$	3.91	—
	$T_1 = 50°C$	3.92	0.00097
2	$T_{ref} = 0°C$	3.90	—
	$T_0 = 25°C$	3.91	—
	$T_1 = 70°C$	3.93	0.00098

Equation 6.6 was used to determine the enthalpy of juice flows. Enthalpy and entropy of condensed water flows were determined from the table of thermodynamic properties of water:

$$h_0 = 104.92\frac{\text{kJ}}{\text{kg}}, s_0 = 0.37\frac{\text{kJ}}{\text{kg.K}}$$

$$h_1 = \left(\frac{C_{P@T_1} + C_{P@T_{ref}}}{2}\right)T_1 + v_1(P_1 - P_0)$$

$$= 3.91(50) + 0.00097(551.325 - 101.325) = 195.86\frac{\text{kJ}}{\text{kg}}$$

$$h_2 = \left(\frac{C_{P@T_2} + C_{P@T_{ref}}}{2}\right)T_2 + v_2(P_2 - P_0)$$

$$= 3.91(70) + 0.00098(451.325 - 101.325) = 274.38\frac{\text{kJ}}{\text{kg}}$$

$$h_3 = 440.54\frac{\text{kJ}}{\text{kg}}, s_3 = 1.36\frac{\text{kJ}}{\text{kg.K}}$$

$$h_4 = 356.23\frac{\text{kJ}}{\text{kg}}, s_4 = 1.13\frac{\text{kJ}}{\text{kg.K}}$$

Mass balance for the steady state:

$$\sum \dot{m}_i = \sum \dot{m}_o \rightarrow \begin{cases} \dot{m}_1 = \dot{m}_2 = \dot{m}_J = 3.5 \,^{\text{kg}}/_{\text{s}} \\ \dot{m}_3 = \dot{m}_4 = \dot{m}_C \end{cases}$$

Energy balance for the steady state:

$$\sum \dot{E}_i = \sum \dot{E}_o$$

$$\rightarrow \dot{m}_1 h_1 + \dot{m}_3 h_3 = \dot{m}_2 h_2 + \dot{m}_4 h_4$$

$$\rightarrow \dot{m}_J(h_2 - h_1) = \dot{m}_C(h_3 - h_4) \rightarrow \dot{m}_C = \dot{m}_J\left(\frac{h_2 - h_1}{h_3 - h_4}\right) = 3.5\left(\frac{274.38 - 195.86}{440.54 - 356.23}\right) = 3.26\frac{\text{kg}}{\text{s}}$$

To calculate the exergy of fruit juice and condensate flows, the following equations were used:

$$ex_1 = \left(\frac{C_{p@T_1} + C_{p@T_0}}{2}\right)\left(T_1 - T_0 - T_0 \ln\left(\frac{T_1}{T_0}\right)\right) + v_{@T_1}(P_1 - P_0)$$

$$= 3.91\left[323.15 - 298.15 - 298.15 \ln\left(\frac{323.15}{298.15}\right)\right]$$

$$+ 0.00097(551.325 - 101.325) = 4.32\frac{\text{kJ}}{\text{kg}}$$

$$ex_2 = \left(\frac{C_{p@T_2} + C_{p@T_0}}{2}\right)\left(T_2 - T_0 - T_0 \ln\left(\frac{T_2}{T_0}\right)\right) + v_{@T_2}(P_2 - P_0)$$
$$= 3.92\left[343.15 - 298.15 - 298.15 \ln\left(\frac{343.15}{298.15}\right)\right]$$
$$+0.00098(451.325 - 101.325) = 12.44\frac{\text{kJ}}{\text{kg}}$$

$$ex_3 = (h_3 - h_0) - T_0(s_3 - s_0)$$
$$= (440.54 - 104.92) - 298.15(1.36 - 0.37) = 38.71\frac{\text{kJ}}{\text{kg}}$$

$$ex_4 = (h_4 - h_0) - T_0(s_4 - s_0)$$
$$= (356.23 - 104.92) - 298.15(1.13 - 0.37) = 22.56\frac{\text{kJ}}{\text{kg}}$$

Calculation of fuel and product exergy:

$$\dot{E}x_F = \dot{E}x_3 - \dot{E}x_4 = \dot{m}_3 ex_3 - \dot{m}_4 ex_4 = \dot{m}_C(ex_3 - ex_4)$$
$$= 3.26(38.71 - 22.56) = 52.62\text{kW}$$

$$\dot{E}x_P = \dot{E}x_2 - \dot{E}x_1 = \dot{m}_2 ex_2 - \dot{m}_1 ex_1 = \dot{m}_J(ex_2 - ex_1)$$
$$= 3.50(12.44 - 4.32) = 28.43\text{kW}$$

Exergy destruction rate:

$$\dot{E}x_{des} = \dot{E}x_F - \dot{E}x_P = 52.62 - 28.43 = 24.19\text{kW}$$

Exergy efficiency:

$$\eta_{ex} = \frac{\dot{E}x_P}{\dot{E}x_F} \times 100 = \frac{28.43}{52.62} \times 100 = 54.03\%$$

✎**Example 6.2: In a cream pasteurization line, hot water used in the process is produced by heating water in a plate heat exchanger (Singh et al., 2019). Table 6.5 presents information about the currents of this heat exchanger. Calculate the rate of entropy generation and the rate of exergy destruction in the process. The surface temperature of the heat exchanger has been measured at 40°C.**

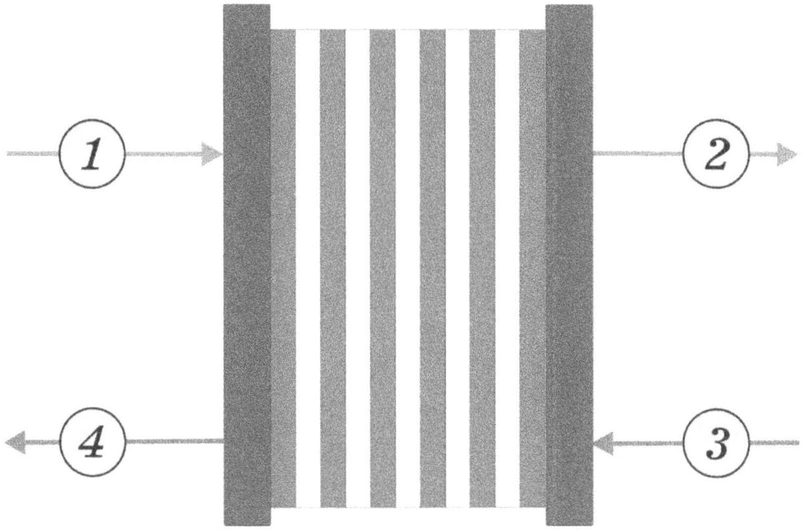

FIGURE 6.13 Schematic of the heat exchanger in Example 6.2.

TABLE 6.5
Details of Example 6.2

State No.	Fluid Type	*T* (°C)	*P* (kPa)	$\dot{m}\left(\frac{kg}{s}\right)$
1	Water	55.62	350	0.2
2	Water	99.00	150	0.2
3	Steam	127.41	350	0.02
4	Condensate	127.00	283	0.02

✍Solution:

Assumptions:

1. The process is considered steady state.
2. Kinetic and potential energy changes are ignored.
3. The dead state temperature and pressure are 25°C and 101.325 kPa.

Mass balance for steady state:

$$\sum \dot{m}_i = \sum \dot{m}_o \rightarrow \begin{cases} \dot{m}_1 = \dot{m}_2 = \dot{m}_w = 0.2 \, {}^{kg}/_{s} \\ \dot{m}_3 = \dot{m}_4 = \dot{m}_s = 0.02 \, {}^{kg}/_{s} \end{cases}$$

Energy balance for steady state:

$$\sum \dot{E}_i = \sum \dot{E}_o \rightarrow \dot{m}_1 h_1 + \dot{m}_3 h_3 = \dot{m}_2 h_2 + \dot{m}_4 h_4 + \dot{Q}_L \rightarrow \dot{Q}_L = \dot{m}_s (h_3 - h_4) - \dot{m}_w (h_2 - h_1)$$

Entropy balance for steady state:

$$\dot{S}_{gen} = \sum_o (\dot{m}s) + \left(\frac{\dot{Q}}{T_b}\right)_o - \sum_i (\dot{m}s) \rightarrow \dot{S}_{gen} = \dot{m}_2 s_2 + \dot{m}_4 s_4 + \frac{\dot{Q}_L}{T_b} - \dot{m}_1 s_1 - \dot{m}_3 s_3$$

$$\rightarrow \dot{S}_{gen} = \dot{m}_w (s_2 - s_1) + \dot{m}_s (s_4 - s_3) + \frac{\dot{Q}_L}{T_b}$$

Exergy balance for steady state:

$$\dot{E}x_{des} = \dot{E}x_F - \dot{E}x_P - \dot{E}x_L$$

$$\dot{E}x_F = \dot{E}x_3 - \dot{E}x_4 = \dot{m}_3 ex_3 - \dot{m}_4 ex_4 = \dot{m}_s (ex_3 - ex_4)$$

$$\dot{E}x_P = \dot{E}x_2 - \dot{E}x_1 = \dot{m}_2 ex_2 - \dot{m}_1 ex_1 = \dot{m}_w (ex_2 - ex_1)$$

$$\dot{E}x_L = \dot{Q}_L \left(1 - \frac{T_0}{T_b}\right)$$

Determination of flow characteristics:

$$@25°C, 101.325 kPa \begin{cases} h_0 = 104.92 \dfrac{kJ}{kg} \\ s_0 = 0.37 \dfrac{kJ}{kg}.K \end{cases}$$

$$@55.62°C, 350 kPa \begin{cases} h_1 = 233.13 \dfrac{kJ}{kg} \\ s_1 = 0.78 \dfrac{kJ}{kg}.K \end{cases}$$

$$@99°C, 150 kPa \begin{cases} h_2 = 414.99 \dfrac{kJ}{kg} \\ s_2 = 1.30 \dfrac{kJ}{kg}.K \end{cases}$$

$$@127.41°C, 350 kPa \begin{cases} h_3 = 2716.69 \dfrac{kJ}{kg} \\ s_3 = 7.05 \dfrac{kJ}{kg}.K \end{cases}$$

$$@127°C, 283kPa \begin{cases} h_4 = 520.82\dfrac{kJ}{kg} \\ s_4 = 1.57\dfrac{kJ}{kg}.K \end{cases}$$

$$ex_1 = (h_1 - h_0) - T_0(s_1 - s_0)$$
$$= (233.13 - 104.92) - 298.15(0.78 - 0.37) = 6.40\frac{kJ}{kg}$$

$$ex_2 = (h_2 - h_0) - T_0(s_2 - s_0)$$
$$= (414.99 - 104.92) - 298.15(1.30 - 0.37) = 33.19\frac{kJ}{kg}$$

$$ex_3 = (h_3 - h_0) - T_0(s_3 - s_0)$$
$$= (2716.69 - 104.92) - 298.15(7.05 - 0.37) = 618.41\frac{kJ}{kg}$$

$$ex_4 = (h_4 - h_0) - T_0(s_4 - s_0)$$
$$= (520.82 - 104.92) - 298.15(1.57 - 0.37) = 57.03\frac{kJ}{kg}$$

Calculation of waste heat rate:

$$\dot{Q}_L = 0.02(2716.69 - 520.82) - 0.2(414.99 - 233.13) = 7.55kW$$

Calculation of entropy generation rate:

$$\dot{S}_{gen} = 0.2(1.30 - 0.78) + 0.02(1.57 - 7.05) + \frac{7.55}{(40 + 273.15)}$$
$$= 0.01848 \cong 0.02\frac{kJ}{s.K}$$

Calculation of exergy destruction rate:

$$\dot{E}x_F = 0.02(618.41 - 57.03) = 11.23kW$$

$$\dot{E}x_P = 0.2(33.19 - 6.40) = 5.36kW$$

$$\dot{E}x_L = 7.55\left[1 - \frac{(25 + 273.15)}{(40 + 273.15)}\right] = 0.36kW$$

$$\dot{E}x_{des} = 5.51kW$$

Exergy destruction rate can also be calculated from the following equation:

$$\dot{E}x_{des} = T_0\dot{S}_{gen} = (25 + 273.15)K\left(0.01848\frac{kJ}{K}\right) = 5.51kW$$

Example 6.3: In a sugar factory, the syrup entering the evaporation line is preheated by a flow of steam in a heat exchanger. Information about the flows of this heat exchanger is presented in Table 6.6. The heat loss rate from the surface of the heat exchanger was calculated by the heat transfer relations and reported as 4 kW; the temperature of the equipment surface was reported as 56°C. Calculate the rate of exergy destruction and exergy efficiency in this process.

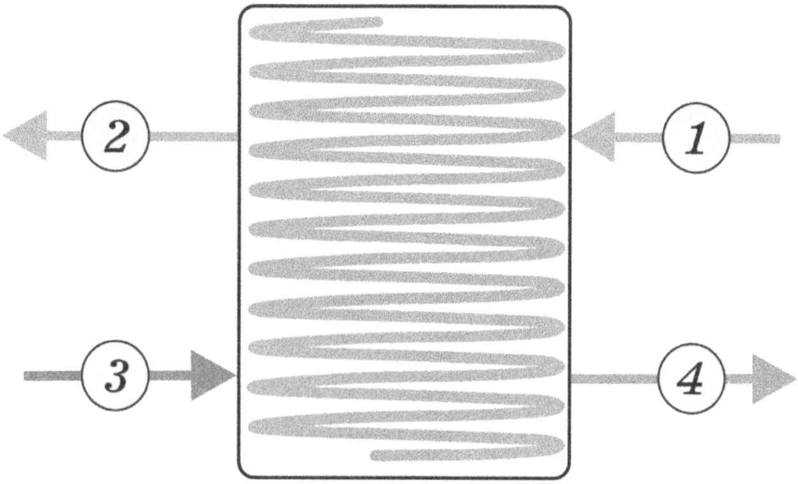

FIGURE 6.14 Schematic of the heat exchanger related to Example 6.3.

TABLE 6.6
Information about the Flows of Example 6.3

State No.	Fluid Type	T (°C)	P (kPa)	Phase description	$\dot{m}\left(\frac{kg}{s}\right)$	Brix (%)
1	Syrup	118	441.325	—	48.60	14.20
2	Syrup	124	391.325	—	—	—
3	Steam	140.80	—	Saturated steam	—	—
4	Condensate	—	—	—	—	—

Solution

Assumptions:

1. The process is considered steady state.
2. The heat loss from the surface of the heat exchanger is insignificant.

3. Kinetic and potential energy changes are ignored.
4. The dead state temperature and pressure are 25°C and 101.325 kPa.
5. The reference temperature and pressure are 0°C and 101.325 kPa.
6. Condensate water leaving the heat exchanger is a saturated liquid.
7. The pressure drop in the steam pipe is negligible.
8. Syrup is composed of water and carbohydrates (sucrose).

The equations in Table 6.1 were used to calculate the specific heat and density of the syrup compounds, and the results are presented in Table 6.7.

TABLE 6.7
Properties of Syrup in Example 6.3

	Carbohydrate		Water	
T (°C)	C_P	ρ	C_P	ρ
$T_{ref} = 0°C$	1.55	—	4.18	—
$T_0 = 25°C$	1.59	—	4.18	—
$T_1 = 118°C$	1.70	1562.47	4.24	954.23
$T_2 = 124°C$	1.70	1560.60	4.25	939.80

Specific heat and specific volume of syrup were determined using Equations 6.4 and 6.7, respectively, and the results are presented in the following table.

TABLE 6.8
Properties of Streams in Example 6.3

State No.		$C_P\left(\frac{kJ}{kg.K}\right)$	$v\left(\frac{m^3}{kg}\right)$
1	$T_{ref} = 0°C$	3.80	—
	$T_0 = 25°C$	3.81	—
	$T_1 = 118°C$	3.88	0.00099
2	$T_{ref} = 0°C$	3.80	—
	$T_0 = 25°C$	3.81	—
	$T_1 = 124°C$	3.89	0.00100

Equation 6.6 was used to determine the enthalpy of syrup flows. Enthalpy and entropy of condensed water flows were determined from the table of thermodynamic properties of water:

$$h_0 = 104.92\frac{kJ}{kg}, \; s_0 = 0.37\frac{kJ}{kg.K}$$

$$h_1 = \left(\frac{C_{P@T_1} + C_{P@T_{ref}}}{2}\right)T_1 + v_1(P_1 - P_0)$$

$$= 3.84(118) + 0.00099(441.325 - 101.325) = 453.67\frac{\text{kJ}}{\text{kg}}$$

$$h_2 = \left(\frac{C_{P@T_2} + C_{P@T_{ref}}}{2}\right)T_2 + v_2(P_2 - P_0)$$

$$= 3.85(124) + 0.00100(391.325 - 101.325) = 477.09\frac{\text{kJ}}{\text{kg}}$$

$$h_3 = h_{g@140.80°C} = 2734.49\frac{\text{kJ}}{\text{kg}},\ s_3 = s_{f@140.80°C} = 6.92\frac{\text{kJ}}{\text{kg.K}}$$

$$h_4 = h_{g@140.80°C} = 592.59\frac{\text{kJ}}{\text{kg}},\ s_4 = s_{f@140.80°C} = 1.75\frac{\text{kJ}}{\text{kg.K}}$$

Mass balance for the steady state:

$$\sum \dot{m}_i = \sum \dot{m}_o \rightarrow \begin{cases} \dot{m}_1 = \dot{m}_2 = \dot{m}_S = 48.60\,{}^{\text{kg}}\!/\!_{\text{s}} \\ \dot{m}_3 = \dot{m}_4 = \dot{m}_V \end{cases}$$

Energy balance for the steady state:

$$\sum \dot{E}_i = \sum \dot{E}_o$$

$$\rightarrow \dot{m}_1 h_1 + \dot{m}_3 h_3 = \dot{m}_2 h_2 + \dot{m}_4 h_4 + \dot{Q}_L$$

$$\rightarrow \dot{m}_S(h_2 - h_1) = \dot{m}_V(h_3 - h_4) + \dot{Q}_L \rightarrow \dot{m}_V = \frac{\dot{m}_S(h_2 - h_1) - \dot{Q}_L}{h_3 - h_4}$$

$$= \frac{48.60(477.09 - 453.67) - 4}{2734.49 - 592.59} = 0.53\frac{\text{kg}}{\text{s}}$$

To calculate the exergy of syrup and water flows, the following equations were used:

$$ex_1 = \left(\frac{C_{p@T_1} + C_{p@T_0}}{2}\right)\left(T_1 - T_0 - T_0 \ln\left(\frac{T_1}{T_0}\right)\right) + v_{@T_1}(P_1 - P_0)$$

$$= 3.85\left[391.15 - 298.15 - 298.15\ln\left(\frac{391.15}{298.15}\right)\right]$$

$$+0.00099(441.325 - 101.325) = 46.69\frac{\text{kJ}}{\text{kg}}$$

$$ex_2 = \left(\frac{C_{p@T_2} + C_{p@T_0}}{2}\right)\left(T_2 - T_0 - T_0 \ln\left(\frac{T_2}{T_0}\right)\right) + v_{@T_2}(P_2 - P_0)$$

$$= 3.85\left[397.15 - 298.15 - 298.15\ln\left(\frac{397.15}{298.15}\right)\right]$$

$$+ + 0.00100(391.325 - 101.325) = 52.31\frac{\text{kJ}}{\text{kg}}$$

$$ex_3 = (h_3 - h_0) - T_0(s_3 - s_0)$$
$$= (2734.49 - 104.92) - 298.15(6.92 - 0.37) = 675.32\frac{kJ}{kg}$$

$$ex_4 = (h_4 - h_0) - T_0(s_4 - s_0)$$
$$= (592.59 - 104.92) - 298.15(1.75 - 0.37) = 76.14\frac{kJ}{kg}$$

Calculation of exergy loss, fuel and product:

$$\dot{E}x_L = \left(1 - \frac{T_0}{T_b}\right)\dot{Q} = \left(1 - \frac{298.15}{329.15}\right)4 = 0.38\text{kW}$$

$$\dot{E}x_F = \dot{E}x_3 - \dot{E}x_4 = \dot{m}_3 ex_3 - \dot{m}_4 ex_4 = \dot{m}_V(ex_3 - ex_4)$$
$$= 0.53(675.32 - 76.14) = 319.58\text{kW}$$

$$\dot{E}x_P = \dot{E}x_2 - \dot{E}x_1 = \dot{m}_2 ex_2 - \dot{m}_1 ex_1 = \dot{m}_S(ex_2 - ex_1) = 48.60$$
$$(52.31 - 46.69) = 272.98\text{kW}$$

Exergy destruction rate:

$$\dot{E}x_{des} = \dot{E}x_F - \dot{E}x_P - \dot{E}x_L = 319.58 - 272.98 - 0.38 = 46.22\text{kW}$$

Exergy efficiency:

$$\eta_{ex} = \frac{\dot{E}x_P}{\dot{E}x_F} \times 100 = \frac{272.98}{319.58} \times 100 = 85.42\%$$

Example 6.4: According to Figure 6.15, a plate pasteurizer is used for cream pasteurization. Table 6.9 presents information about the flows of this pasteurizer.

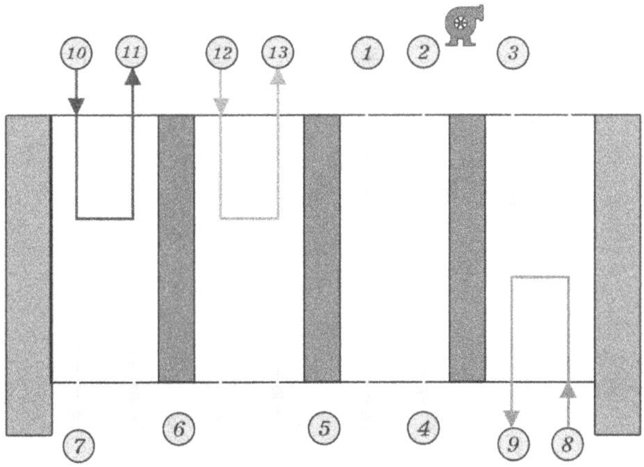

FIGURE 6.15 Schematic of the heat exchanger related to Example 6.4.

For cream, the mass fraction of water is 0.5465, protein is 0.0202, carbohydrate is 0.0285, fat is 0.4, and ash is 0.0048. Calculate the rate of exergy destruction and exergy efficiency in this process. The surface temperature of the heat exchanger has been measured at 40°C.

TABLE 6.9
Details of Example 6.4

State No.	Fluid Type	T (°C)	P (kPa)	$\dot{m}\left(\frac{kg}{s}\right)$
1	Cream	54.00	300.00	0.67
2	Cream	78.12	235.00	0.67
3	Cream	78.12	450.00	0.67
4	Cream	90.00	425.00	0.67
5	Cream	62.00	400.00	0.67
6	Cream	35.50	350.00	0.67
7	Cream	7.00	325.00	0.67
8	Hot water	99.00	150.00	0.20
9	Water	56.12	125.00	0.20
10	Chilled water	1.00	250.00	2.02
11	Cold water	8.50	200.00	2.02
12	Cold water	27.00	300.00	2.02
13	Cold water	32.00	250.00	2.02

✍Solution

Assumptions:

1. The process is considered steady state.
2. Kinetic and potential energy changes are ignored.
3. The dead state temperature and pressure are 25°C and 101.325 kPa.

Energy balance for steady state:

$$\sum \dot{E}_i = \sum \dot{E}_o \rightarrow \dot{m}_1 h_1 + \dot{m}_3 h_3 + \dot{m}_8 h_8 + \dot{m}_{10} h_{10} + \dot{m}_{12} h_{12} = \dot{m}_2 h_2 + \dot{m}_7 h_7$$
$$+ \dot{m}_9 h_9 + \dot{m}_{11} h_{11} + \dot{m}_{13} h_{13} + \dot{Q}_L$$

Exergy balance for steady state:

$$\dot{E}x_{des} = \dot{E}x_F - \dot{E}x_P - \dot{E}x_L$$

The temperature of flows 7, 10, and 11 is lower than the temperature of the dead state, and they have cold exergy; therefore, in the calculation of fuel and product exergy, the exergy of flows 7, 10, and 11 is considered negatively:

$$\dot{E}x_F = \left(\dot{E}x_8 - \dot{E}x_9\right) + \left(\dot{E}x_4 - \dot{E}x_5\right) + \left(\dot{E}x_5 - \dot{E}x_6\right) + \left[\dot{E}x_6 - \left(-\dot{E}x_7\right)\right] =$$

$$[\dot{m}_8 ex_8 - \dot{m}_9 ex_9] + [\dot{m}_4 ex_4 - \dot{m}_5 ex_5] + [\dot{m}_5 ex_5 - \dot{m}_6 ex_6] + [\dot{m}_6 ex_6 - (-\dot{m}_7 ex_7)]$$

$$\dot{E}x_P = (\dot{E}x_2 - \dot{E}x_1) + (\dot{E}x_4 - \dot{E}x_3) + [(-\dot{E}x_{11}) - (-\dot{E}x_{10})] + (\dot{E}x_{13} - \dot{E}x_{12}) =$$

$$[\dot{m}_2 ex_2 - \dot{m}_1 ex_1] + [\dot{m}_4 ex_4 - \dot{m}_3 ex_3] + [(-\dot{m}_{11} ex_{11}) - (-\dot{m}_{10} ex_{10})] + [\dot{m}_{13} ex_{13} - \dot{m}_{12} ex_{12}]$$

$$\dot{E}x_L = \dot{Q}_L \left(1 - \frac{T_0}{T_b}\right)$$

To calculate the specific heat and density of cream compounds, the equations in Table 6.1 were used, and the results are presented in Table 6.10.

TABLE 6.10
Specific Heat and Density of Cream Compounds for Example 6.4

	Ash		Carbohydrate		Fat		Protein		Water	
T (°C)	C_P	ρ	C_P	ρ	C_P	ρ	C_P	ρ	C_P	ρ
$T_{ref} = 0$	1.09	—	1.55	—	1.98	—	2.01	—	4.18	—
$T_0 = 25$	1.14	—	1.59	—	2.02	—	2.04	—	4.18	—
$T_1 = 54$	1.18	2408.65	1.64	1582.34	2.05	903.04	2.07	1301.91	4.19	986.39
$T_2 = 78.12$	1.22	2401.88	1.67	1574.85	2.07	892.97	2.09	1289.40	4.20	974.50
$T_3 = 78.12$	1.22	2401.88	1.67	1574.85	2.07	892.97	2.09	1289.40	4.20	974.50
$T_4 = 90$	1.23	2398.54	1.68	1571.16	2.08	888.01	2.11	1283.24	4.21	967.03
$T_5 = 62$	1.20	2406.40	1.65	1579.85	2.06	899.70	2.08	1297.76	4.19	982.93
$T_6 = 35.5$	1.16	2413.84	1.61	1588.08	2.03	910.77	2.05	1311.50	4.18	992.56
$T_7 = 7$	1.11	2421.84	1.56	1596.93	1.99	922.67	2.02	1326.27	4.18	997.02

Specific heat and specific volume of cream were determined using Equations 6.4 and 6.7, respectively, and the results are presented in Table 6.11.

TABLE 6.11
Specific Heat and Density of Streams in Example 6.4

T (°C)	$C_P\left(\frac{\text{kJ}}{\text{kg.K}}\right)$	$v\left(\frac{\text{m}^3}{\text{kg}}\right)$
$T_{ref} = 0$	3.17	—
$T_0 = 25$	3.18	—
$T_1 = 54$	3.20	0.00103251
$T_2 = 78.12$	3.22	0.00104451
$T_3 = 78.12$	3.22	0.00104451
$T_4 = 90$	3.23	0.00105146
$T_5 = 62$	3.21	0.00103618
$T_6 = 35.5$	3.19	0.00102513
$T_7 = 7$	3.17	0.00101672

Equation 6.6 was used to determine the enthalpy of cream flows. Enthalpy and entropy of water flows were determined from the table of thermodynamic properties of water:

$$h_1 = \left(\frac{C_{P@T_1} + C_{P@T_{ref}}}{2}\right) T_1 + v_1 (P_1 - P_0)$$

$$= 3.18(54) + 0.00103251(300 - 101.33) = 172.15 \frac{kJ}{kg}$$

$$h_2 = \left(\frac{C_{P@T_2} + C_{P@T_{ref}}}{2}\right) T_2 + v_2 (P_2 - P_0)$$

$$= 3.19(78.12) + 0.00104451(235 - 101.33) = 249.59 \frac{kJ}{kg}$$

$$h_3 = \left(\frac{C_{P@T_3} + C_{P@T_{ref}}}{2}\right) T_3 + v_3 (P_3 - P_0)$$

$$= 3.19(78.12) + 0.00104451(450 - 101.33) = 249.81 \frac{kJ}{kg}$$

$$h_4 = \left(\frac{C_{P@T_4} + C_{P@T_{ref}}}{2}\right) T_4 + v_4 (P_4 - P_0)$$

$$= 3.20(90) + 0.00105146(425 - 101.33) = 288.13 \frac{kJ}{kg}$$

$$h_5 = \left(\frac{C_{P@T_5} + C_{P@T_{ref}}}{2}\right) T_5 + v_5 (P_5 - P_0)$$

$$= 3.19(62) + 0.00103618(400 - 101.33) = 197.91 \frac{kJ}{kg}$$

$$h_6 = \left(\frac{C_{P@T_6} + C_{P@T_{ref}}}{2}\right) T_6 + v_6 (P_6 - P_0)$$

$$= 3.18(35.5) + 0.00102513(350 - 101.33) = 113.06 \frac{kJ}{kg}$$

$$h_7 = \left(\frac{C_{P@T_7} + C_{P@T_{ref}}}{2}\right) T_7 + v_7 (P_7 - P_0)$$

$$= 3.17(7) + 0.00101672(325 - 101.33) = 22.40 \frac{kJ}{kg}$$

$$@\,25°C, 101.325 kPa \begin{cases} h_0 = 104.92 \dfrac{kJ}{kg} \\ s_0 = 0.37 \dfrac{kJ}{kg}.K \end{cases}$$

$$@55.62°C, 350kPa \begin{cases} h_8 = 414.99 \dfrac{kJ}{kg} \\ s_8 = 1.30 \dfrac{kJ}{kg}.K \end{cases}$$

$$@99°C, 150kPa \begin{cases} h_9 = 235.03 \dfrac{kJ}{kg} \\ s_9 = 0.78 \dfrac{kJ}{kg}.K \end{cases}$$

$$@127.41°C, 350kPa \begin{cases} h_{10} = 4.43 \dfrac{kJ}{kg} \\ s_{10} = 0.02 \dfrac{kJ}{kg}.K \end{cases}$$

$$@127°C, 283kPa \begin{cases} h_{11} = 35.92 \dfrac{kJ}{kg} \\ s_{11} = 0.13 \dfrac{kJ}{kg}.K \end{cases}$$

$$@127°C, 283kPa \begin{cases} h_{12} = 113.46 \dfrac{kJ}{kg} \\ s_{12} = 0.40 \dfrac{kJ}{kg}.K \end{cases}$$

$$@127°C, 283kPa \begin{cases} h_{13} = 134.32 \dfrac{kJ}{kg} \\ s_{13} = 0.46 \dfrac{kJ}{kg}.K \end{cases}$$

Equations 6.8 and 5.10 were used to calculate the exergies of the cream and water streams:

$$ex_1 = \left(\frac{C_{p@T_1} + C_{p@T_0}}{2}\right)\left(T_1 - T_0 - T_0 \ln\left(\frac{T_1}{T_0}\right)\right) + v_{@T_1}(P_1 - P_0)$$
$$= 3.19\left[327.15 - 298.15 - 298.15 \ln\left(\frac{327.15}{298.15}\right)\right]$$
$$+ 0.00103251(300 - 101.325) = 4.44 \frac{kJ}{kg}$$

$$ex_2 = \left(\frac{C_{p@T_2} + C_{p@T_0}}{2}\right)\left(T_2 - T_0 - T_0 \ln\left(\frac{T_2}{T_0}\right)\right) + v_{@T_2}(P_2 - P_0)$$
$$= 3.20\left[351.27 - 298.15 - 298.15 \ln\left(\frac{351.27}{298.15}\right)\right]$$
$$+ 0.00104451(235 - 101.325) = 13.70 \frac{kJ}{kg}$$

$$ex_3 = \left(\frac{C_{p@T_3} + C_{p@T_0}}{2}\right)\left(T_3 - T_0 - T_0 \ln\left(\frac{T_3}{T_0}\right)\right) + v_{@T_3}\left(P_3 - P_0\right)$$
$$= 3.20\left[351.27 - 298.15 - 298.15 \ln\left(\frac{351.27}{298.15}\right)\right]$$
$$+ 0.00104451(450 - 101.325) = 13.92 \frac{\text{kJ}}{\text{kg}}$$

$$ex_4 = \left(\frac{C_{p@T_4} + C_{p@T_0}}{2}\right)\left(T_4 - T_0 - T_0 \ln\left(\frac{T_4}{T_0}\right)\right) + v_{@T_4}\left(P_4 - P_0\right)$$
$$= 3.21\left[363.15 - 298.15 - 298.15 \ln\left(\frac{363.15}{298.15}\right)\right]$$
$$+ 0.00105146(425 - 101.325) = 20.21 \frac{\text{kJ}}{\text{kg}}$$

$$ex_5 = \left(\frac{C_{p@T_5} + C_{p@T_0}}{2}\right)\left(T_5 - T_0 - T_0 \ln\left(\frac{T_5}{T_0}\right)\right) + v_{@T_5}\left(P_5 - P_0\right) = 3.20\left[335.15\right.$$
$$\left. - 298.15 \ln\left(\frac{335.15}{298.15}\right)\right] + 0.00103618(400 - 101.325) = 7.09 \frac{\text{kJ}}{\text{kg}}$$

$$ex_6 = \left(\frac{C_{p@T_6} + C_{p@T_0}}{2}\right)\left(T_6 - T_0 - T_0 \ln\left(\frac{T_6}{T_0}\right)\right) + v_{@T_6}\left(P_6 - P_0\right)$$
$$= 3.19\left[308.65 - 298.15 - 298.15 \ln\left(\frac{308.65}{298.15}\right)\right]$$
$$+ 0.00102513(350 - 101.325) = 0.83 \frac{\text{kJ}}{\text{kg}}$$

$$ex_7 = \left(\frac{C_{p@T_7} + C_{p@T_0}}{2}\right)\left(T_7 - T_0 - T_0 \ln\left(\frac{T_7}{T_0}\right)\right) + v_{@T_7}\left(P_7 - P_0\right)$$
$$= 3.18\left[280.15 - 298.15 - 298.15 \ln\left(\frac{280.15}{298.15}\right)\right]$$
$$+ 0.00101673(325 - 101.325) = 2.03 \frac{\text{kJ}}{\text{kg}}$$

$$ex_8 = \left(h_8 - h_0\right) - T_0\left(s_8 - s_0\right)$$
$$= (414.99 - 104.92) - 298.15(1.30 - 0.37) = 33.19 \frac{\text{kJ}}{\text{kg}}$$

$$ex_9 = \left(h_9 - h_0\right) - T_0\left(s_9 - s_0\right)$$
$$= (235.03 - 104.92) - 298.15(0.78 - 0.37) = 6.38 \frac{\text{kJ}}{\text{kg}}$$

$$ex_{10} = (h_{10} - h_0) - T_0(s_{10} - s_0)$$
$$= (4.43 - 104.92) - 298.15(0.02 - 0.37) = 4.44\frac{\text{kJ}}{\text{kg}}$$

$$ex_{11} = (h_{11} - h_0) - T_0(s_{11} - s_0)$$
$$= (35.92 - 104.92) - 298.15(0.13 - 0.37) = 2.09\frac{\text{kJ}}{\text{kg}}$$

$$ex_{12} = (h_{12} - h_0) - T_0(s_{12} - s_0)$$
$$= (113.46 - 104.92) - 298.15(0.4 - 0.37) = 0.23\frac{\text{kJ}}{\text{kg}}$$

$$ex_{13} = (h_{13} - h_0) - T_0(s_{13} - s_0)$$
$$= (134.32 - 104.92) - 298.15(0.46 - 0.37) = 0.49\frac{\text{kJ}}{\text{kg}}$$

Calculation of waste heat rate:

$$\dot{Q}_L = (\dot{m}_1h_1 + \dot{m}_3h_3 + \dot{m}_8h_8 + \dot{m}_{10}h_{10} + \dot{m}_{12}h_{12})$$
$$-(\dot{m}_2h_2 + \dot{m}_7h_7 + \dot{m}_9h_9 + \dot{m}_{11}h_{11} + \dot{m}_{13}h_{13}) =$$

$$[0.67(172.15) + 0.67(249.81) + 0.2(414.99) + 2.02(4.43) + 2.02(113.46)]$$
$$-[0.67(249.59) + 0.67(22.40) + 0.20(235.03) + 2.02(35.92) + 2.02(134.32)]$$
$$= 30.74kW$$

Calculation of exergy destruction rate:

$$\dot{E}x_F = [0.2(33.19) - 0.2(6.38)] + [0.67(20.21) - 0.67(7.09)]$$
$$+[0.67(7.09) - 0.67(0.83)] + [0.67(0.83) - 0.67(-2.03)]$$
$$= 20.26\text{kW}$$

$$\dot{E}x_P = [0.67(13.70) - 0.67(4.44)] + [0.67(20.21) - 0.67(13.92)]$$
$$+[2.02(-2.09) - 2.02(-4.44)]$$
$$+[2.02(0.49) - 2.02(0.23)] = 15.69\text{kW}$$

$$\dot{E}x_L = 30.74\left(1 - \frac{298.15}{313.15}\right) = 1.47\text{kW}$$

$$\dot{E}x_{des} = 20.26 - 15.69 - 1.47 = 3.10\text{kW}$$

6.7.2 Pumps

The function of the pump is to add energy to the liquid, and as a result, the pressure of the liquid increases, and it moves. Pumps are placed in two large groups, each of which includes different types of pumps, such as discrete flow pumps and continuous flow pumps.

6.7.2.1 Discrete Flow Pumps

The main characteristic of discrete flow pumps is that the flow can always be under pressure, so if the pressure exceeds a certain value, it causes damage to the piping system. This feature is caused by the fact that the inlet and outlet flow are separated by two one-way valves; that is, the flow is discrete and not continuous. These pumps are mostly used to increase the pressure in the fluid. They transfer fluid with low flow rate and high pressure, and during fluid transfer, the inlet and outlet parts of the pump are separated by a piece. Discrete flow pumps are divided into two categories, reciprocating pumps and circulating (or rotating) pumps.

In rotary pumps, the moving part of the pump rotates around the axis or axes of the pump, but in reciprocating pumps, the moving part of the pump moves back and forth in a certain direction. Based on the shape of its moving part, rotary pumps exist in gear, lobe, double screw, and peristaltic types, among which the lobe type is the most used in the food industry, and its material is stainless steel.

There are reciprocating pumps in the piston and diaphragm types, of which the piston type is mostly used in the food industry. Rotary pumps are widely used in the food industry compared to reciprocating pumps. The piston pump consists of a piston that moves forward and backward inside the cylinder and consists of two valves, one at the inlet and the other at the outlet of the pump. As the piston moves backward, the inlet valve opens and the outlet valve closes, and the fluid is sucked into the pump, and then the inlet valve closes and the outlet valve opens, and due to the forward movement of the piston, the fluid comes out of the cylinder. The forward movement of the piston transfers energy to the fluid trapped inside the cylinder. A piston pump is used to transfer a small volume of liquid with low viscosity and high pressure. A diaphragm pump consists of two plastic or rubber membranes that are connected to a piston and move forward and backward by it. It also has two chambers; when the membrane moves forward, pressure is formed in one chamber and a vacuum is formed in the other chamber, as a result, the food material is pressurized and pumped in one chamber, and in the other chamber, due to the vacuum, the food material enters this chamber, and then due to the movement of the piston backward, the opposite of this action happens. Compared to piston pumps, diaphragm pumps transfer a larger volume of liquid with higher viscosity and lower pressure.

6.7.2.2 Continuous Flow Pumps

The characteristic of continuous flow pumps is that the input and output flow is connected; therefore, if the outlet channel of the pump is blocked, the fluid inside the pump rotates idle. The pump flow becomes zero, but there is no damage to the system. These pumps are mostly used to transfer fluid from one point to another. They transfer fluid with a high flow rate and low pressure. In other words, they are used in places where the only resistance to the movement of the fluid is the weight of the fluid and the friction of the movement path. These pumps are divided into three categories, radial flow pumps (centrifuge), axial flow pumps, and mixed flow pumps.

Axial centrifugal pumps are not used much in the food industry and only used to transfer large volumes at low heights, such as canning factory wastewater. The radial centrifugal pump is widely used in the food industry. The basic working principles

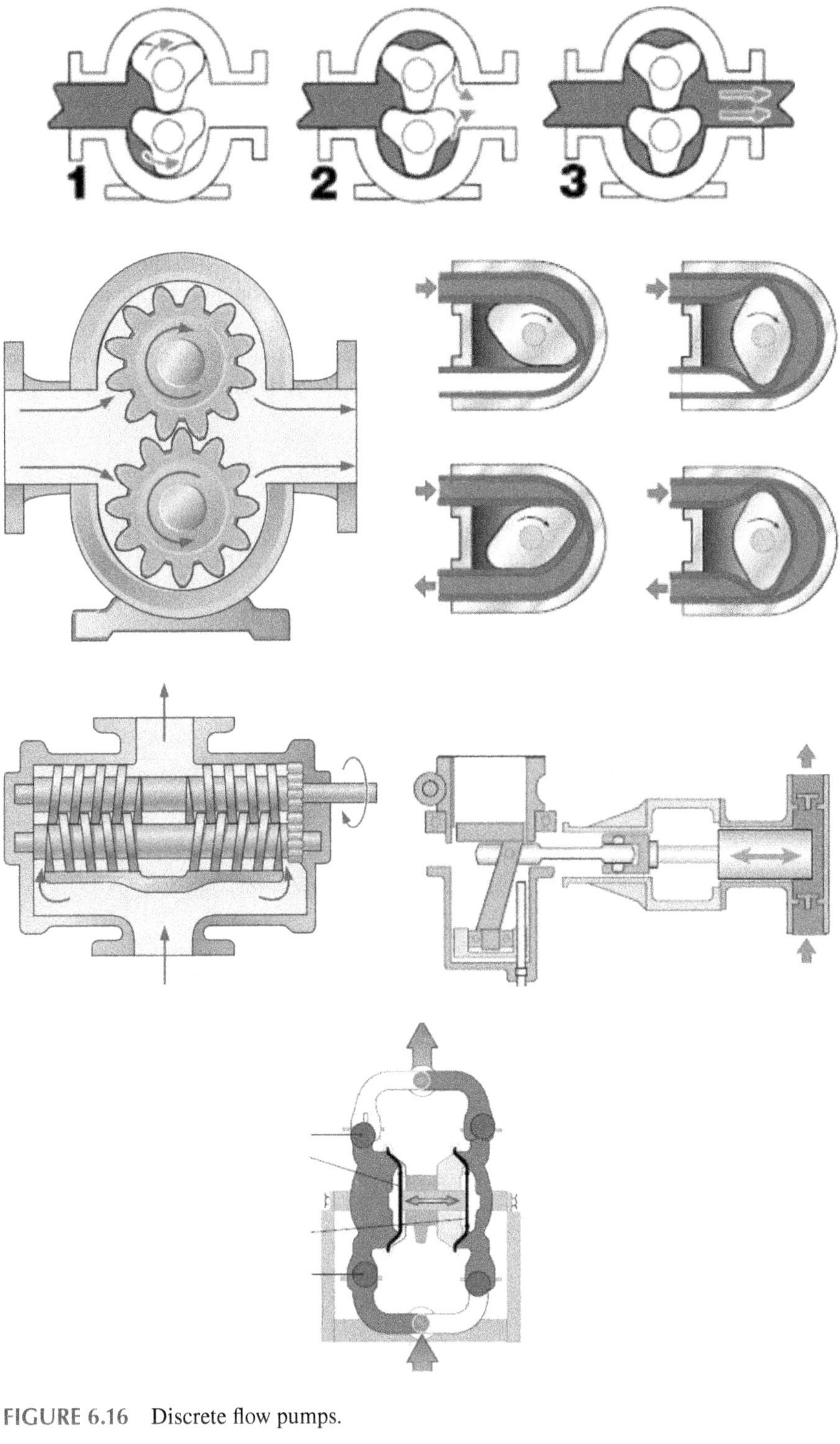

FIGURE 6.16 Discrete flow pumps.

of all centrifugal pumps are based on the use of centrifugal force. Any object that moves in a circular path is affected by centrifugal force. The direction of the centrifugal force is such that it always tends to move the object away from the axis or center of rotation. In centrifugal pumps, the fluid enters from the center of the impeller due to the suction force, and due to the rotation of the impeller, under the influence of the centrifugal force, it leaves the impeller and enters the pump chamber. The chamber's significant volume and its specific geometric attributes lead to a reduction in fluid velocity. As a result, a portion of its kinetic energy transforms into pressure energy, compelling it to exit through the outlet pipe under the exerted pressure. Radial centrifugal pumps employ impellers in three variations based on their application: open, semi-open, and closed. The open type is preferred for transferring Newtonian fluids with substantial particles, while semi-closed types are suitable for fluids with small insoluble particles. On the other hand, the closed type is employed for Newtonian fluids lacking insoluble particles.

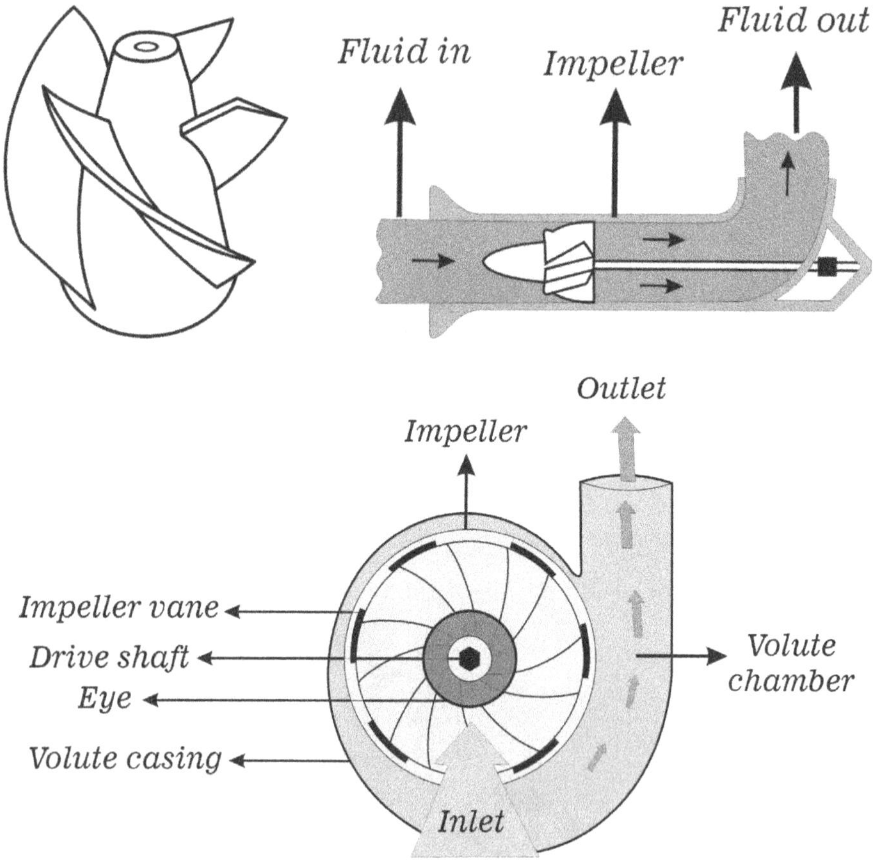

FIGURE 6.17 Continuous flow pumps. A) Mixed flow pump. B) Axial flow pump. C) Centrifugal pump.

6.7.2.3 Balance Equations

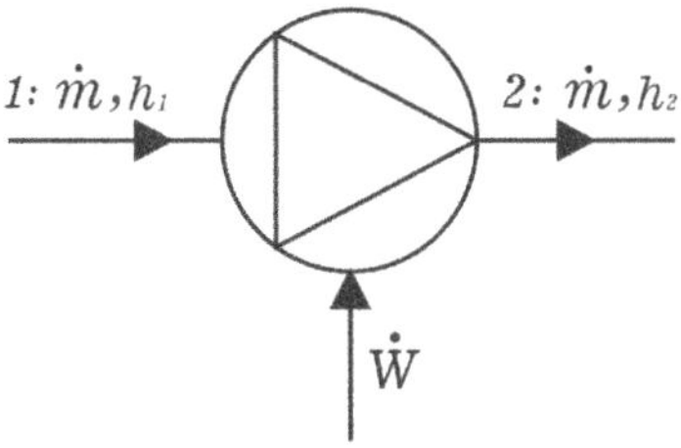

FIGURE 6.18 Symbolic representation of a pump.

Mass balance:

$$\dot{m}_1 = \dot{m}_2 \tag{6.26}$$

Energy balance:

$$\dot{m}_1 h_1 + \dot{W} = \dot{m}_2 h_2 \tag{6.27}$$

Entropy balance:

$$\dot{m}_1 s_1 + \dot{S}_{gen} = \dot{m}_2 s_2 \tag{6.28}$$

Exergy balance:

$$\dot{E}x_F = \dot{W} \tag{6.29}$$

$$\dot{E}x_P = \dot{E}x_2 - \dot{E}x_1 \tag{6.30}$$

$$\dot{E}x_{des} = \dot{W} - \left(\dot{E}x_2 - \dot{E}x_1\right) \tag{6.31}$$

Exergy efficiency:

$$\eta_{ex} = \frac{\dot{E}x_2 - \dot{E}x_1}{\dot{W}} \tag{6.32}$$

6.7.2.4 Mechanical Energy and Pump Efficiency

A pump gives mechanical energy to the fluid by increasing its pressure; therefore, the pressure of fluid flow is related to its mechanical energy. The change in mechanical energy of a fluid during incompressible flow and ignoring the changes in kinetic and potential energy is as follows:

$$\Delta e_{mech} = v\left(P_2 - P_1\right) \tag{6.33}$$

Mechanical energy transfer is usually done by a rotating shaft, and mechanical energy is often called shaft work. A pump receives shaft work (usually from an electric motor) and transfers it to the fluid as mechanical energy. In the absence of any irreversibility, including friction, mechanical energy can be completely converted from one mechanical form to another. The mechanical efficiency of a device or a process can be defined as follows:

$$\eta_{mech} = \frac{E_{mech,out}}{E_{mech,in}} = 1 - \frac{E_{mech,loss}}{E_{mech,in}} \tag{6.34}$$

Mechanical efficiency of less than 100% indicates that the conversion has not been completed and losses have occurred during this conversion. A mechanical efficiency of 97% shows that 3% of the input mechanical energy has been converted into thermal energy (internal energy) as a result of thermal waste (due to friction), which will cause a slight increase in fluid temperature. The rate of conversion of mechanical work consumed into fluid mechanical energy is expressed by pump efficiency, which is as follows:

$$\eta_{mech,Pump} = \frac{\Delta \dot{E}_{mech,Fluid}}{\dot{W}_{shaft,in}} \tag{6.35}$$

which means

$$\Delta \dot{E}_{mech,Fluid} = \dot{E}_{mech,out} - \dot{E}_{mech,in} = \dot{m}v(P_2 - P_1) \tag{6.36}$$

Mechanical efficiency should not be confused with engine efficiency, which is defined as follows:

$$\eta_{engine} = \frac{\dot{W}_{shaft}}{\dot{W}_{elect}} \tag{6.37}$$

A pump is coupled with its motor. Therefore, the combined or overall efficiency is considered the pump-motor set and is defined as follows:

$$\eta_{Pump-motor} = \eta_{Pump} \times \eta_{motor} = \frac{\Delta \dot{E}_{mech,Fluid}}{\dot{W}_{shaft,in}} \times \frac{\dot{W}_{shaft}}{\dot{W}_{elect}} = \frac{\Delta \dot{E}_{mech,Fluid}}{\dot{W}_{elect}} \tag{6.38}$$

Therefore, we can write:

$$\eta_{Pump-motor} = \frac{\dot{m}v_1\left(P_2 - P_1\right)}{\dot{W}_{elect}} \tag{6.39}$$

✐Example 6.5: The pump shown in Figure 6.19 is used to transfer syrup in a sugar factory. The pump-motor combined efficiency is 75% and the information on input and output flows is presented in Table 6.12. Calculate the rate of exergy destruction and exergy efficiency.

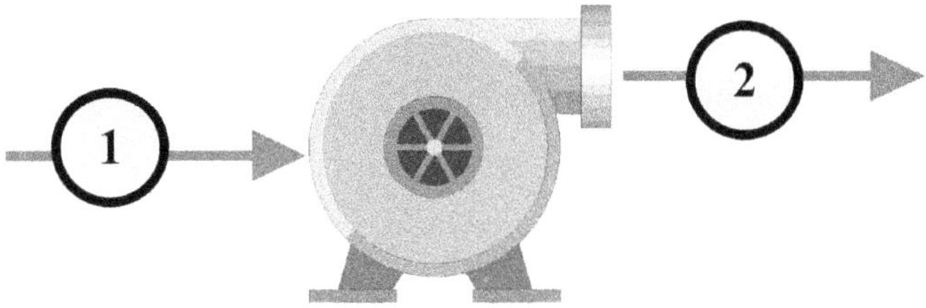

FIGURE 6.19 Diagram of the pump related to Example 6.5.

TABLE 6.12
Details of Example 6.5

State No.	T (°C)	P (kPa)	$\dot{m}\left(\frac{kg}{s}\right)$	Brix (%)
1	90.00	101.325	48.60	14.20
2	90.10	551.325	—	—

✍Solution

Assumptions:

1. The process is considered steady state.
2. The heat loss from the surface of the pump is insignificant.
3. Kinetic and potential energy changes are ignored.
4. The dead state temperature and pressure are 25°C and 101.325 kPa.
5. The syrup is composed of water and carbohydrates (sucrose).

The equations in Table 6.1 were used to calculate the specific heat and density of the syrup compounds, and the results are presented in Table 6.13.

TABLE 6.13
Properties of Juice in Example 6.5

	Carbohydrate		Water	
T (°C)	C_p	ρ	C_p	ρ
$T_0 = 25°C$	1.59	—	4.18	—
$T_1 = 90°C$	1.68	1571.16	4.21	967.03
$T_2 = 90.1°C$	1.68	1571.13	4.21	966.96

Specific heat and volume of syrup were determined using Equations 6.4 and 6.7, respectively, and the results are presented in Table 6.14.

TABLE 6.14
properties of streams in example 6.5

State no.		$C_P\left(\frac{kJ}{kg.K}\right)$	$v\left(\frac{m^3}{kg}\right)$
0	$T_0 = 25$	3.81	—
1	$T_1 = 90$	3.85	0.00098
2	$T_1 = 90.1$	3.85	0.00098

Mass balance for the steady state conditions:

$$\sum \dot{m}_i = \sum \dot{m}_o \rightarrow \dot{m}_1 = \dot{m}_2 = \dot{m}_S = 48.60\,{}^{kg}\!/_{s}$$

Equation 6.39 was used to calculate the electric power consumption of the pump:

$$\dot{W}_{elect} = \frac{\dot{m}v_1(P_2 - P_1)}{\eta_{Pump-motor}} = \frac{48.60(0.00098)(551.325 - 101.325)}{0.75} = 28.51\text{kW}$$

Equation 6.8 was used to calculate the exergy of syrup flows:

$$ex_1 = \left(\frac{C_{p@T_1} + C_{p@T_0}}{2}\right)\left(T_1 - T_0 - T_0 \ln\left(\frac{T_1}{T_0}\right)\right) + v_{@T_1}(P_1 - P_0)$$
$$= 3.83\left[363.15 - 298.15 - 298.15\ln\left(\frac{363.15}{298.15}\right)\right]$$
$$+ (0.00098)(101.325 - 101.325) = 23.75\frac{kJ}{kg}$$

$$ex_2 = \left(\frac{C_{p@T_2} + C_{p@T_0}}{2}\right)\left(T_2 - T_0 - T_0 \ln\left(\frac{T_2}{T_0}\right)\right) + v_{@T_2}(P_2 - P_0)$$
$$= 3.83\left[363.25 - 298.15 - 298.15\ln\left(\frac{363.25}{298.15}\right)\right]$$
$$+ (0.00098)(551.325 - 101.325) = 24.26\frac{kJ}{kg}$$

Calculation of fuel and product exergy:

$$\dot{E}x_F = \dot{W}_{elect} = 28.51\text{kW}$$
$$\dot{E}x_P = \dot{E}x_2 - \dot{E}x_1 = \dot{m}_2 ex_2 - \dot{m}_1 ex_1 = \dot{m}_S(ex_2 - ex_1) = 48.60$$
$$(24.26 - 23.75) = 24.73\text{kW}$$

Exergy destruction rate:

$$\dot{E}x_{des} = \dot{E}x_F - \dot{E}x_P = 28.51 - 24.73 = 3.78\text{kW}$$

Exergy efficiency:

$$\eta_{ex} = \frac{\dot{E}x_P}{\dot{E}x_F} \times 100 = \frac{24.73}{28.51} \times 100 = 86.75\%$$

6.7.3 Mixing

Blending multiple fluid streams is a routine procedure within the food industry, often carried out in a mixing chamber. This chamber need not be distinctively designed; any equipment that combines multiple flows and produces an output flow can be treated as and scrutinized as a mixing chamber. The fundamental principle of mass conservation dictates that the total incoming flow rate must equate to the exiting mass flow rate of the mixture within a mixing chamber. Typically, mixing chambers are not characterized by active processes. Additionally, the effects of kinetic and potential energies can be disregarded within this equipment.

6.7.3.1 Direct Contact Heating Equipment

Direct contact heat exchangers function by effectively blending steam with the food product, resulting in a notably heightened heat transfer intensity. This category

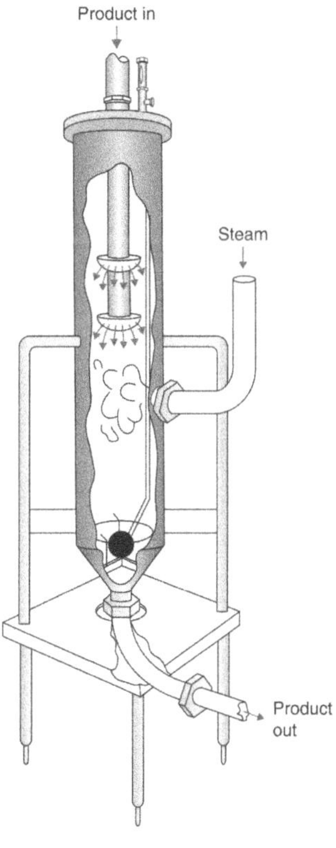

FIGURE 6.20 Direct contact heat exchangers. A) Steam infusion (Singh & Heldman, 2001). B) Steam injection.

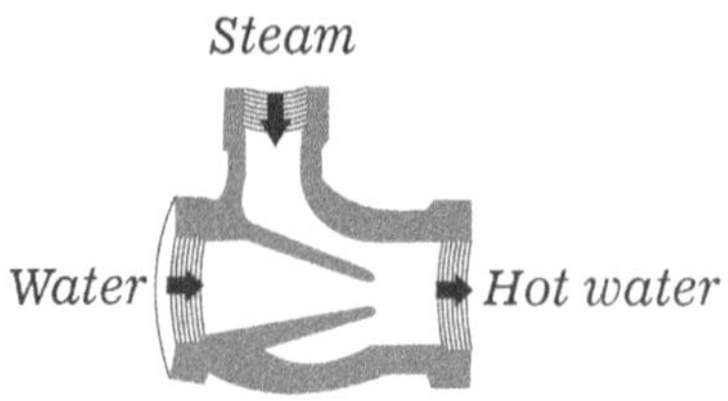

FIGURE 6.20 (Continued)

encompasses two distinct types: steam injection and steam infusion. In the injection variant, steam is introduced directly into the product, while in the stream infusion type, the food is sprayed into a tank containing the steam. The steam used must be tasteless, odorless, and free of toxic compounds and dissolved gas. The water entering the boiler to produce steam used in direct heat exchangers must have the characteristics of drinking water and not contain any additional chemicals. In these heat exchangers, the temperature of the food material increases suddenly due to mixing with steam to the desired temperature; therefore, the nutritional and sensory value of food is better preserved. The steam mixed with the food becomes a liquid due to the loss of its latent heat, which causes the dilution of the product. To return the concentration to the original state, the diluted product is passed through a tank under a vacuum (flash tank), which causes evaporation of excess water in the product and its sudden cooling. The amount of vacuum and, as a result, the temperature of the tank should be such that the same amount of water is removed from the product as steam is injected into the product.

6.7.3.2 Balance Equations

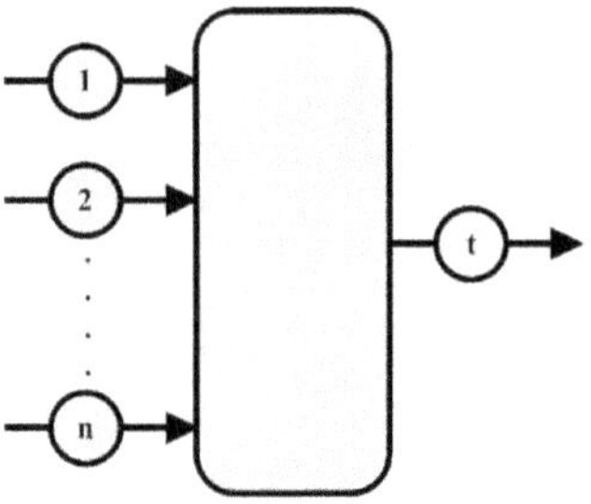

FIGURE 6.21 Schematic of a mixing chamber.

Mass balance:

$$\dot{m}_1 + \dot{m}_2 + \ldots + \dot{m}_n = \dot{m}_t \tag{6.40}$$

Energy balance:

$$\dot{m}_1 h_1 + \dot{m}_2 h_2 + \ldots + \dot{m}_n h_n = \dot{m}_t h_t + \dot{Q}\,. \tag{6.41}$$

Entropy balance:

$$\dot{m}_1 s_1 + \dot{m}_2 s_2 + \ldots + \dot{m}_n s_n + \dot{S}_{\text{gen}} = \dot{m}_t s_t + \frac{\dot{Q}}{T_b} \tag{6.42}$$

Exergy balance:

$$\dot{E}x_F = \dot{E}x_1 + \dot{E}x_2 + \ldots + \dot{E}x_n \tag{6.43}$$

$$\dot{E}x_P = \dot{E}x_t \tag{6.44}$$

$$\dot{E}x_L = \left(1 - \frac{T_0}{T_b}\right)\dot{Q} \tag{6.45}$$

$$\dot{E}x_{des} = \left(\dot{E}x_1 + \dot{E}x_2 + \ldots + \dot{E}x_n\right) - \dot{E}x_t - \left(1 - \frac{T_0}{T_b}\right)\dot{Q} \tag{6.46}$$

Exergy efficiency:

$$\eta_{ex} = \frac{\dot{E}x_t}{\dot{E}x_1 + \dot{E}x_2 + \ldots + \dot{E}x_n} \tag{6.47}$$

✐Example 6.6: In a concentrate processing plant, vapor from the last stage is converted into condensate by a barometric condenser as a result of mixing with water. Using the information in Table 6.15, calculate the flow of water entering the condenser and the condensate leaving it. Obtain the rate of exergy destruction in this process.

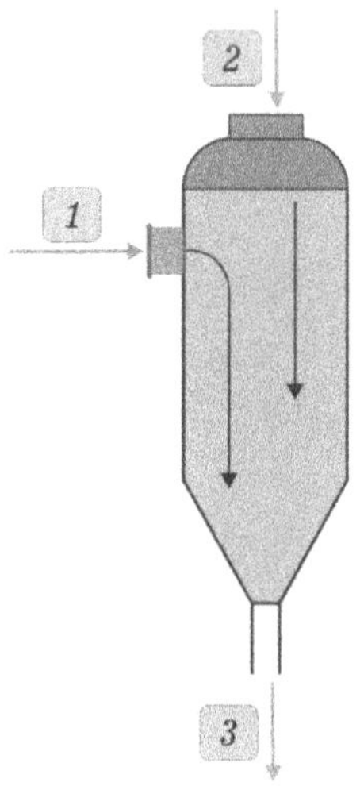

FIGURE 6.22 Schematic of Example 6.6.

TABLE 6.15
Details of Example 6.6

State No.	T (°C)	P (kPa)	Phase Description	$\dot{m}\left(\frac{kg}{s}\right)$
1	—	81.325	Saturated steam	0.82
2	25	601.325	—	—
3	40	121.325	—	—

✍Solution

Assumptions:

1. The process is considered steady state.
2. Heat loss from the surface of the condenser is negligible.
3. Ignore changes in kinetic and potential energy.
4. Consider the dead state temperature and pressure 25°C and 101.325 kPa, respectively.

Mass and energy balance:

$$\sum_i \dot{m} = \sum_o \dot{m} \rightarrow \dot{m}_1 + \dot{m}_2 = \dot{m}_3$$

$$\sum_i \dot{E} = \sum_o \dot{E} \rightarrow \dot{m}_1 h_1 + \dot{m}_2 h_2 = \dot{m}_3 h_3$$

Determining the properties of flows:

$$h_0 = 104.92\frac{kJ}{kg}, \; s_0 = 0.37\frac{kJ}{kg.K}$$

$$h_1 = h_{g@81.325kPa} = 2665.91\frac{kJ}{kg}, \; s_1 = s_{g@81.325kPa} = 7.43\frac{kJ}{kg.K}$$

$$h_2 = 105.38\frac{kJ}{kg}, \; s_2 = 0.37\frac{kJ}{kg.K}$$

$$h_3 = 167.63\frac{kJ}{kg}, \; s_3 = 0.57\frac{kJ}{kg.K}$$

Simultaneous solution of mass and energy balance:

$$\begin{cases} 0.82 + \dot{m}_2 = \dot{m}_3 \\ 0.82(2666.33) + \dot{m}_2(271.96) = \dot{m}_3(317.62) \end{cases} \rightarrow \begin{array}{l} \dot{m}_2 = 32.91\frac{kg}{s} \\ \dot{m}_3 = 33.73\frac{kg}{s} \end{array}$$

Determining the exergy of flows:

$$\begin{cases} ex_1 = (h_1 - h_0) - T_0(s_1 - s_0) = 455.68\dfrac{kJ}{kg} \\ ex_2 = (h_2 - h_0) - T_0(s_2 - s_0) = 0.50\dfrac{kJ}{kg} \\ ex_3 = (h_3 - h_0) - T_0(s_3 - s_0) = 1.55\dfrac{kJ}{kg} \end{cases}$$

Calculation of fuel and product exergy:

$$\dot{E}x_F = \dot{E}x_1 + \dot{E}x_2 = \dot{m}_1 ex_1 + \dot{m}_2 ex_2 = 0.82(455.68) + 32.91(0.5) = 390.16\text{kW}$$

$$\dot{E}x_P = \dot{E}x_3 = \dot{m}_3 ex_3 = 33.73(1.55) = 52.15\text{kW}$$

Exergy destruction rate:

$$\dot{E}x_{des} = \dot{E}x_F - \dot{E}x_P = 390.16 - 52.15 = 338.01\text{kW}$$

Exergy efficiency:

$$\eta_{ex} = \frac{\dot{E}x_P}{\dot{E}x_F} \times 100 = \frac{52.15}{390.16} \times 100 = 13.37\%$$

✐Example 6.7: Superheated steam is not as suitable for process heating as saturated steam because it provides a lower rate of heat transfer and requires a larger heat transfer area. Therefore, superheated steam must be desuperheated. A common method of desuperheating is mixing superheated steam with cooling water to produce saturated steam in a direct-contact heat exchanger known as a desuperheater. The steam produced in the steam production unit of sugar factories is saturated by a desuperheater before entering the evaporation line. Table 6.16 presents information about the desuperheater flows used in a sugar factory. Calculate the rate of exergy destruction and the rate of entropy generation in this process.

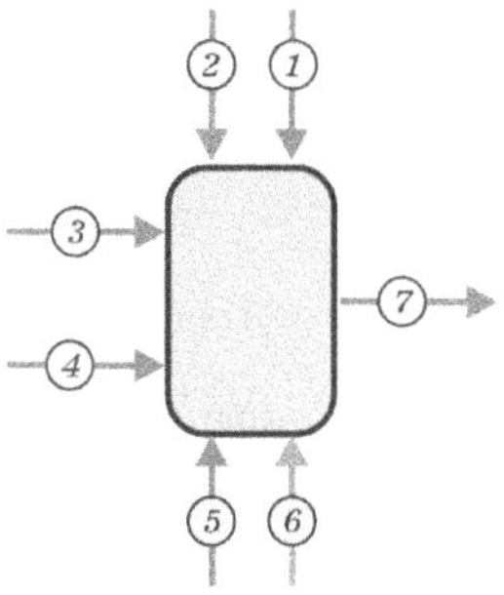

FIGURE 6.23 Schematic of Example 6.7.

TABLE 6.16
Details of Example 6.7

State No.	Fluid Type	*T* (°C)	*P* (kPa)	$\dot{m}\left(\frac{kg}{s}\right)$
1	Superheated steam	420.00	2801.33	3.06
2	Superheated steam	190.00	369.50	5.97
3	Superheated steam	190.00	369.50	5.97
4	Superheated steam	190.00	369.50	8.33
5	Saturated steam	—	1001.33	1.94
6	Water	95.10	369.50	—
7	Saturated steam	—	369.45	—

✍Solution

Assumptions:

1. The process is considered steady state.
2. Heat loss from the surface is negligible.
3. Kinetic and potential energy changes are ignored.
4. The dead state temperature and pressure are 25°C and 101.325 kPa, respectively.

Mass balance for steady state:

$$\sum \dot{m}_i = \sum \dot{m}_o \rightarrow \dot{m}_1 + \dot{m}_2 + \dot{m}_3 + \dot{m}_4 + \dot{m}_5 + \dot{m}_6 = \dot{m}_7$$

Energy balance for steady state:

$$\sum \dot{E}_i = \sum \dot{E}_o \rightarrow \dot{m}_1 h_1 + \dot{m}_2 h_2 + \dot{m}_3 h_3 + \dot{m}_4 h_4 + \dot{m}_5 h_5 + \dot{m}_6 h_6 = \dot{m}_7 h_7$$

Entropy balance for steady state:

$$\dot{S}_{gen} = \sum_o (\dot{m}s) + \left(\frac{\dot{Q}}{T_b}\right)_o - \sum_i (\dot{m}s) \rightarrow \dot{S}_{gen} = \dot{m}_7 s_7 - (\dot{m}_1 s_1 + \dot{m}_2 s_2 + \dot{m}_3 s_3 + \dot{m}_4 s_4 + \dot{m}_5 s_5 + \dot{m}_6 s_6)$$

Exergy balance for steady state:

$$\dot{E}x_{des} = \dot{E}x_F - \dot{E}x_P - \dot{E}x_L$$

$$\dot{E}x_F = \dot{E}x_1 + \dot{E}x_2 + \dot{E}x_3 + \dot{E}x_4 + \dot{E}x_5 + \dot{E}x_6$$
$$= \dot{m}_1 ex_1 + \dot{m}_2 ex_2 + \dot{m}_3 ex_3 + \dot{m}_4 ex_4 + \dot{m}_5 ex_5 + \dot{m}_6 x_6$$

$$\dot{E}x_P = \dot{E}x_7 = \dot{m}_7 ex_7$$

$$\dot{E}x_L = 0$$

Determination of flow characteristics:

$$@25°C,101.325kPa\begin{cases} h_0 = 104.92\frac{kJ}{kg} \\ s_0 = 0.37\frac{kJ}{kg}.K \end{cases}$$

$$@420°C,2801.33kPa\begin{cases} h_1 = 3280.18\frac{kJ}{kg} \\ s_1 = 7.02\frac{kJ}{kg}.K \end{cases}$$

$$@190°C,369.50kPa\begin{cases} h_2 = h_3 = h_4 = 2841.54\frac{kJ}{kg} \\ s_2 = s_3 = s_4 = 7.17\frac{kJ}{kg}.K \end{cases}$$

$$@1001.33kPa\begin{cases} h_5 = h_{f@1001.33kPa} = 762.77\frac{kJ}{kg} \\ s_5 = s_{f@1001.33kPa} = 2.14\frac{kJ}{kg}.K \end{cases}$$

$$@95.10°C,369.5kPa\begin{cases} h_6 = 398.73\frac{kJ}{kg} \\ s_6 = 1.25\frac{kJ}{kg}.K \end{cases}$$

$$@369.45kPa\begin{cases} h_7 = h_{g@369.45kPa} = 2734.45\frac{kJ}{kg} \\ s_7 = s_{g@369.45kPa} = 6.92\frac{kJ}{kg}.K \end{cases}$$

$$ex_1 = (h_1 - h_0) - T_0(s_1 - s_0)$$
$$= (3280.18 - 104.92) - 298.15(7.02 - 0.37) = 1190.26\frac{kJ}{kg}$$

$$ex_2 = ex_3 = ex_3 = (h_2 - h_0) - T_0(s_2 - s_0)$$
$$= (2841.54 - 104.92) - 298.15(7.17 - 0.37) = 709.35\frac{kJ}{kg}$$

$$ex_5 = (h_5 - h_0) - T_0(s_5 - s_0)$$
$$= (762.77 - 104.92) - 298.15(2.14 - 0.37) = 129.70\frac{kJ}{kg}$$

$$ex_6 = (h_6 - h_0) - T_0(s_6 - s_0)$$
$$= (398.73 - 104.92) - 298.15(1.25 - 0.37) = 30.21\frac{kJ}{kg}$$

$$ex_7 = (h_7 - h_0) - T_0(s_7 - s_0)$$
$$= (2734.45 - 104.92) - 298.15(6.92 - 0.37) = 675.18\frac{kJ}{kg}$$

Calculation of flow rates of streams 6 and 7 from mass and energy balance:

$$\begin{cases} 3.06 + 5.97 + 5.97 + 8.33 + 1.94 + \dot{m}_6 = \dot{m}_7 \\ 3.06(3280.18) + 5.97(2841.54) + 5.97(2841.54) + 8.33(2841.54) + 1.94(762.77) \\ +\dot{m}_6(398.73) = \dot{m}_7(2734.45) \end{cases}$$

$$\rightarrow \begin{cases} \dot{m}_6 = 0.01\frac{kg}{s} \\ \dot{m}_7 = 25.28\frac{kg}{s} \end{cases}$$

Calculation of entropy generation rate:

$$\begin{aligned}\dot{S}_{gen} &= 25.28(6.92) \\ &-[3.06(7.02) + 5.97(7.17) + 5.97(7.17) \\ &+8.33(7.17) + 1.94(2.14) + 0.01(1.25)] \\ &= 4.0451\frac{kJ}{s.K}\end{aligned}$$

Calculation of exergy destruction rate:

$$\dot{E}x_F = 3.06(1190.26) + 5.97(709.35) + 5.97(709.35) + 8.33(709.35) + 1.94(129.70) + 0.01(30.21) = 18272.50\text{kW}$$

$$\dot{E}x_P = 25.28(675.18) = 17066.45\text{kW}$$

$$\dot{E}x_{des} = 18272.50 - 17066.45 = 1206.05\text{kW}$$

Example 6.8: Hot water with a temperature of 85°C is produced by mixing saturated steam at atmospheric pressure and tap water with a temperature of 15°C. Determine the ratio of the mass of steam consumed to tap water.

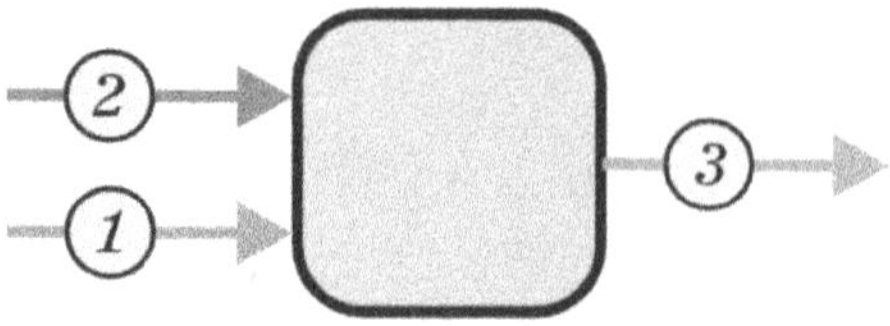

FIGURE 6.24 Schematic of Example 6.8.

✍Solution

Assumptions:

1. The process is considered steady state.
2. Heat loss from the surface is negligible.
3. Kinetic and potential energy changes are ignored.

Mass balance for steady state:

$$\sum \dot{m}_i = \sum \dot{m}_o \rightarrow \dot{m}_1 + \dot{m}_2 = \dot{m}_3$$

Energy balance for steady state:

$$\sum \dot{E}_i = \sum \dot{E}_o \rightarrow \dot{m}_1 h_1 + \dot{m}_2 h_2 = \dot{m}_3 h_3$$

Determination of flow characteristics:

$$h_1 = h_{@15^\circ C, 101.325 kPa} = 63.08 \frac{kJ}{kg}$$

$$h_2 = h_{g @ 101.325 kPa} = 2675.55 \frac{kJ}{kg}$$

$$h_3 = h_{@85^\circ C, 101.325 kPa} = 356.05 \frac{kJ}{kg}$$

Calculation of the mass flow rate ratio of flow 2 to flow 1:

$$\dot{m}_1 h_1 + \dot{m}_2 h_2 = (\dot{m}_1 + \dot{m}_2) h_3 \overset{\div \dot{m}_1}{\Rightarrow} h_1 + \frac{\dot{m}_2}{\dot{m}_1} h_2 = \left(1 + \frac{\dot{m}_2}{\dot{m}_1}\right) h_3$$

$$\rightarrow \frac{\dot{m}_2}{\dot{m}_1} = \frac{h_3 - h_1}{h_2 - h_3} = \frac{356.05 - 63.08}{2675.55 - 356.05} = 0.126$$

6.7.4 Evaporators

Water is the main component of food, and, along with other compounds (carbohydrates, proteins, salts, and fat), it plays an essential role in the life of living organisms. The water in the food affects the speed of chemical reactions, the activity of microorganisms, and the shelf life of food. Therefore, high moisture content reduces the shelf life of food. It should be noted that the amount of water in food alone cannot be used as a criterion for measuring spoilage, but the amount of available water is essential and significant to determine the spoilage of food. Water activity is a measure of the usability of water to participate in chemical, microbiological, and enzymatic reactions. In the process of concentration, a major part of the water content of the liquid food is physically removed.

Food concentration is done to achieve the following goals:

- Reducing the water activity of the product to increase shelf life
- Minimizing the costs of packaging, storage, and transportation
- Facilitating the control and management of the final product

Food concentration methods:

- Thermal concentration (evaporation)
- Membrane concentration
- Freeze concentration

Employing heat allows for the extraction of water from food, forming the foundation of the evaporation process. This procedure involves raising the liquid's temperature to its boiling point using a heating agent, resulting in the removal of a portion of its water content. Evaporation leads to an augmentation in the concentration of dissolved solids within the food while concurrently lowering the product's water activity.

Consider a solution that contains a solvent and a non-volatile solute component. The increase in concentration is done by removing part of the solvent, and the necessary energy for the evaporation of the solvent is provided by heat. In most systems, the solvent is water, and the goal is to obtain a concentrated solution from a dilute solution. Among the common examples of evaporation, we can mention the increase in the concentration of dilute solutions of sugar, salt, milk, and juices. In these samples, the concentrated solution is considered a desirable and valuable product, and the evaporated water (solvent) is worthless. The only thing that is produced from seawater is drinking water; water is a valuable product. It should be noted that the purpose of distillation is to separate components that have close boiling points and cannot be separated using the evaporation process, such as water and ethanol; however, in the process of evaporation, the boiling points of the constituents have a significant difference from each other, and in fact, it is considered a non-volatile soluble component. The drying process is also different from the evaporation process; in the drying process, the result of the solid process is almost dry, but the result of the evaporation process is a concentrated solution.

Evaporators are employed to thermally concentrate liquids. Fundamentally, an evaporator comprises a heat exchanger situated within a spacious chamber. This heat exchanger operates as a non-contact type, enabling efficient heat transfer from steam to the product. Evaporation commonly occurs under atmospheric pressure or in a vacuum, which effectively lowers the product's boiling point and subsequently curtails energy usage. This approach additionally prevents excessive heating of the product.

6.7.4.1 Non-Continuous Evaporators (Pan)

These evaporators consist of a tank with an agitator such that heating and evaporating water from food liquids is done using its jacketed thermal surface. The presence of an agitator in this evaporator increases the intensity of heat transfer (reduces operation time) and prevents burns, adhesion, and deposition of food on the inner wall of the device (thermal surface). The operation in this type of evaporator is non-continuous or batch, and the residence time of the food in this evaporator is long (between 30 minutes and several hours) due to the low ratio of the thermal surface to the volume of the food.

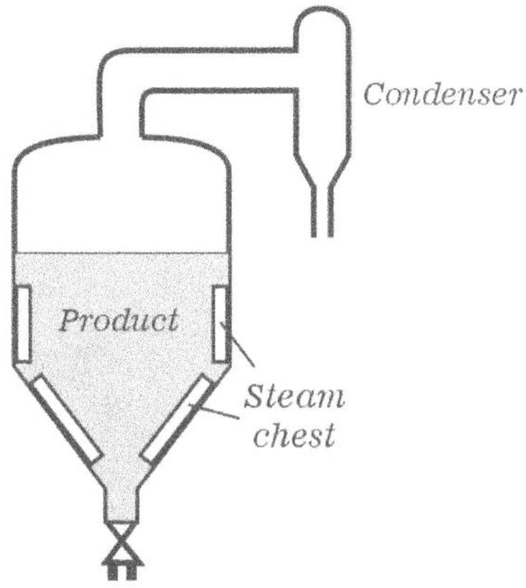

FIGURE 6.25 Pan evaporator.

6.7.4.2 Forced Circulation Evaporators

In forced circulation evaporators, the dilute liquid enters the evaporator exchanger from the bottom and by pump pressures and is heated while passing through the heat exchanger, and when leaving the exchanger, it suddenly boils inside the separator, and as a result of evaporation, it releases its vapor. The liquid coming out of the separator, which has become thicker, circulates inside the system until it reaches the desired concentration; that is, it passes through the heat exchanger and separator several times by the pump. Due to the pressure created by the pump, the food liquid does not boil in the heat exchanger and evaporates suddenly after leaving the exchanger.

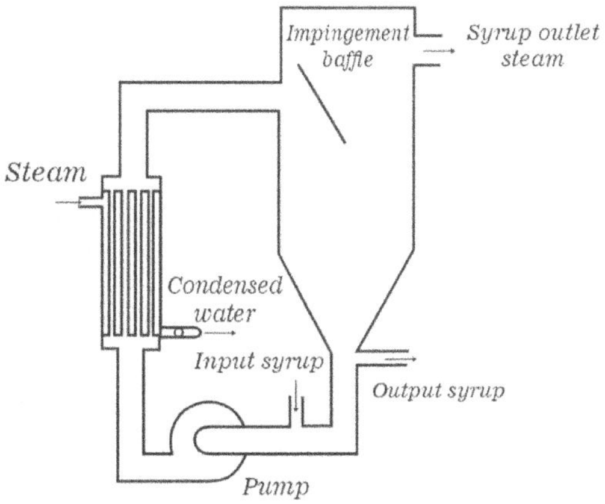

FIGURE 6.26 Forced circulation evaporator.

6.7.4.3 Natural Circulation and Horizontal Tube Evaporators

There is always vapor in the tubes of this evaporator, and the boiling process is done outside the tubes. The vapor coming out of these evaporators contains liquid droplets. By using plates called demister pads, exiting liquid drops can be prevented.

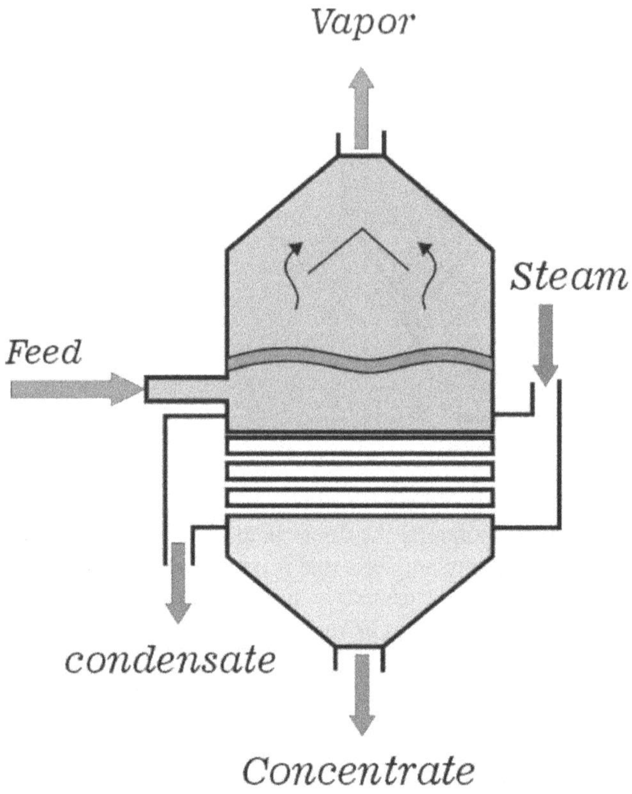

FIGURE 6.27 Natural circulation and horizontal-tube evaporator.

6.7.4.4 Natural Circulation and Vertical Tube Evaporators

In natural circulation and vertical tube evaporators, the process involves steam condensation occurring outside the tubes, while liquid boiling takes place within the tubes. As the fluid in proximity to the heat transfer surfaces (inside the tubes) is heated, its density decreases, prompting it to move upward within the tube as it undergoes boiling. Upon exiting the tubes, this fluid cools due to its distance from the inner wall's thermal surfaces, causing it to descend through a central, thicker tube. This cooling-driven increase in density propels the fluid downward, ultimately emerging as a concentrated product through the lower tube. Robert's evaporator is a type of natural circulation and horizontal-tube evaporator used in the sugar industry.

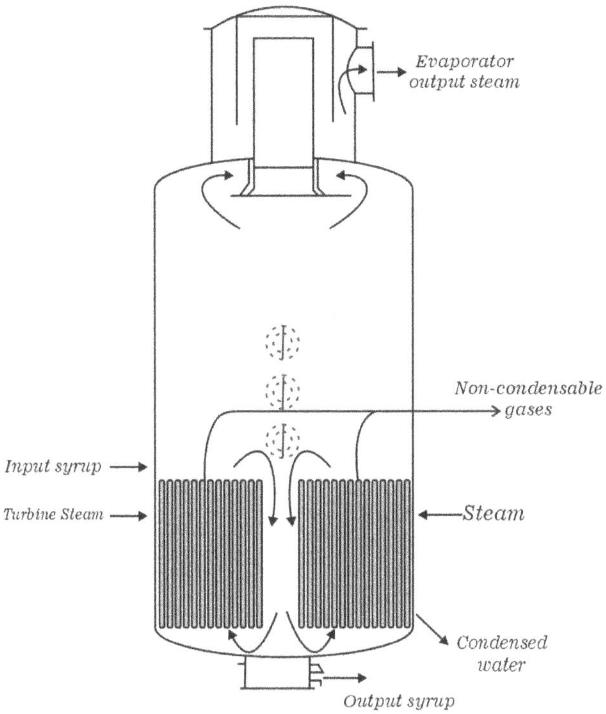

FIGURE 6.28 Robert-type evaporator.

In another type of natural circulation and vertical tube evaporators, the heat exchanger tubes are outside the separator tank, which causes the size of the tubes not to be a function of the size and shape of the tank; therefore, a higher capacity can be achieved.

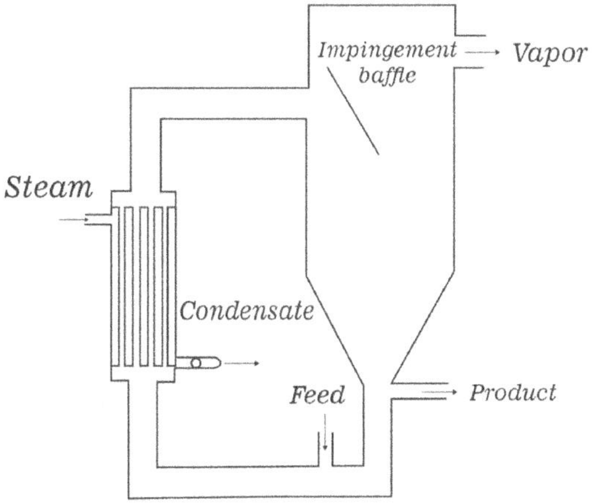

FIGURE 6.29 Natural circulation and vertical tube evaporator.

6.7.4.5 Long-Tube Falling-Film Evaporators

To retain properties and prevent undesired alterations in temperature-sensitive food, the concentration process should be executed swiftly and at the lowest feasible temperature. For this purpose, falling-film evaporators are usually used. In this evaporator, the feed enters from the top, is divided between the tubes by a distributor system, flows in the form of a thin film on the walls of the tubes, and moves downward due to the force of weight. In this type of evaporator, it is possible to concentrate foods with high viscosity.

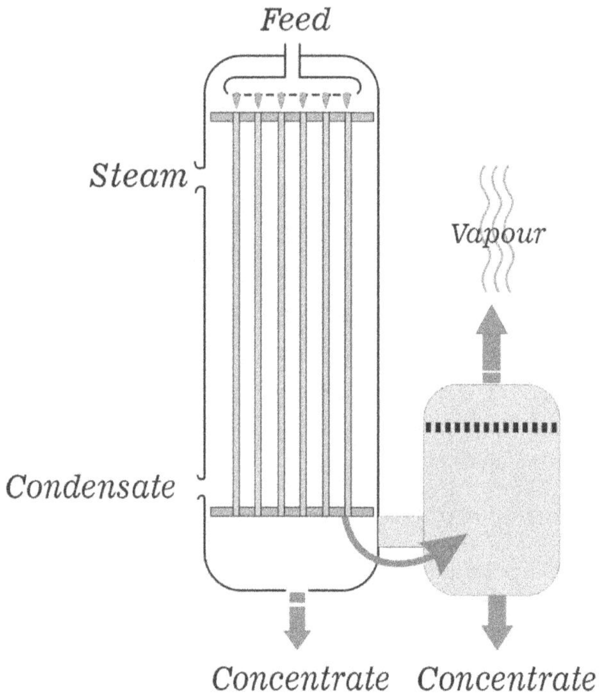

FIGURE 6.30 Falling film evaporator.

6.7.4.6 Long-Tube Rising-Film Evaporators

In this evaporator design, the food, which has already reached boiling temperature due to passing through a heat exchanger, enters the evaporating heat exchanger from the lower part with boiling temperature. While passing through this heat exchanger, steam bubbles are created inside the tubes of the heat exchanger due to receiving the heat of the heating medium. With the upward movement of the liquid, the size of these bubbles becomes larger, and finally, with the increase in the volume of steam, the created steam fills the inner part of the tube, and as a result of increasing steam pressure, a thin layer of liquid (film) is formed on the thermal surfaces, and this steam pressure causes the film to move upwards. The formation of a thin layer near the tube (heat transfer surface) increases the heat transfer efficiency and evaporation rate.

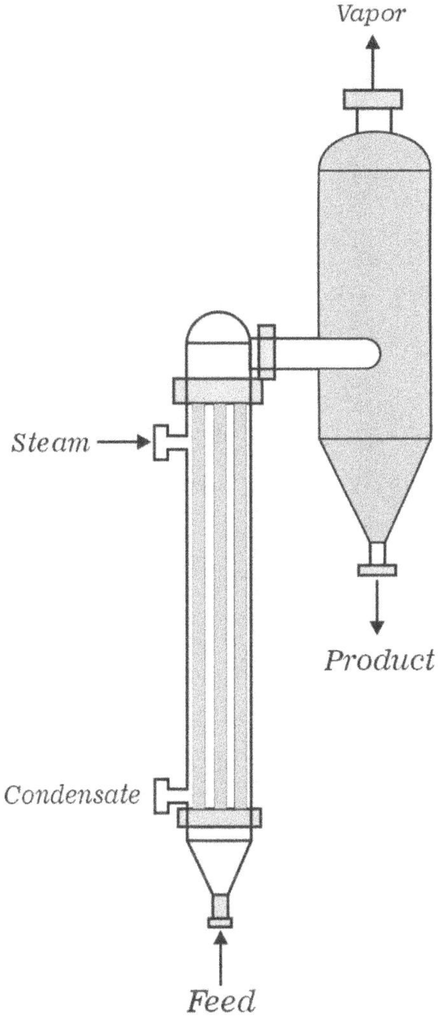

FIGURE 6.31 Rising-film evaporator.

6.7.4.7 Rising-Falling Film Evaporators

In this evaporator design, first, the dilute liquid enters the rising part, and then, with the increase in concentration and as a result of the increase in viscosity, it enters the falling part. Since the liquid enters the rising part first, there is no need to use distribution devices in this system; therefore, its price is lower than the falling type, and because the liquid enters the falling part after the rising part when its viscosity increases, it can be used for liquids with high viscosity like the falling film type. Due to the lower temperature difference between the heating medium and the food, the sediment formation rate is lower in this rising type. Therefore, this system has the advantages of both rising and falling film systems.

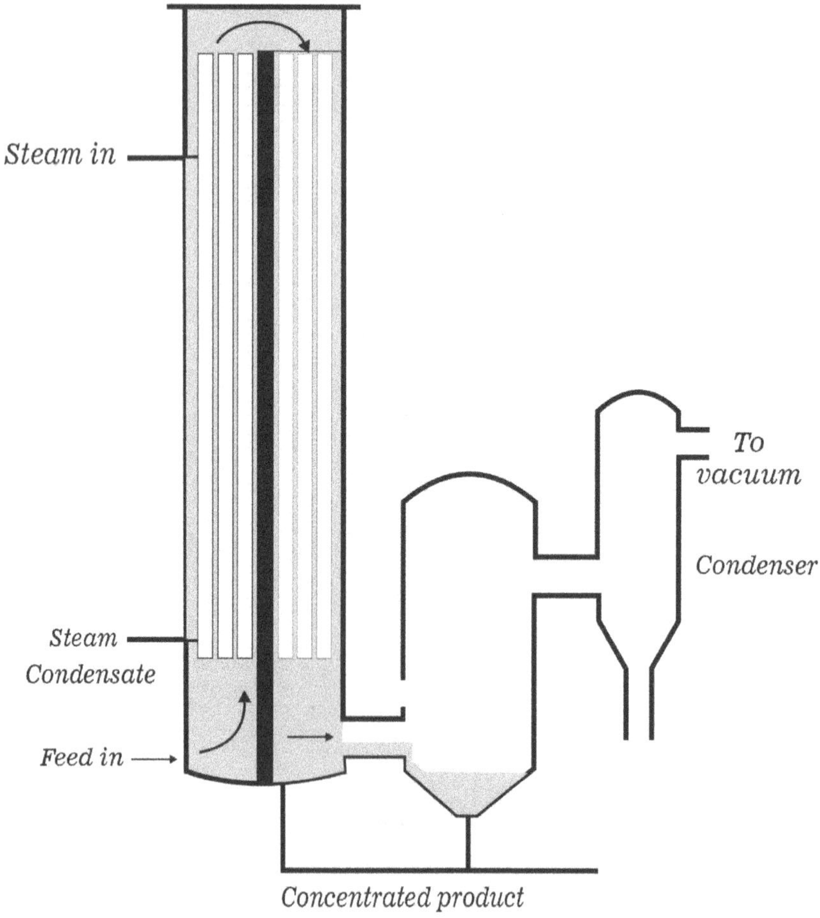

FIGURE 6.32 Rising-falling film evaporator.

6.7.4.8 Agitated Evaporators

In evaporators, the main resistance to heat transfer is in the liquid phase. One of the ways to increase the heat transfer rate is to increase the turbulence (increase the heat transfer coefficient) by using mechanical agitators (scraping blades) in the liquid phase. The food liquid from the upper part of this heat exchanger, while being disturbed by the scraping blade, flows from the upper part of the heat exchanger to the lower part as a film between the inner surface of the heat exchanger and the scraping blade. The rotation of the scraping blade improves the heat transfer and prevents sticking, sedimentation, and burning of the food to the inner wall of the heat exchanger. This evaporator is suitable for quick heat transfer to highly viscous liquids, such as gelatin. In other evaporators, with the increase in viscosity, the heat transfer coefficient decreases, but in this type of evaporator, the decrease in the heat transfer coefficient is less.

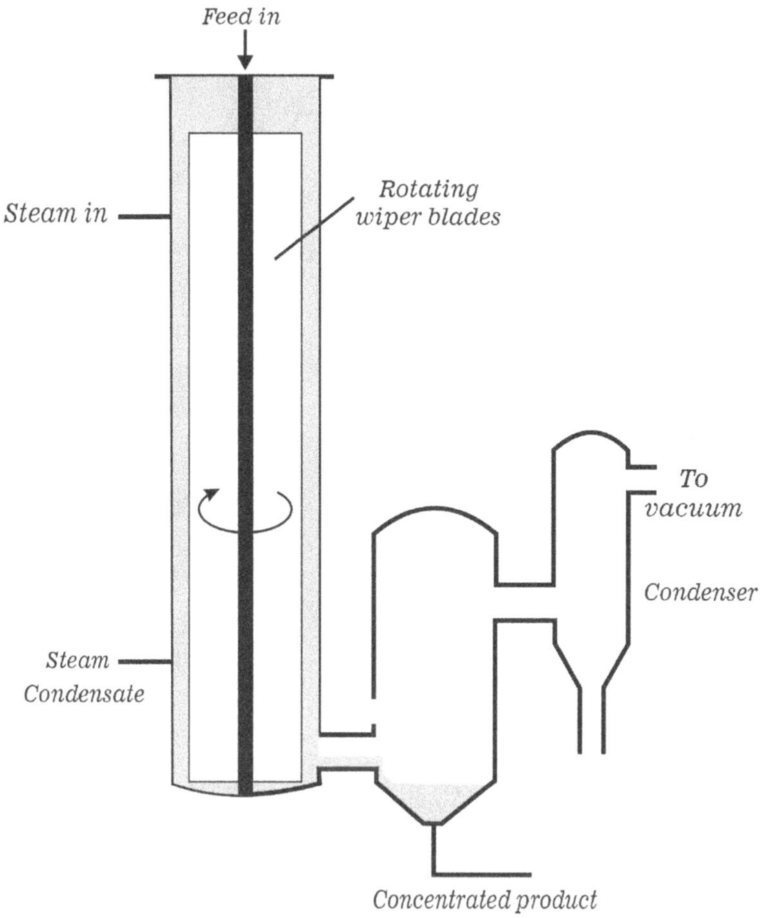

FIGURE 6.33 Agitated evaporator.

6.7.4.9 Plate Evaporators

The plate evaporator is the same as the falling film evaporator, whose heat exchanger is a plate type instead of a tubular shell. In the following section, the long tube evaporator is compared with the plate evaporator:

1. The height of plate evaporators is less than that of long tube evaporators, but the area occupied by this evaporator is greater.
2. Plate evaporators are easier to open and close compared to long tube evaporators, and their maintenance is also simpler.
3. The heat transfer coefficient is higher in the plate type, and as a result, the heat transfer speed is also higher and the residence time of the product is lower.
4. The capacity of the plate evaporator is less than that of the long tube type.

6.7.4.10 Balance Equations

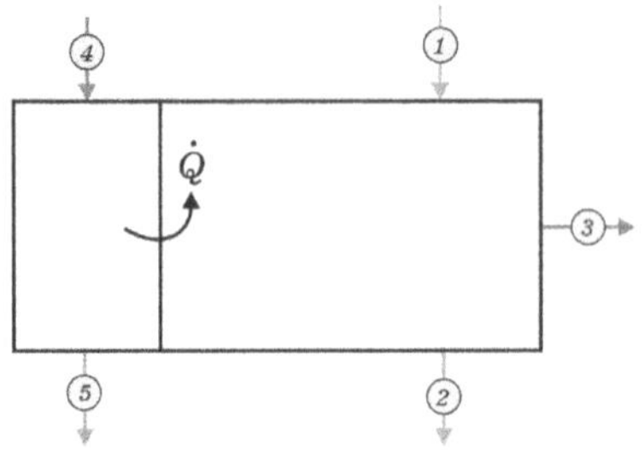

FIGURE 6.34 Schematic of inflow and outflow streams in a typical evaporator.

Mass balance:

$$\dot{m}_1 = \dot{m}_2 + \dot{m}_3 \qquad (6.48)$$
$$\dot{m}_4 = \dot{m}_5$$

Material balance:

$$TS_1\dot{m}_1 = TS_2\dot{m}_2 \qquad (6.49)$$

Energy balance:

$$\dot{m}_1 h_1 + \dot{m}_4 h_4 = \dot{m}_2 h_2 + \dot{m}_3 h_3 + \dot{m}_5 h_5 + \dot{Q} \qquad (6.50)$$

Entropy balance:

$$\dot{m}_1 s_1 + \dot{m}_4 s_4 + \dot{S}_{gen} = \dot{m}_2 s_2 + \dot{m}_3 s_3 + \dot{m}_5 s_5 + \frac{\dot{Q}}{T_b} \qquad (6.51)$$

Exergy balance:

$$\dot{E}x_F = \left(\dot{E}x_4 - \dot{E}x_5\right) + \dot{E}x_1 \qquad (6.52)$$

$$\dot{E}x_P = \dot{E}x_2 + \dot{E}x_3 \qquad (6.53)$$

$$\dot{E}x_L = \left(1 - \frac{T_0}{T_b}\right)\dot{Q} \qquad (6.54)$$

$$\dot{E}x_{des} = \left(\dot{E}x_4 - \dot{E}x_5\right) + \dot{E}x_1 - \left(\dot{E}x_2 + \dot{E}x_3\right) - \left(1 - \frac{T_0}{T_b}\right)\dot{Q} \qquad (6.55)$$

Exergy efficiency:

$$\eta_{ex} = \frac{\dot{E}x_2 + \dot{E}x_3}{\left(\dot{E}x_4 - \dot{E}x_5\right) + \dot{E}x_1} \qquad (6.56)$$

✐Example 6.9: An evaporator is used to concentrate juice from 31% of solid material to 70%. The information about the flows is presented in Table 6.17.

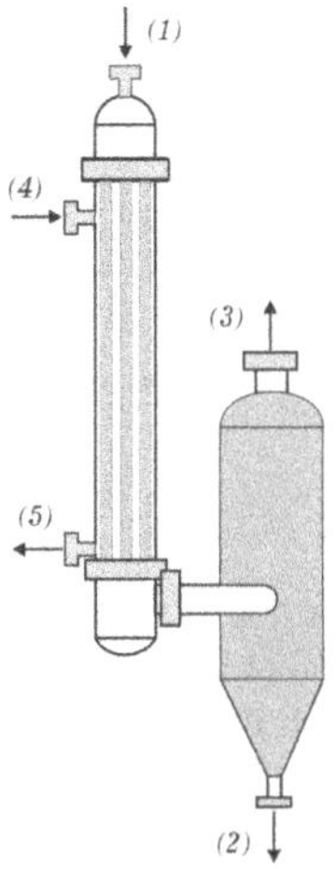

FIGURE 6.35 Schematic of Example 6.9.

TABLE 6.17
Details of Example 6.9

State No.	Fluid Type	T (°C)	P (kPa)	$\dot{m}\left(\frac{kg}{s}\right)$	Phase Description
1	Juice	107	401.325	1.45	—
2	Juice	95.8	81.325	—	—
3	Steam	—	81.325	—	Saturated steam
4	Steam	—	126.325	0.86	Saturated steam
5	Condensate	—	—	—	—

The evaporator surface temperature is reported as 45°C. Also, the composition of juice flows is in Table 6.18.

TABLE 6.18
Mass Fraction of Components of Juice in Example 6.9

State No.	X_{Water}	$X_{Protein}$	X_{Fat}	$X_{Carbohydrate}$	X_{Ash}	TS (%)
1	0.69	0.00946	0.00027	0.28983	0.01044	31
2	—	—	—	—	—	70

Calculate the rate of exergy destruction and the exergy efficiency of the process.

✍Solution

Assumptions:

1. The process is considered steady state.
2. Changes in kinetic energy and potential can be ignored.
3. The dead state temperature and pressure are 25°C and 101.325 kPa.
4. The reference temperature and pressure are 0°C and 101.325 kPa.
5. Condensate water leaving the heat exchanger is saturated liquid.
6. The pressure drop in the steam tube is insignificant.

Determining the composition of juice leaving from the evaporator:

$$X_{water,2} = 1 - TS_2 = 1 - 0.7 = 0.3$$

While the total quantity of each compound remains constant in both the input and output juices, the evaporation of water from the juice introduces a shift in the mass fraction of each component. Interestingly, the ratio of the mass fraction for each compound in the input and output juices mirrors the ratio of solid substances in both scenarios. This intriguing relationship underscores the intricate interplay between composition and evaporation within the juice processing dynamics.

$$\frac{X_{1,2}}{X_{i,1}} = \frac{TS_2}{TS_1}$$

$$\rightarrow \begin{cases} X_{Protein,2} = X_{Protein,1} \times \dfrac{TS_2}{TS_1} = 0.00946 \times \dfrac{70}{31} = 0.02136 \\ X_{Fat,2} = X_{Fat,1} \times \dfrac{TS_2}{TS_1} = 0.00027 \times \dfrac{70}{31} = 0.00061 \\ X_{Ash,2} = X_{Ash,1} \times \dfrac{TS_2}{TS_1} = 0.01044 \times \dfrac{70}{31} = 0.02357 \\ X_{Carbohydrate,2} = X_{Carbohydrate,1} \times \dfrac{TS_2}{TS_1} = 0.28983 \times \dfrac{70}{31} = 0.65445 \end{cases}$$

The equations in Table 6.1 were used to calculate the specific heat and density of the juice compounds, and the results are presented in Table 6.19.

TABLE 6.19
Properties of Juice in Example 6.9

	Protein		Fat		Carbohydrate		Ash		Water	
T (°C)	C_P	ρ	C_P	ρ	C_P	ρ	C_P	ρ	C_P	ρ
$T_{ref} = 0°C$	2.01	—	1.98	—	1.55	—	1.09	—	4.18	—
$T_0 = 25°C$	2.04	—	2.02	—	1.59	—	1.14	—	4.18	—
$T_1 = 107°C$	2.12	1274.43	2.09	880.91	1.69	1565.88	1.25	2393.77	4.23	954.50
$T_2 = 95.8°C$	2.11	1280.24	2.08	885.59	1.68	1569.36	1.24	2396.92	4.22	963.00

Specific heat and specific volume of fruit juice were determined using Equations 6.4 and 6.7. The results are presented in Table 6.20.

TABLE 6.20
Properties of Streams in Example 6.9

State No.		$C_P\left(\frac{kJ}{kg.K}\right)$	$v\left(\frac{m^3}{kg}\right)$
1	$T_{ref}=0°C$	3.36	—
	$T_0=25°C$	3.38	—
	$T_1=107°C$	3.44	0.00092
2	$T_{ref}=0°C$	3.34	—
	$T_0=25°C$	3.37	—
	$T_1=95.8°C$	2.44	0.00076

Equation 6.6 was used to determine the enthalpy of juice flows. Enthalpy and entropy of condensed water flows were determined from the table of thermodynamic properties of water:

$$h_0=104.92\frac{kJ}{kg},\ s_0=0.37\frac{kJ}{kg.K}$$

$$h_1=\left(\frac{C_{P@T_1}+C_{P@T_{ref}}}{2}\right)T_1+v_1(P_1-P_0)$$

$$=3.40(107)+0.00092(401.325-101.325)=364.25\frac{kJ}{kg}$$

$$h_2=\left(\frac{C_{P@T_2}+C_{P@T_{ref}}}{2}\right)T_2+v_2(P_2-P_0)$$

$$=2.39(95.8)+0.00076(81.325-101.325)=228.86\frac{kJ}{kg}$$

$$h_3=h_{g@P_3}=2668.91\frac{kJ}{kg},\ s_3=s_{g@P_3}=7.43\frac{kJ}{kg.K}$$

$$h_4=h_{g@P_4}=2685.38\frac{kJ}{kg},\ s_4=s_{g@P_4}=7.28\frac{kJ}{kg.K}$$

$$h_5=h_{f@P_5}=445.65\frac{kJ}{kg},\ s_5=s_{f@P_5}=1.38\frac{kJ}{kg.K}$$

Mass balance for the steady state:

$$\sum\dot{m}_i=\sum\dot{m}_o\rightarrow\begin{cases}TS_1\dot{m}_1=TS_2\dot{m}_2\\ \dot{m}_1=\dot{m}_2+\dot{m}_3\\ \dot{m}_4=\dot{m}_5\end{cases}\rightarrow\begin{cases}0.31(1.45)=0.7\dot{m}_2\\ 1.45=\dot{m}_2+\dot{m}_3\\ 0.8=\dot{m}_5\end{cases}\rightarrow\begin{cases}\dot{m}_2=0.64\,{}^{kg}/_{s}\\ \dot{m}_3=0.81\,{}^{kg}/_{s}\\ \dot{m}_5=0.80\,{}^{kg}/_{s}\end{cases}$$

Energy balance for the steady state:

$$\sum \dot{E}_i = \sum \dot{E}_o$$

$$\rightarrow \dot{m}_1 h_1 + \dot{m}_4 h_4 = \dot{m}_2 h_2 + \dot{m}_3 h_3 + \dot{m}_5 h_5 + \dot{Q}_L$$

$$\rightarrow 1.45(364.25) + 0.8(2685.38) = 0.64(228.86) + 0.81(2668.91) + 0.80(445.65) + \dot{Q}_L$$

$$\rightarrow \dot{Q}_L = 19.31\text{kW}$$

To calculate the exergy of juice and water flows, the following equations were used:

$$ex_1 = \left(\frac{C_{p@T_1} + C_{p@T_0}}{2}\right)\left(T_1 - T_0 - T_0 \ln\left(\frac{T_1}{T_0}\right)\right) + v_{@T_1}(P_1 - P_0)$$

$$= 3.41\left[380.15 - 298.15 - 298.15\ln\left(\frac{380.15}{298.15}\right)\right] + 0.00092(401.325 - 101.325) = 32.86\frac{\text{kJ}}{\text{kg}}$$

$$ex_2 = \left(\frac{C_{p@T_2} + C_{p@T_0}}{2}\right)\left(T_2 - T_0 - T_0 \ln\left(\frac{T_2}{T_0}\right)\right) + v_{@T_2}(P_2 - P_0)$$

$$= 2.40\left[368.95 - 298.15 - 298.15\ln\left(\frac{368.95}{298.15}\right)\right] + 0.00076(81.325 - 101.325) = 17.48\frac{\text{kJ}}{\text{kg}}$$

$$ex_3 = (h_3 - h_0) - T_0(s_3 - s_0)$$

$$= (2668.91 - 104.92) - 298.15(7.43 - 0.37) = 455.68\frac{\text{kJ}}{\text{kg}}$$

$$ex_4 = (h_4 - h_0) - T_0(s_4 - s_0)$$

$$= (2685.38 - 104.92) - 298.15(7.28 - 0.37) = 519.25\frac{\text{kJ}}{\text{kg}}$$

$$ex_5 = (h_5 - h_0) - T_0(s_5 - s_0)$$

$$= (445.65 - 104.92) - 298.15(1.38 - 0.37) = 39.5\frac{\text{kJ}}{\text{kg}}$$

Calculation of exergy of fuel, product, and loss:

$$\dot{E}x_F = \left(\dot{E}x_4 - \dot{E}x_5\right) + \dot{E}x_1 = \dot{m}_4 ex_4 - \dot{m}_5 ex_5 + \dot{m}_1 ex_1$$

$$= \dot{m}_4(ex_4 - ex_5) + \dot{m}_1 ex_1$$

$$= 0.8(519.25 - 39.5) + 1.45(32.83) = 431.45\text{kW}$$

$$\dot{E}x_P = \dot{E}x_2 + \dot{E}x_3 = \dot{m}_2 ex_2 + \dot{m}_3 ex_3 = 0.64(17.48) + 0.81(455.68) = 379.35\text{kW}$$

$$\dot{E}x_L = \left(1 - \frac{T_0}{T_b}\right)\dot{Q} = \left(1 - \frac{298.15}{318.15}\right)19.31 = 1.21\text{kW}$$

Exergy destruction rate:

$$\dot{E}x_{des} = \dot{E}x_F - \dot{E}x_P - \dot{E}x_L = 431.45 - 379.35 - 1.21 = 50.89\text{kW}$$

Exergy efficiency:

$$\eta_{ex} = \frac{\dot{E}x_P}{\dot{E}x_F} \times 100 = \frac{379.35}{431.45} \times 100 = 87.92\%$$

✎Example 6.10: In a sugar factory, Brix of the syrup is increased from 14.2% to 21.9% by a Robert-type evaporator (Piri et al., 2021). The rate of heat loss from the surface of the evaporator is calculated by the equations of heat transfer and reported as 30 kW, and the temperature of the surface of the equipment is 60°C. Calculate the rate of exergy destruction and exergy efficiency in this process.

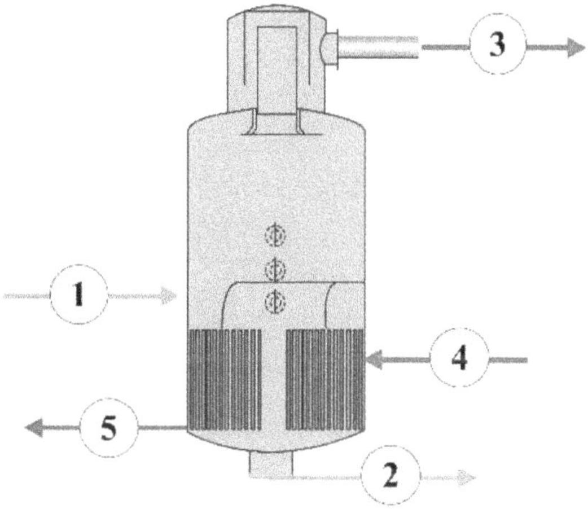

FIGURE 6.36 Schematic of the evaporator related to Example 6.10.

TABLE 6.21
Details of Example 6.10

State No.	Fluid Type	T (°C)	P (kPa)	$\dot{m}\left(\frac{\text{kg}}{\text{s}}\right)$	Phase Description	Brix (%)
1	Syrup	124	391.325	36.63	—	14.2
2	Syrup	126.60	241.325	—	—	21.9
3	Steam	—	241.325	—	Saturated steam	—
4	Steam	140.80	—	—	Saturated steam	—
5	Condensate	—	—	—	—	—

✍Solution

Assumptions:

1. The process is considered steady state.
2. Kinetic and potential energy changes are ignored.
3. The dead state temperature and pressure are 25°C and 101.325 kPa.
4. The reference temperature and pressure are 0°C and 101.325 kPa.
5. Condensate water leaving the heat exchanger is a saturated liquid.
6. The pressure drop in the steam pipe is negligible.
7. Syrup is composed of water and carbohydrates (sucrose).

The equations in Table 6.1 were used to calculate the specific heat and density of the syrup compounds, and the results are presented in Table 6.22.

TABLE 6.22
Properties of Juice in Example 6.10

	Carbohydrate		Water	
T **(°C)**	C_P	ρ	C_P	ρ
$T_{ref} = 0°C$	1.55	—	4.18	—
$T_0 = 25°C$	1.59	—	4.18	—
$T_1 = 124°C$	1.70	1560.60	4.25	939.80
$T_2 = 126.6°C$	1.70	1559.80	4.25	937.36

The specific heat and volume of fruit syrup were determined using Equations 6.4 and 6.7. The results are presented in Table 6.23.

TABLE 6.23
Properties of Streams in Example 6.10

State No.		$C_P\left(\frac{kJ}{kg.K}\right)$	$v\left(\frac{m^3}{kg}\right)$
1	$T_{ref} = 0°C$	3.80	—
	$T_0 = 25°C$	3.81	—
	$T_1 = 124°C$	3.89	0.00100
2	$T_{ref} = 0°C$	3.60	—
	$T_0 = 25°C$	3.61	—
	$T_1 = 126.6°C$	3.69	0.00097

Equation 6.6 was used to determine the enthalpy of syrup flows. Enthalpy and entropy of condensed water flows were determined from the table of thermodynamic properties of water:

$$h_0 = 104.92\frac{\text{kJ}}{\text{kg}}, s_0 = 0.37\frac{\text{kJ}}{\text{kg.K}}$$

$$h_1 = \left(\frac{C_{P@T_1} + C_{P@T_{ref}}}{2}\right)T_1 + v_1(P_1 - P_0)$$

$$= 3.85(124) + 0.001(391.325 - 101.325) = 477.09\frac{\text{kJ}}{\text{kg}}$$

$$h_2 = \left(\frac{C_{P@T_2} + C_{P@T_{ref}}}{2}\right)T_2 + v_2(P_2 - P_0)$$

$$= 3.65(126.6) + 0.00097(241.325 - 101.325) = 461.89\frac{\text{kJ}}{\text{kg}}$$

$$h_3 = h_{g@P_3} = 2714.89\frac{\text{kJ}}{\text{kg}}, s_3 = s_{g@P_3} = 7.06\frac{\text{kJ}}{\text{kg.K}}$$

$$h_4 = h_{g@T_4} = 2734.49\frac{\text{kJ}}{\text{kg}}, s_4 = s_{g@T_4} = 6.92\frac{\text{kJ}}{\text{kg.K}}$$

$$h_5 = h_{f@T_5} = 592.59\frac{\text{kJ}}{\text{kg}}, s_5 = s_{f@T_5} = 1.75\frac{\text{kJ}}{\text{kg.K}}$$

Mass balance for the steady state:

$$\sum \dot{m}_i = \sum \dot{m}_o \rightarrow \begin{cases} \text{Brix}_1\dot{m}_1 = \text{Brix}_2\dot{m}_2 \\ \dot{m}_1 = \dot{m}_2 + \dot{m}_3 \\ \dot{m}_4 = \dot{m}_5 \end{cases} \rightarrow \begin{cases} 0.142(36.63) = 0.219\dot{m}_2 \\ 36.63 = \dot{m}_2 + \dot{m}_3 \\ \dot{m}_4 = \dot{m}_5 \end{cases} \rightarrow \begin{cases} \dot{m}_2 = 23.75\,\text{kg}/\text{s} \\ \dot{m}_3 = 12.88\,\text{kg}/\text{s} \\ \dot{m}_4 = \dot{m}_5 \end{cases}$$

Energy balance for the steady state:

$$\sum \dot{E}_i = \sum \dot{E}_o$$

$$\rightarrow \dot{m}_1h_1 + \dot{m}_4h_4 = \dot{m}_2h_2 + \dot{m}_3h_3 + \dot{m}_5h_5 + \dot{Q}_L$$

$$\rightarrow 36.63(477.09) + \dot{m}_4(2734.49) = 23.75(461.89)$$

$$+12.88(2714.89) + \dot{m}_5(592.59) + 30$$

$$\rightarrow \dot{m}_4 = \dot{m}_5 = 13.30\frac{\text{kg}}{\text{s}}$$

To calculate the exergy of syrup and water flows, the following equations were used:

$$ex_1 = \left(\frac{C_{p@T_1} + C_{p@T_0}}{2}\right)\left(T_1 - T_0 - T_0\ln\left(\frac{T_1}{T_0}\right)\right) + v_{@T_1}(P_1 - P_0)$$

$$= 3.85\left[397.15 - 298.15 - 298.15\ln\left(\frac{397.15}{298.15}\right)\right]$$

$$+0.001(391.325 - 101.325) = 52.31\frac{\text{kJ}}{\text{kg}}$$

$$ex_2 = \left(\frac{C_{p@T_2} + C_{p@T_0}}{2}\right)\left(T_2 - T_0 - T_0 \ln\left(\frac{T_2}{T_0}\right)\right) + v_{@T_2}(P_2 - P_0)$$

$$= 3.65\left[399.75 - 298.15 - 298.15\ln\left(\frac{399.75}{298.15}\right)\right]$$

$$+0.00097(241.325 - 101.325) = 51.89\frac{\text{kJ}}{\text{kg}}$$

$$ex_3 = (h_3 - h_0) - T_0(s_3 - s_0)$$

$$= (2714.89 - 104.92) - 298.15(7.06 - 0.37) = 613.23\frac{\text{kJ}}{\text{kg}}$$

$$ex_4 = (h_4 - h_0) - T_0(s_4 - s_0)$$

$$= (2734.49 - 104.92) - 298.15(6.92 - 0.37) = 675.32\frac{\text{kJ}}{\text{kg}}$$

$$ex_5 = (h_5 - h_0) - T_0(s_5 - s_0)$$

$$= (592.59 - 104.92) - 298.15(1.75 - 0.37) = 76.14\frac{\text{kJ}}{\text{kg}}$$

Calculation of exergy of fuel, product, and loss:

$$\dot{E}x_F = \left(\dot{E}x_4 - \dot{E}x_5\right) + \dot{E}x_1 = \dot{m}_4 ex_4 - \dot{m}_5 ex_5 + \dot{m}_1 ex_1$$

$$= \dot{m}_4(ex_4 - ex_5) + \dot{m}_1 ex_1$$

$$= 13.30(675.32 - 76.14) + 36.63(52.31) = 9885.86\text{kW}$$

$$\dot{E}x_P = \dot{E}x_2 + \dot{E}x_3 = \dot{m}_2 ex_2 + \dot{m}_3 ex_3 = 23.75(51.86) + 12.88(613.23)$$

$$= 9130.36\text{kW}$$

$$\dot{E}x_L = \left(1 - \frac{T_0}{T_b}\right)\dot{Q} = \left(1 - \frac{298.15}{333.15}\right)30 = 3.15\text{kW}$$

Exergy destruction rate:

$$\dot{E}x_{des} = \dot{E}x_F - \dot{E}x_P - \dot{E}x_L = 9885.86 - 9130.36 - 3.15 = 752.35\text{kW}$$

Exergy efficiency:

$$\eta_{ex} = \frac{\dot{E}x_P}{\dot{E}x_F} \times 100 = \frac{9130.36}{9885.86} \times 100 = 92.36\%$$

6.7.4.11 Batch Evaporators

The working method of the batch evaporators is that first the valve related to the feed flow (F) is opened, and the evaporator tank is filled with a certain amount of product. After the loading is completed, the valve is closed. Then the valve related to the steam flow is opened, and the heating of the product starts. As a result of heat transfer, the product boils and starts to evaporate. After the product

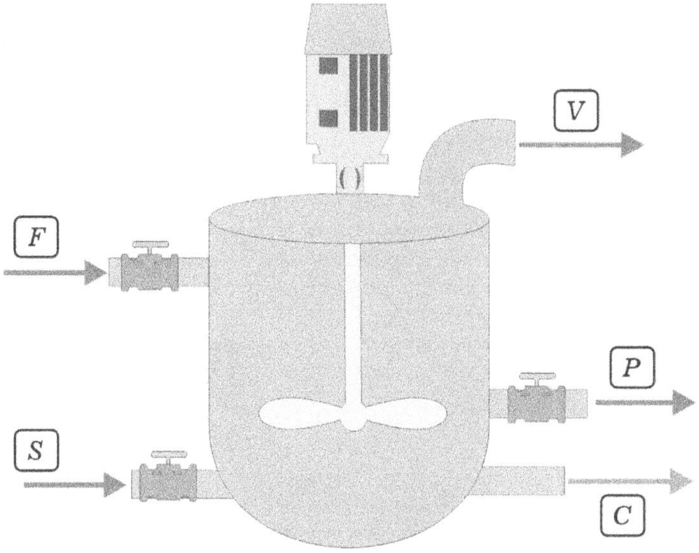

FIGURE 6.37 Batch evaporator.

reaches the desired concentration, the heating is stopped, the valve related to the product flow (P) is returned, and the concentrated product is removed from the evaporator. During the evaporation process, the product is continuously stirred by an agitator to cause uniform heat transfer and prevent the product from burning in the walls.

Example 6.11: In a fruit concentration factory, a batch evaporator is used to concentrate tomato juice. Tomato juice is evaporated at a temperature of 75°C, its solid substance is 6.5%, it is 1000 kg, and it comes out at a temperature of 69.5°C and solid substance of 20%. The pressure of the evaporator is 30 kPa, and saturated steam with a pressure of 300 kPa is used to heat the tomato juice. The shaft power of the agitator is 0.5 kW. If the process takes 1 hour, calculate the amount of exergy destruction in the process. The composition of tomato juice at the beginning and end of the process is shown in Table 6.24.

TABLE 6.24
Mass Fraction of Components in Example 6.11.

X_{Water}	$X_{Protein}$	X_{Fat}	$XC_{arbohydrate}$	X_{Ash}	TS (%)
0.935	0.011	0.002	0.047	0.005	6.5

✍Solution

Assumptions:

1. Heat loss from the surface of the evaporator is insignificant.
2. Kinetic and potential energy changes are ignored.
3. After transferring heat to the tomato juice, the steam leaves the evaporator in a saturated form and at the same pressure.
4. The dead state temperature and pressure are 25°C and 101.325 kPa.
5. The reference temperature and pressure are 0°C and 101.325 kPa.

The following represents determining the composition of tomato juice leaving the evaporator:

$$X_{water,2} = 1 - TS_2 = 1 - 0.2 = 0.8$$

The mass fraction ratio of each component of tomato juice in the primary and secondary states is equal to the ratio of solid substance in both states:

$$\frac{X_{i,2}}{X_{i,1}} = \frac{TS_2}{TS_1}$$

$$\rightarrow \begin{cases} X_{Protein,2} = X_{Protein,1} \times \dfrac{TS_2}{TS_1} = 0.0011 \times \dfrac{20}{6.5} = 0.33846 \\ X_{Fat,2} = X_{Fat,1} \times \dfrac{TS_2}{TS_1} = 0.002 \times \dfrac{20}{6.5} = 0.00645 \\ X_{Ash,2} = X_{Ash,1} \times \dfrac{TS_2}{TS_1} = 0.005 \times \dfrac{20}{6.5} = 0.01538 \\ X_{Carbohydrate,2} = X_{Carbohydrate,1} \times \dfrac{TS_2}{TS_1} = 0.047 \times \dfrac{20}{6.5} = 0.14462 \end{cases}$$

The equations in Table 6.1 were used to calculate the specific heat and density of the tomato juice compounds, and the results are presented in Table 6.25.

TABLE 6.25
Properties of Tomato Juice in Example 6.11

	Protein		Fat		Carbohydrate		Ash		Water	
T (°C)	C_P	ρ	C_P	ρ	C_P	ρ	C_P	ρ	C_P	ρ
$T_{ref} = 0°C$	2.01	—	1.98	—	1.55	—	1.09	—	4.18	—
$T_0 = 25°C$	2.04	—	2.02	—	1.59	—	1.14	—	4.18	—
$T_1 = 75°C$	2.09	1291.02	2.07	894.27	1.66	1575.82	1.21	2402.75	4.20	976.28
$T_2 = 69.5°C$	20.9	1293.87	2.06	896.57	1.66	1577.52	1.21	2404.30	4.20	979.25

Like Example 6.10, the specific heat and specific volume of tomato juice were determined using Equations 6.4 and 6.7. The results are presented in Table 6.26.

TABLE 6.26
Properties of Example 6.11

State No.		$C_P\left(\frac{\text{kJ}}{\text{kg.K}}\right)$	$v\left(\frac{\text{m}^3}{\text{kg}}\right)$
1	$T_{ref} = 0°C$	4.01	—
	$T_0 = 25°C$	4.01	0.00098
	$T_1 = 75°C$	4.04	0.00100
2	$T_{ref} = 0°C$	3.66	—
	$T_0 = 25°C$	3.67	0.00093
	$T_1 = 69.5°C$	3.70	0.00095

Equation 3.26 was used to determine the internal energy of the tomato juice. The enthalpy and entropy of condensed water flows are determined from the table of the thermodynamic properties of water as follows:

$$h_0 = 104.92\frac{\text{kJ}}{\text{kg}}, s_0 = 0.37\frac{\text{kJ}}{\text{kg.K}}$$

$$u_1 = \left(\frac{C_{P@T_1} + C_{P@T_{ref}}}{2}\right)T_1 = 4.02(75) = 301.78\frac{\text{kJ}}{\text{kg}}$$

$$u_2 = \left(\frac{C_{P@T_2} + C_{P@T_{ref}}}{2}\right)T_2 = 3.68(69.5) = 255.77\frac{\text{kJ}}{\text{kg}}$$

$$h_V = h_{g@P_3} = 2624.55\frac{\text{kJ}}{\text{kg}}, s_V = s_{g@P_3} = 7.77\frac{\text{kJ}}{\text{kg.K}}$$

$$h_S = h_{g@P_4} = 2724.90\frac{\text{kJ}}{\text{kg}}, s_S = s_{g@P_4} = 6.99\frac{\text{kJ}}{\text{kg.K}}$$

$$h_C = h_{f@P_5} = 561.43\frac{\text{kJ}}{\text{kg}}, s_C = s_{f@P_5} = 1.67\frac{\text{kJ}}{\text{kg.K}}$$

Mass balance:

$$\begin{cases} m_2 = m_1 - m_V \\ m_2\text{TS}_2 = m_1\text{TS}_1 \\ m_S = m_C \end{cases} \rightarrow \begin{cases} m_2 = 325\text{kg} \\ m_V = 675\text{kg} \end{cases}$$

Energy balance:

$$E_{in} - E_{out} = \Delta E_{system} = E_2 - E_1$$

$$\rightarrow m_S h_S + W - m_V h_V - m_C h_C = m_2 u_2 - m_1 u_1$$

$$\rightarrow m_S(h_S - h_C) + W - m_V h_V = m_2 u_2 - m_1 u_1$$

$$\rightarrow m_S = \frac{m_2u_2 - m_1u_1 + m_Vh_V - W}{(h_S - h_C)}$$

$$= \frac{325\text{kg}(255.77) - 1000\text{kg}(301.78) + 675\text{kg}(2624.55) - 0.5(3600)}{2724.90 - 561.43}$$

$$= 995.94\text{kg}$$

To calculate the exergy of tomato juice and water, the following equations were used:

$$ex_1 = \left(\frac{C_{p@T_1} + C_{p@T_0}}{2}\right)\left(T_1 - T_0 - T_0\ln\left(\frac{T_1}{T_0}\right)\right) + P_0(v_1 - v_0)$$

$$= 4.03\left[348.15 - 298.15 - 298.15\ \ln\left(\frac{348.15}{298.15}\right)\right]$$

$$+101.33(0.001 - 0.00098) = 15.20\frac{\text{kJ}}{\text{kg}}$$

$$ex_2 = \left(\frac{C_{p@T_2} + C_{p@T_0}}{2}\right)\left(T_2 - T_0 - T_0\ln\left(\frac{T_2}{T_0}\right)\right) + P_0(v_2 - v_0)$$

$$= 3.68\left[342.65 - 298.15 - 298.15\ln\left(\frac{342.65}{298.15}\right)\right]$$

$$+101.33(0.00095 - 0.00093) = 11.14\frac{\text{kJ}}{\text{kg}}$$

$$ex_V = (h_V - h_0) - T_0(s_V - s_0)$$

$$= (2624.55 - 104.92) - 298.15(7.77 - 0.37) = 313.24\frac{\text{kJ}}{\text{kg}}$$

$$ex_S = (h_S - h_0) - T_0(s_S - s_0)$$

$$= (2724.90 - 104.92) - 298.15(6.99 - 0.37) = 644.90\frac{\text{kJ}}{\text{kg}}$$

$$ex_C = (h_C - h_0) - T_0(s_C - s_0)$$

$$= (561.43 - 104.92) - 298.15(1.67 - 0.37) = 67.56\frac{\text{kJ}}{\text{kg}}$$

Exergy balance:

$$Ex_S + W - Ex_C - Ex_V - Ex_{des} = Ex_2 - Ex_1$$

$$\rightarrow m_Sex_S + W - m_Cex_C - m_Vex_V - Ex_{des} = m_2ex_2 - m_1ex_2$$

$$\rightarrow Ex_{des} = m_S(ex_S - ex_C) + W - m_Vex_V - m_2ex_2 + m_1ex_2$$

$$= 995.94(644.90 - 67.56) + 0.5(3600) - 675(313.24)$$

$$-325(11.14) - 1000(15.20) = 376932.93\text{kJ}$$

6.7.5 Separators

Separation of vapor from the fluid flow is done by separators. For this purpose, the separator is under lower pressure than the fluid flow. In the separators used in the concentration process, after leaving the heat exchanger, the food enters the separator under a vacuum, and the vapor phase is separated from the liquid. The vapor separated from the product exits from the top of the separator and goes to the condenser. Along with the vapor coming out of the product, there are also drops of thick liquid, which cause the product to drop and reduce the heating value of the vapor, which is economically undesirable; therefore, the separator tank is designed in such a way that the liquid droplets are separated from the vapor and returned to the product. If the food is boiled in the heat exchanger, the speed of the vapor entering the separator is so high that it creates a centrifugal force and the vapor spirals upwards inside the separator; therefore, due to having a higher density than vapor, the droplets trapped in the vapor are thrown on the inner wall of the separator by the centrifugal force and return to the product after losing their kinetic energy due to the force of weight.

To effectively separate the product droplets from the outgoing steam vapor and generate centrifugal force, it is essential for the incoming vapor to come into contact with the separator's body. This interaction facilitates the generation of the necessary centrifugal forces to accomplish the successful separation process.

6.7.5.1 Balance Equations

Mass balance:

$$\dot{m}_1 + \dot{m}_2 + \ldots + \dot{m}_n = \dot{m}_v + \dot{m}_l \tag{6.57}$$

Material balance:

$$x_1\dot{m}_1 + x_2\dot{m}_2 + \ldots + x_n\dot{m}_n = \dot{m}_v + x_l\dot{m}_l \tag{6.58}$$

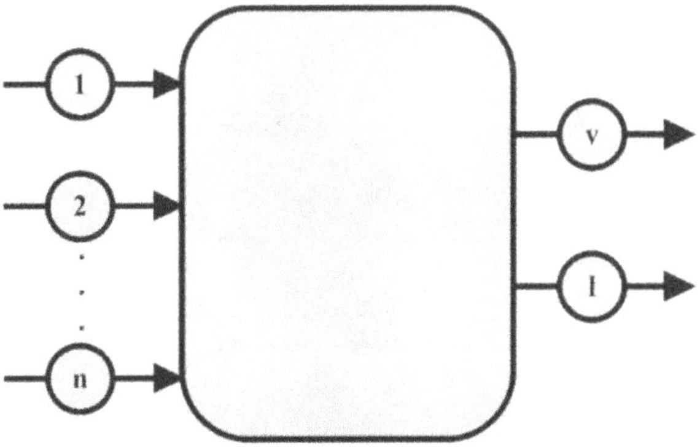

FIGURE 6.38 Schematic of inflow and outflow streams in a typical separator.

Energy balance:

$$h_1\dot{m}_1 + h_2\dot{m}_2 + \ldots + h_n\dot{m}_n = h_v\dot{m}_v + h_l\dot{m}_l \tag{6.59}$$

Entropy balance:

$$\dot{m}_1 s_1 + \dot{m}_2 s_2 + \ldots + \dot{m}_n s_n + \dot{S}_{gen} = \dot{m}_v s_v + \dot{m}_l s_l \tag{6.60}$$

Exergy balance:

$$\dot{E}x_F = \dot{E}x_1 + \dot{E}x_2 + \ldots + \dot{E}x_n \tag{6.61}$$

$$\dot{E}x_P = \dot{E}x_v + \dot{E}x_l \tag{6.62}$$

$$\dot{E}x_{des} = \left(\dot{E}x_1 + \dot{E}x_2 + \ldots + \dot{E}x_n\right) - \left(\dot{E}x_v + \dot{E}x_l\right) \tag{6.63}$$

Exergy efficiency:

$$\eta_{ex} = \frac{\dot{E}x_v + \dot{E}x_l}{\dot{E}x_1 + \dot{E}x_2 + \ldots + \dot{E}x_n} \tag{6.64}$$

✐Example 6.12: In a sugar factory, a flash drum is used to produce steam from condensate. The characteristics of the input and output flows of the flash drum are given in the table. Calculate the efficiency and rate of exergy destruction in this process.

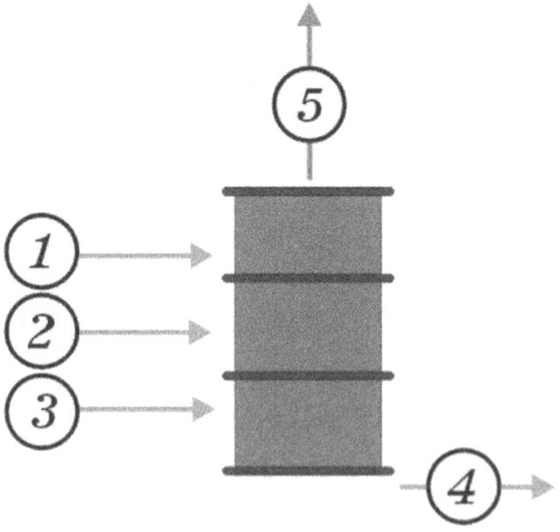

FIGURE 6.39 Schematic of Example 6.12.

TABLE 6.27
Properties of Flows of Example 6.12

State No.	Fluid Type	P (kPa)	$\dot{m}\left(\frac{kg}{s}\right)$	Phase Description
1	Water	375	13.30	Saturated liquid
2	Water	375	4.36	Saturated liquid
3	Water	375	0.52	Saturated liquid
4	Water	225	—	Saturated liquid
5	steam	225	—	Saturated steam

✍Solution

Assumptions:

1. The process is considered steady state.
2. Heat loss from the surface of the flash drum is negligible.
3. Changes in kinetic energy and potential can be ignored.
4. The dead state temperature and pressure are 25°C and 101.325 kPa.

Determining the enthalpy of flows:

$$h_0 = 104.92\frac{kJ}{kg}, s_0 = 0.37\frac{kJ}{kg.K}$$

$$h_1 = h_2 = h_3 = h_{f@P=375kPa} = 594.73\frac{kJ}{kg},$$

$$s_1 = s_2 = s_3 = s_{f@P=375kPa} = 1.75\frac{kJ}{kg.K}$$

$$h_4 = h_{f@P=225kPa} = 520.71\frac{kJ}{kg}, s_4 = s_{f@P=225kPa} = 1.57\frac{kJ}{kg.K}$$

$$h_5 = h_{g@225kPa} = 2711.66\frac{kJ}{kg}, s_5 = s_{g@225kPa} = 7.09\frac{kJ}{kg.K}$$

Mass and energy balance:

$$\dot{m}_1 + \dot{m}_2 + \dot{m}_3 = \dot{m}_4 + \dot{m}_5$$

$$\dot{m}_1 h_1 + \dot{m}_2 h_2 + \dot{m}_3 h_3 = \dot{m}_4 h_4 + \dot{m}_5 h_5$$

Simultaneous solution of mass and energy balance:

$$\begin{cases} 13.30 + 4.36 + 0.52 = \dot{m}_4 + \dot{m}_5 \\ 13.3(594.73) + 4.36(594.73) + 0.52(594.73) = \dot{m}_4(520.71) + \dot{m}_5(2711.66) \end{cases}$$

$$\rightarrow \dot{m}_4 = 17.57\frac{\text{kg}}{\text{s}}$$

$$\rightarrow \dot{m}_5 = 0.61\frac{\text{kg}}{\text{s}}$$

Equation 5.10 was used to calculate the exergy of water:

$$ex_1 = ex_2 = ex_3 = (h_1 - h_0) - T_0(s_1 - s_0)$$
$$= (594.73 - 104.92) - 298.15(1.75 - 0.37) = 76.74\frac{\text{kJ}}{\text{kg}}$$

$$ex_4 = (h_4 - h_0) - T_0(s_4 - s_0)$$
$$= (520.71 - 104.92) - 298.15(1.57 - 0.37) = 56.99\frac{\text{kJ}}{\text{kg}}$$

$$ex_5 = (h_5 - h_0) - T_0(s_5 - s_0)$$
$$= (2711.66 - 104.92) - 298.15(7.09 - 0.37) = 603.04\frac{\text{kJ}}{\text{kg}}$$

Fuel and product exergy:

$$\dot{E}x_F = \dot{E}x_1 + \dot{E}x_2 + \dot{E}x_3 = ex_1(\dot{m}_1 + \dot{m}_2 + \dot{m}_3)$$
$$= 76.74(13.30 + 4.36 + 0.52) = 1395.12\text{kW}$$

$$\dot{E}x_P = \dot{E}x_1 + \dot{E}x_2 = \dot{m}_4 ex_4 + \dot{m}_5 ex_5$$
$$= 17.57(56.99) + 0.61(603.04) = 1371.53\text{kW}$$

Exergy destruction rate:

$$\dot{E}x_{des} = \dot{E}x_F - \dot{E}x_P = 1395.12 - 1371.53 = 23.59\text{kW}$$

Exergy efficiency:

$$\eta_{ex} = \frac{\dot{E}x_P}{\dot{E}x_F} \times 100 = \frac{1371.53}{1395.12} \times 100 = 98.31\%$$

6.7.6 Throttling Valves

The throttling process occurs when the fluid flow abruptly encounters an obstruction in its course. This obstacle can be a plate with a hole, a half-open valve, or a sudden decrease in the diameter of the tube. In this case, the fluid must find a way to pass through the limited path, which causes its pressure to drop. The main difference between this tool and the nozzle is the abruptness of the rate of change in the cross-sectional area. Usually, there is a slight increase in kinetic energy in this process, but the amount of kinetic energy in the input and output is so small that it can be ignored. In this process, no work is done, and the amount of potential energy change is either exactly zero or very low. The short time of this process causes no opportunity for

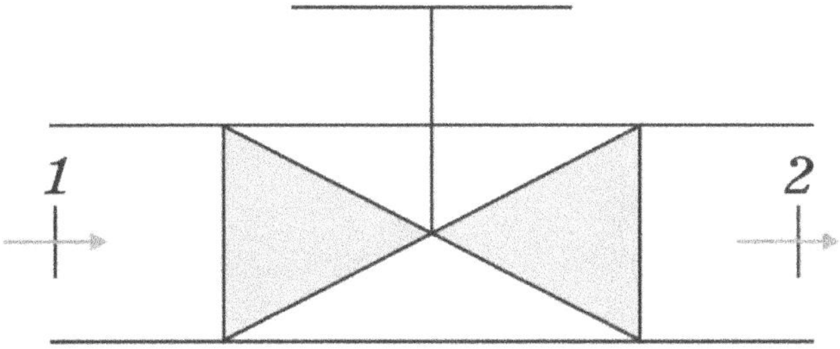

FIGURE 6.40 Schematic of a throttling valve.

heat transfer. Given this issue, the only remaining term in the first law is enthalpy. The throttling process is a constant enthalpy process.

6.7.6.1 Balance Equations

Mass balance:

$$\dot{m}_1 = \dot{m}_2 \tag{6.65}$$

Energy balance:

$$h_1 = h_2 \tag{6.66}$$

Entropy balance:

$$\dot{m}_1 s_1 + \dot{S}_{gen} = \dot{m}_2 s_2 \tag{6.67}$$

Exergy balance:

$$\dot{E}x_F = \dot{E}x_1 \tag{6.68}$$

$$\dot{E}x_P = \dot{E}x_2 \tag{6.69}$$

$$\dot{E}x_{des} = \dot{E}x_1 - \dot{E}x_2 \tag{6.70}$$

Exergy efficiency:

$$\eta_{ex} = \frac{\text{ex}_2}{\text{ex}_1} \tag{6.71}$$

✐**Example 6.13: A pressure-reducing valve is used to reduce the pressure of steam entering the evaporation line of a concentrate processing. The steam enters the pressure-reducing valve with a pressure of 800 kPa and a temperature of 180°C and leaves it with a pressure of 300 kPa. Calculate exergy destruction per unit mass flow.**

FIGURE 6.41 Schematic of the pressure reducing valve related to Example 6.13.

✍Solution

Assumptions:

1. The process is considered steady state.
2. Ignore changes in kinetic and potential energy.
3. Consider dead state temperature and pressure 25°C and 101.325 kPa, respectively.

The throttling process in the pressure-reducing valve is constant enthalpy:

$$h_2 = h_1$$

Determining flow characteristics:

$$\begin{matrix} P_1 = 800\text{kPa} \\ T_1 = 180 \end{matrix} \rightarrow \begin{cases} h_1 = 2792.44\dfrac{\text{kJ}}{\text{kg}} \\ s_1 = 6.72\dfrac{\text{kJ}}{\text{kg.K}} \end{cases}$$

$$\begin{matrix} P_2 = 300\text{kPa} \\ h_2 = h_1 = 2792.44\dfrac{\text{kJ}}{\text{kg}} \end{matrix} \rightarrow s_2 = 7.15\dfrac{\text{kJ}}{\text{kg.K}}$$

Exergy balance:

$$\dot{E}x_1 = \dot{E}x_2 + \dot{E}x_{des} \rightarrow \dot{E}x_{des} = \dot{E}x_1 - \dot{E}x_2 = \dot{m}\left(ex_1 - ex_2\right)$$

$$\rightarrow \frac{\dot{E}x_{des}}{\dot{m}} = ex_1 - ex_2 = \left(h_1 - h_2\right) - T_0\left(s_1 - s_2\right)$$

$$\rightarrow \frac{\dot{E}x_{des}}{\dot{m}} = 0 - 298.15(6.72 - 7.15) = 130.08\frac{\text{kJ}}{\text{kg}}$$

6.7.7 Compressors

A compressor is used to increase the pressure of gases. The mechanical energy input to the compressor is used to increase the gas enthalpy and heat loss from the compressor.

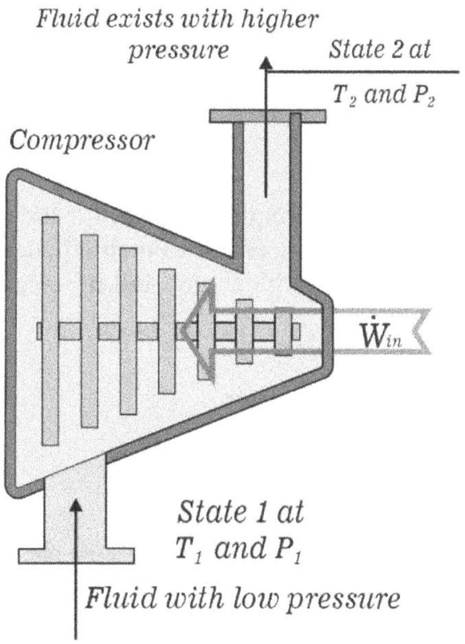

FIGURE 6.42 Schematic of a typical compressor.

6.7.7.1 Balance Equations

Mass balance:

$$\dot{m}_1 = \dot{m}_2 \tag{6.72}$$

Energy balance:

$$\dot{m}_1 h_1 + \dot{W} = \dot{m}_2 h_1 + \dot{Q} \tag{6.73}$$

Entropy balance:

$$\dot{m}_1 s_1 + \dot{S}_{gen} = \dot{m}_2 s_2 + \frac{\dot{Q}}{T_b} \tag{6.74}$$

Exergy balance:

$$\dot{E}x_F = \dot{W} \tag{6.75}$$

$$\dot{E}x_P = \dot{E}x_2 - \dot{E}x_1 \tag{6.76}$$

$$\dot{E}x_L = \left(1 - \frac{T_0}{T_b}\right)\dot{Q} \tag{6.77}$$

$$\dot{E}x_{des} = \dot{W} - \left(\dot{E}x_2 - \dot{E}x_1\right) - \left(1 - \frac{T_0}{T_b}\right)\dot{Q} \tag{6.78}$$

Exergy efficiency:

$$\eta_{ex} = \frac{\dot{E}x_2 - \dot{E}x_1}{\dot{W}} \tag{6.79}$$

Example 6.14: Calculate the required power for the compressor of the figure. The compressor works stably. Potential energy changes are negligible. Consider air an ideal gas, and its specific heat is 1.01 kJ/(kg.K).

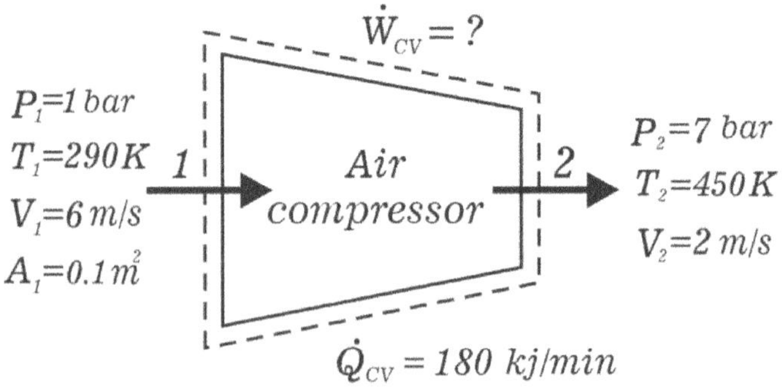

FIGURE 6.43 Schematic of Example 6.14.

Solution

1. The process is considered steady state.
2. Air is an ideal gas.
3. Potential energy changes are ignored.

Mass balance for steady state:

$$\sum \dot{m}_i = \sum \dot{m}_o \rightarrow \dot{m}_1 = \dot{m}_2 = \dot{m}$$

Energy balance for steady state:

$$\sum \dot{E}_i = \sum \dot{E}_o \rightarrow \dot{m}_1\left(h_1 + \frac{V_1^2}{2}\right) + \dot{W} = \dot{m}_2\left(h_2 + \frac{V_2^2}{2}\right) + \dot{Q} \rightarrow \dot{W} = \dot{m}\left[h_2 - h_1 + \frac{V_2^2 - V_1^2}{2}\right] + \dot{Q}$$

Air is an ideal gas, so the energy balance is written as follows:

$$\dot{W} = \dot{m}\left[C_p(T_2 - T_1) + \frac{V_2^2 - V_1^2}{2}\right] + \dot{Q}$$

The volume flow of air can be calculated as follows:

$$\begin{cases} \dot{m} = \rho_1 V_1 A_1 \\ \rho_1 = \dfrac{P_1}{RT_1} \end{cases} \rightarrow \dot{m} = \frac{P_1 V_1 A_1}{RT_1} = \frac{(100)(6)(0.1)}{(0.287)(290)} = 0.73\frac{\text{kg}}{\text{s}}$$

By considering the energy balance relationship, the amount of input power to the compressor is calculated:

$$\dot{W} = 0.73\left[1.1(450-290) + \frac{2^2-6^2}{2}\left|\frac{1\frac{\text{kJ}}{\text{kg}}}{\frac{1000\text{m}^2}{s^2}}\right|\right] + \frac{180}{60} \rightarrow \dot{W} = 131.47\text{kW}$$

Example 6.15: The compressor used in a spray dryer system converts dry air from state 1 to state 2 during an adiabatic process. According to the information given in the figure, calculate the destruction rate and exergy efficiency of the compressor. Consider air an ideal gas with a constant specific heat of 1.005 kJ/(kg°C).

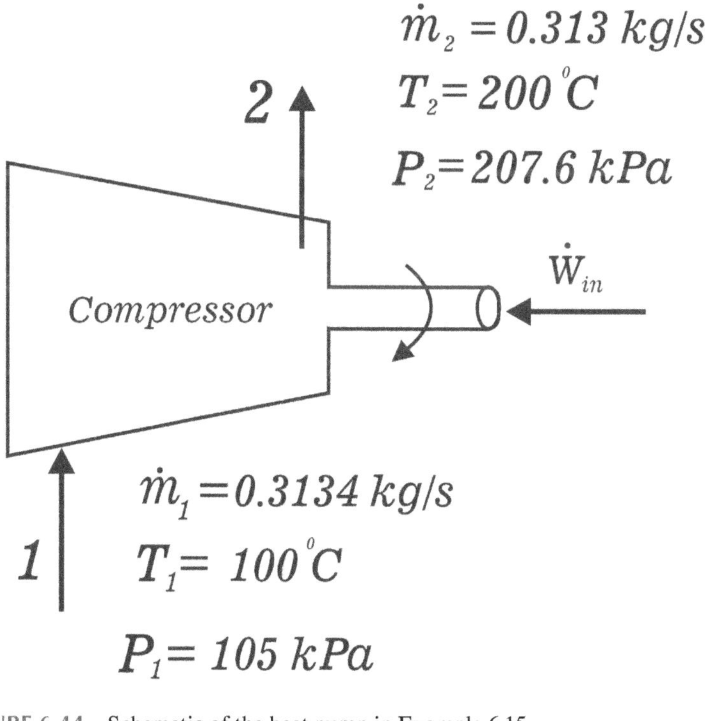

FIGURE 6.44 Schematic of the heat pump in Example 6.15.

✍Solution

1. The process is considered steady state.
2. Air is an ideal gas.
3. Kinetic and potential energy changes are ignored.
4. Heat loss from the surface of the compressor is negligible.

Mass balance for steady state:

$$\sum \dot{m}_i = \sum \dot{m}_o \rightarrow \dot{m}_1 = \dot{m}_2 = \dot{m}$$

Energy balance for steady state:

$$\sum \dot{E}_i = \sum \dot{E}_o \rightarrow \dot{m}_1 h_1 + \dot{W} = \dot{m}_2 h_2 \rightarrow \dot{W} = \dot{m}\left[h_2 - h_1\right]$$

Air is an ideal gas; therefore, the energy balance is written as follows:

$$\dot{W} = \dot{m}\left[C_p\left(T_2 - T_1\right)\right] = \left(0.3134\frac{\text{kg}}{\text{s}}\right)\left(1.005\frac{\text{kJ}}{\text{kg°C}}\right)(200-100)\text{°C} = 31.50\text{kW}$$

Exergy balance:

$$\dot{E}x_{des} = \dot{E}x_F - \dot{E}x_P - \dot{E}x_L$$

$$\dot{E}x_L = 0$$

$$\dot{E}x_F = \dot{W} = 31.50\text{kW}$$

$$\dot{E}x_P = \dot{E}x_2 - \dot{E}x_1 = \dot{m}\left(ex_2 - ex_1\right) = \dot{m}\left[C_{p,air}\left(T_2 - T_1\right) - T_0\left(C_{p,air}\ln\frac{T_2}{T_1} - R_{air}\ln\frac{P_2}{P_1}\right)\right]$$

$$\rightarrow \dot{E}x_P = 0.3134\Big[(1.005)(200-100)$$

$$-298.15\left((1.005)\ln\frac{473.15}{373.15} - (0.287)\ln\frac{207.6}{105}\right)\Big]$$

$$= 27.48\text{kW}$$

Calculation of exergy destruction rate:

$$\dot{E}x_{des} = 31.50 - 27.48 = 4.02\text{kW}$$

Exergy efficiency:

$$\eta_{ex} = \frac{\dot{E}x_2 - \dot{E}x_1}{\dot{W}} = \frac{27.48\text{kW}}{31.5\text{kW}} = 0.8723 \text{ or } 87.23\%$$

6.7.8 Steam Boilers

A boiler is a set of closed cylindrical chamber and pipes in which heat is given to water or other fluids, and finally, steam is produced. This process requires heat, which can be provided through the combustion of fuels, electricity, or nuclear energy. The steam and hot water produced are used in various industries such as food, pharmaceutical, papermaking, heating and cooling systems, and power plants. The produced steam can be used in the food industry for thermal processes of food (such as sterilization, pasteurization, and evaporation).

6.7.8.1 Calorific Value of Fuel in Boilers

The calorific value of a fuel is defined as the amount of energy released per unit mass of the fuel when it burns completely with oxygen. Low heat value (LHV) and high heat value (HHV) are determined from the state of water obtained at the end of complete combustion of a fuel with oxygen. The remaining water can be liquid or steam. The following equation shows the general formula for the complete combustion of organic matter with oxygen:

$$C_xH_y + \left(x + \frac{y}{4}\right)O_2 \rightarrow xCO_2 + \left(\frac{y}{2}\right)H_2O + \text{Heat} \tag{6.80}$$

The total heat resulting from complete combustion per unit mass of combustible material is called the HHV or gross heat value (GHV). The high calorific value includes the total heat released from the fuel and also includes the latent heat of the amount of water steam in the combustion products. If the steam produced during burning is not condensed by the boiling device, some of the heat produced will be wasted along with the uncondensed steam. The heat resulting from this combustion is called the LHV or net heat value (NHV). The LHV includes the net heat released from the fuel and does not include the latent heat of the water in the fuel. To calculate

FIGURE 6.45 An industrial boiler.

fuel energy in combustion processes, an LHV of fuel has been proposed (Dinçer et al., 2016):

$$\dot{E}_{Fuel} = \dot{m}_{Fuel}\mathrm{LHV}_{Fuel} \quad (6.81)$$

6.7.8.2 Mass and Energy Balance

In general, the energy analysis of steam boilers is carried out along with the blower attached to it, which has the role of providing the necessary air for combustion:

Mass balance:

$$\dot{m}_5 = \dot{m}_2 + \dot{m}_3 \quad (6.82)$$

$$\dot{m}_1 = \dot{m}_4$$

Energy balance:

$$\dot{m}_1 h_1 + \dot{m}_2\mathrm{LHV}_{fuel} + \dot{m}_3 h_3 + \dot{W} = \dot{m}_4 h_4 + \dot{m}_5 h_5 + \dot{Q}_{loss} \quad (6.83)$$

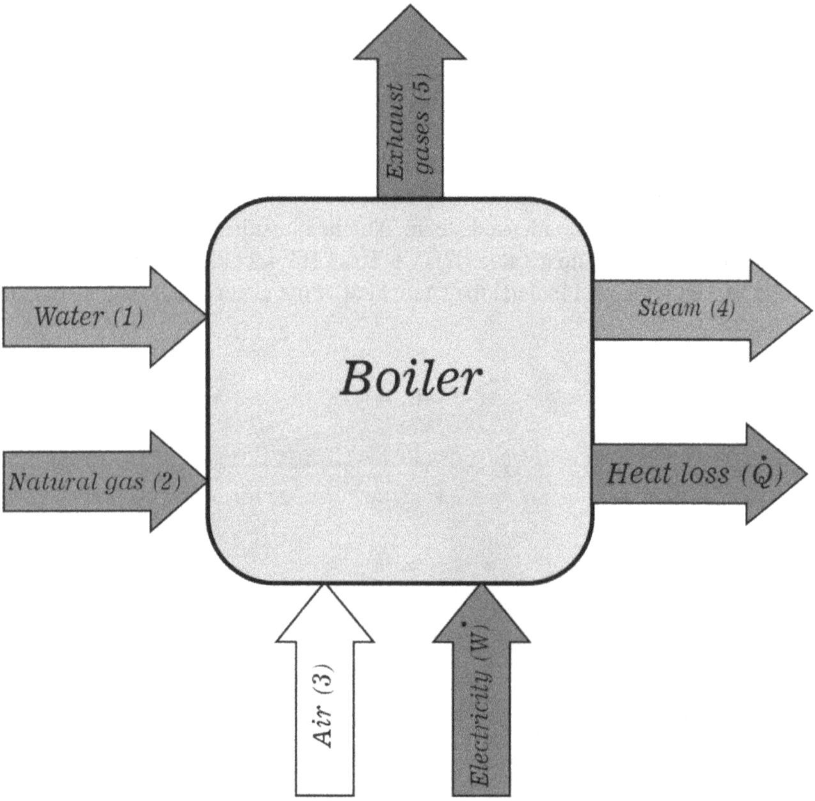

FIGURE 6.46 Energy flows in the boiler.

The thermal efficiency of the boiler is defined as the ratio of the enthalpy increase of the water flow in the boiler to the fuel energy:

$$\eta_{Boiler} = \frac{\Delta H_{water}}{\dot{m}_{fuel} LHV_{fuel}} = \frac{\dot{m}_1 (h_4 - h_1)}{\dot{m}_2 LHV_{fuel}} \tag{6.84}$$

✐Example 6.16: Natural gas flow is used to produce saturated steam at a pressure of 500 kPa in a steam boiler. Water enters the boiler at a temperature of 70°C and a volumetric flow rate of 58 L/min. The amount of natural gas consumed in a steam boiler with a temperature of 25°C and a pressure of 101.323 kPa, for a period of 2 hours, has been measured as 500 m³. If the LHV of fuel is 55,500 kJ/kg and its density is 0.6548kg/m³, calculate the thermal efficiency of the boiler.

✍Solution

The characteristics of water at the inlet and outlet of the boiler are determined from the saturated water table:

$$h_{inlet\ water} = h_{f@70°C} = 293.07 \frac{kJ}{kg}$$

$$v_{inlet\ water} = v_{f@70°C} = 0.001023 \frac{m^3}{kg}$$

$$h_{outlet\ water} = h_{g@500kPa} = 2748.1 \frac{kJ}{kg}$$

Considering the volume flow rate and the specific volume of the inlet water, it is possible to calculate the mass flow rate of water:

$$\dot{m}_{water} = \frac{\dot{V}}{v_{inlet\ water}} = \frac{\left(\frac{0.058}{60}\right)\frac{m^3}{s}}{0.001023\frac{m^3}{kg}} = 0.945 \frac{kg}{s}$$

Determining the mass flow rate of fuel:

$$\dot{m}_{fuel} = \rho_{fuel}\dot{V} = \left(0.6548 \frac{kg}{m^3}\right)\frac{500m^3}{2hr} \times \frac{1hr}{3600s} = 0.04547 \frac{kg}{s}$$

The thermal efficiency of the steam boiler is equal to:

$$\eta_{boiler} = \frac{\dot{m}_{water}\left(h_{outlet\ water} - h_{inlet\ water}\right)}{\dot{m}_{fuel}LHV_{fuel}} = \frac{0.945(2748.1 - 293.07)}{(0.04547)(55500)} = 0.9193 = 91.93\%$$

Example 6.17: The information about the input and output flows of the steam boiler is presented in Table 6.28. Calculate the waste heat and thermal efficiency of the boiler. Use the following information for analysis:

1. The low calorific value of the fuel is 44,661 kJ/kg.
2. The specific heat of air and steam are constant and 1.005 kJ/kg.°C and 1.155 kJ/kg.°C, respectively.
3. The power consumption of the blower is 1 kW.

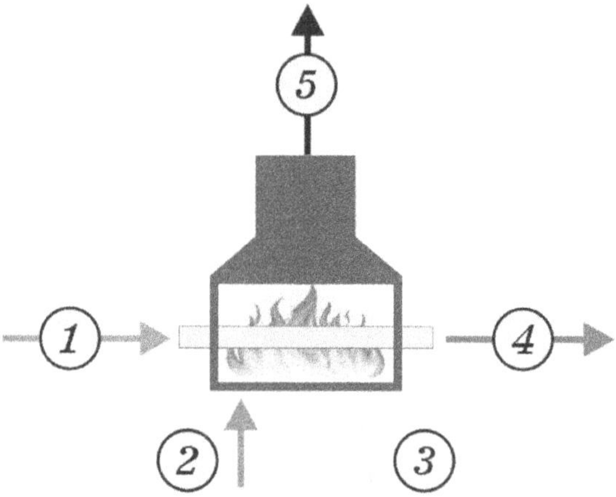

FIGURE 6.47 Schematic of flows in Example 6.17.

TABLE 6.28
Details of Example 6.17

State No.	Fluid Type	T (°C)	P (kPa)	$\dot{m}\left(\frac{kg}{s}\right)$
1	Water	70	950	3.24
2	Fuel	25	100	0.2
3	Air	25	100	5.02
4	Steam	180	800	3.24
5	Exhaust gases	130	170	5.22

✍Solution

Assumptions:

1. The air entering the boiler and the steam coming out of the boiler are ideal gases.
2. The process is considered steady state.
3. Kinetic and potential energy changes are ignored.

Determining the enthalpy of flows:

$$h_1 = h_{f\,@70°C} = 293.07\frac{\text{kJ}}{\text{kg}}$$

$$h_3 = C_pT_3 = 1.005(25) = 25.125\frac{\text{kJ}}{\text{kg}}$$

$$h_4 = h_{@800\text{kPa},18°C} = 2791.47\frac{\text{kJ}}{\text{kg}}$$

$$h_5 = C_pT_5 = 1.155(130) = 150.15\frac{\text{kJ}}{\text{kg}}$$

The energy balance for the boiler is as follows:

$$\dot{m}_1h_1 + \dot{m}_2LHV_{fuel} + \dot{m}_3h_3 + \dot{W} = \dot{m}_4h_4 + \dot{m}_5h_5 + \dot{Q}_{loss}$$

$$3.24(293.07) + 0.2(44661) + 5.02(25.125) + 1 = 3.24(2791.47) + 5.22(150.15) + \dot{Q}_{loss}$$

$$\rightarrow \dot{Q}_{loss} = 180.59\text{kW}$$

Boiler efficiency:

$$\eta_{boiler} = \frac{\dot{m}_1(h_4 - h_1)}{\dot{m}_2LHV_{fuel}} = \frac{3.24(2791.4724 - 293.07)}{0.2(44661)} = 0.9062 \text{ or } 90.62\%$$

6.8 REVERSIBLE WORK OF STEADY FLOW

The energy conservation equation for a steady flow device undergoing an internally reversible process can be expressed in differential form as:

$$\delta q_{rev} - \delta w_{rev} = dh + de_K + de_P \tag{6.85}$$

For an internal reversible process:

$$\delta q_{int\ rev} = Tds \tag{6.86}$$

Now, using Equation 7.4, the work of reversible steady flow can be calculated as follows:

$$-\delta w_{rev} = v\,dP + de_K + de_P \tag{6.87}$$

And by integration:

$$w_{rev} = -\int_1^2 v\,dP - \Delta e_K - \Delta e_P \tag{6.88}$$

If kinetic and potential changes can be ignored:

$$w_{rev} = -\int_1^2 v\,dP \tag{6.89}$$

✐Example 6.18: An adiabatic pump compresses saturated water at a pressure of 10 kPa to a pressure of 15 MPa during an internal reversible process. Determine the required power of the pump per unit mass of water.

✍Solution

Characteristics of flows:

$$\begin{matrix} P_1 = 10\text{kPa} \\ x_1 = 0 \end{matrix} \rightarrow \begin{cases} h_1 = 191.81\dfrac{\text{kJ}}{\text{kg}} \\ s_1 = 0.6492\dfrac{\text{kJ}}{\text{kg.K}} \\ v_1 = 0.001010\dfrac{\text{m}^3}{\text{kg}} \end{cases}$$

$$\begin{matrix} P_2 = 15\text{MPa} \\ s_2 = s_1 \end{matrix} \rightarrow \begin{cases} h_2 = 206.9\dfrac{\text{kJ}}{\text{kg}} \\ v_2 = 0.001004\dfrac{\text{m}^3}{\text{kg}} \end{cases}$$

Calculation of power consumption using energy balance:

$$w = h_2 - h_1 = 206.9 - 191.81 = 15.1\frac{\text{kJ}}{\text{kg}}$$

Calculation of power consumption using the flow reversible work relationship:

$$w = v_1(P_2 - P_1) = \left(0.001010\frac{\text{m}^3}{\text{kg}}\right)(15000 - 10)\text{kPa} = 15.14\frac{\text{kJ}}{\text{kg}}$$

6.9 ISENTROPIC EFFICIENCY OF STEADY FLOW EQUIPMENT

Any type of heat dissipation is undesirable in steady flow devices, and devices should ideally be adiabatic. In addition, an ideal process should not include irreversible factors because these factors reduce the efficiency of the device; therefore, the desirable ideal process is isentropic. Bringing a real process closer to an ideal isentropic process will result in better performance of the device. Isentropic efficiency expresses the degree of deviation of real processes from ideal isentropic processes. In steady state equipment, the inlet fluid state (T_1, P_1, h_1, and s_1) is the same in both real and ideal processes, and the output pressure is also the same in both real and ideal processes ($P_2 = P_{2s}$); because the pressure is affected by the environment in which the fluid is discharged.

It should be noted that the definitions of isentropic efficiency and second law efficiency are different. In isentropic efficiency, it is assumed that the process between the real initial state and the hypothetical output state is an ideal isentropic state. The exergy efficiency of a process, assessed between the actual input and output states, is assumed to be ideal and reversible. As a result, the isentropic and exergy efficiencies yield values that are closely related, yet distinct.

6.9.1 Isentropic Efficiency of the Pump

When the changes in kinetic energy and fluid potential can be ignored, the isentropic efficiency of a pump is defined as follows (Cimbala and Cengel, 2006):

$$\eta_P = \frac{w_{is}}{w_{act}} = \frac{h_{2s} - h_1}{h_{2a} - h_1} = \frac{v(P_2 - P_1)}{h_{2a} - h_1} \tag{6.90}$$

✐Example 6.19: Water enters a pump with a pressure of 100 kPa and a temperature of 15°C. Calculate the power consumption of the pump of 1.5 kW, the water mass flow rate of 1.2 kg/s, and the isentropic efficiency of 75%. Determine the output water pressure.

✍Solution

$$w_{act} = \frac{\dot{W}_{act}}{\dot{m}} = \frac{1.5\text{kW}}{1.2\frac{\text{kg}}{\text{s}}} = 1.25\frac{\text{kJ}}{\text{kg}}$$

$$\eta_P = \frac{w_{is}}{w_{act}} \rightarrow w_{is} = w_{act}\eta_P = 1.25(0.75) = 0.9375\frac{\text{kJ}}{\text{kg}}$$

$$w_{is} = v(P_2 - P_1) \rightarrow 0.9375 = 0.001001(P_2 - 100) \rightarrow P_2 = 1036.56\text{kPa}$$

6.9.2 Compressor Isentropic Efficiency

$$\eta_C = \frac{w_{is}}{w_{act}} \tag{6.91}$$

When the changes in kinetic and potential energy of the gas being compressed are negligible, the inlet work to a direct compressor will be equal to the enthalpy change of the fluid:

$$\eta_C = \frac{h_{2s} - h_1}{h_{2a} - h_1} \tag{6.92}$$

h_{2a} and h_{2s} are the enthalpy values in the outlet state for the real process and the isentropic process, respectively.

Example 6.20: Air enters a compressor at a pressure of 100 kPa and 300 K and is compressed to a pressure of 150 kPa. The compressor efficiency is 70%. Determine the work required per kilogram of air.

Solution

Assumptions:

1. Air is an ideal gas.
2. Heat loss from the surface of the compressor is negligible.
3. Kinetic and potential energy changes are neglected.

For an isentropic ideal gas process:

$$\left(\frac{T_2}{T_1}\right)_{s=const} = \left(\frac{P_2}{P_1}\right)^{\frac{k-1}{k}} \rightarrow T_{2s} = T_1\left(\frac{P_2}{P_1}\right)^{\frac{k-1}{k}} \rightarrow T_{2s} = 300\left(\frac{150}{100}\right)^{\frac{1.4-1}{1.4}} = 336.9 \text{ K}$$

$$C_{p,300K} = 1.005, C_{p,336.9K} = 1.007 \rightarrow C_{p,ave} = 1.006$$

Energy balance per mass flow unit in an ideal state:

$$w_{is} = C_{p,ave}\left(T_{2s} - T_1\right) = 1.006\left(336.9 - 300\right) = 37.12\frac{\text{kJ}}{\text{kg}}$$

Using the definition of isentropic efficiency:

$$w_{act} = \frac{w_{is}}{\eta_T} = \frac{37.12}{0.7} = 53.03\frac{\text{kJ}}{\text{kg}}$$

6.10 CASE EXAMPLES FROM THE FOOD INDUSTRY

6.10.1 Evaporation of Raw Juice in the Concentrate Production Line

✐Example 6.21: A falling film evaporator is used for two-stage concentration of raw sour cherry juice in the concentrate production line (Herfeh et al., 2022). This evaporator includes three heating coils, one of which is used to preheat the raw sour cherry juice after the first stage of concentration, and the other two are used to preheat the clear fruit juice. The heat required for these processes is supplied through live steam entering the evaporator. Table 6.29 presents the juice compositions, and Table 6.30 presents the thermophysical properties of the streams. If the power consumption of the pump is 1 kW, calculate the destruction rate and exergy efficiency in the process.

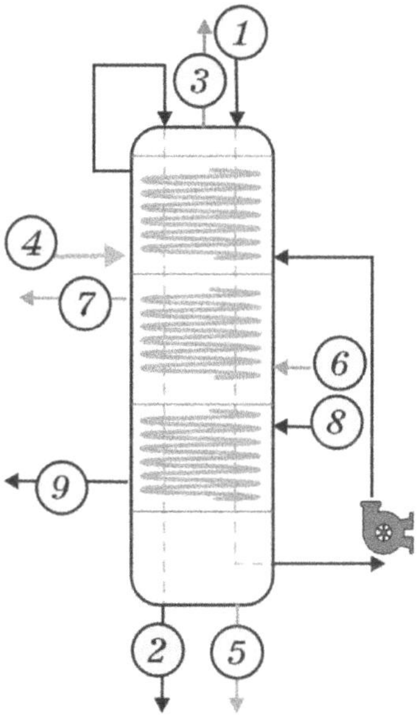

FIGURE 6.48 Diagram of falling film evaporator related to Example 6.21.

TABLE 6.29

Compositions of Juice Flows in Example 6.21

State No.	X_{ash}	$X_{carbohydrate}$	X_{water}	$X_{protein}$	X_{fat}	TS
1, 8, 9	0.00320	0.09892	0.89490	0.00290	0.00008	0.1051
2	0.00428	0.13215	0.85958	0.00388	0.00011	0.1404
6, 7	0.00433	0.12029	0.87134	0.00393	0.00011	0.1287

TABLE 6.30
Details of Example 6.21

State No.	Fluid Type	T (°C)	P (kPa)	$\dot{m}\left(\frac{kg}{s}\right)$	Phase Description
1	Raw juice	87	360	3.5	—
2	Raw juice	115.4	170	—	—
3	Steam	—	170	—	Saturated steam
4	Steam	—	240	—	Saturated steam
5	Condensate	—	240	—	Saturated water
6	Filtered juice	74	400	3.5	—
7	Filtered juice	79	350	—	—
8	Raw juice	62	600	3.5	—
9	Raw juice	67	550	—	—

✍Solution

Assumptions:

1. The process is considered steady state.
2. Kinetic and potential energy changes are ignored.
3. The dead state temperature and pressure are 25°C and 101.325 kPa.
4. The reference temperature and pressure are 0°C and 101.325 kPa.
5. Condensate water coming out of the evaporator is a saturated liquid.
6. The pressure drop in the steam pipe is insignificant.
7. Heat loss from the surface of the evaporator is insignificant.

To calculate the specific heat and density of juice compounds, the equations in Table 6.1 were used, and the results are presented in Table 6.31.

TABLE 6.31
Specific Heat and Density of Juice Compounds for Example 6.21

	Ash		Carbohydrate		Fat		Protein		Water	
T (°C)	C_P	ρ	C_P	ρ	C_P	ρ	C_P	ρ	C_P	ρ
$T_{ref}=0$	1.0926	—	1.5488	—	1.9842	—	2.0082	—	4.1762	—
$T_0=25$	1.1375	—	1.5942	—	2.0180	—	2.0376	—	4.1773	—
$T_1=87$	1.2291	2399.09	1.6746	1572.09	2.0760	889.26	2.1034	1284.80	4.2097	969.01
$T_2=115.4$	1.2616	2391.42	1.6962	1563.27	2.0903	877.40	2.1302	1270.08	4.2386	947.50
$T_6=74$	1.2123	2403.03	1.6615	1567.13	2.0669	894.69	2.0905	1291.54	4.1994	976.84
$T_7=79$	1.2189	2401.63	1.6668	1574.57	2.0706	892.60	2.0955	1288.95	4.2032	973.98
$T_8=62$	1.1956	2406.40	1.6476	1579.85	2.0571	899.70	2.0781	1297.76	4.1916	982.93
$T_9=67$	1.2027	2405.00	1.6536	1578.30	2.0614	897.61	2.0833	1295.17	4.1947	980.52

The specific heat and volume of syrup were determined using Equations 6.4 and 6.7, respectively, and the results are presented in Table 6.32.

TABLE 6.32
Specific Heat and Density of Juice Streams for Example 6.21

State No.		$C_P\left(\frac{kJ}{kg.K}\right)$	$v\left(\frac{m^3}{kg}\right)$
1	$T_{ref}=0°C$	3.9000	—
	$T_0=25°C$	3.9057	—
	$T_1=87°C$	3.9431	0.00099012
2	$T_{ref}=0°C$	3.8071	—
	$T_0=25°C$	3.8144	—
	$T_2=115.4°C$	3.8815	0.00099671
6	$T_{ref}=0°C$	3.8380	—
	$T_0=25°C$	3.8448	—
	$T_6=74°C$	3.8727	0.00097329
7	$T_{ref}=0°C$	3.8380	—
	$T_0=25°C$	3.8448	—
	$T_7=79°C$	3.8766	0.00097599
8	$T_{ref}=0°C$	3.9000	—
	$T_0=25°C$	3.9057	—
	$T_8=62°C$	3.9241	0.00097671
9	$T_{ref}=0°C$	3.9000	—
	$T_0=25°C$	3.9057	—
	$T_9=67°C$	3.9275	0.00097901

Equation 6.6 was used to determine the entropy of juice flows. The enthalpy and entropy of water flows were determined from the table of the thermodynamic properties of water:

$$h_0 = 104.92\frac{kJ}{kg}, s_0 = 0.37\frac{kJ}{kg.K}$$

$$h_1 = \left(\frac{C_{P@T_1} + C_{P@T_{ref}}}{2}\right)T_1 + v_1(P_1 - P_0)$$

$$= 3.9215(87) + 0.00099012(360 - 101.325) = 341.43\frac{kJ}{kg}$$

$$h_2 = \left(\frac{C_{P@T_2} + C_{P@T_{ref}}}{2}\right)T_2 + v_2(P_2 - P_0)$$

$$= 3.8443(115.4) + 0.00099671(170 - 101.325) = 443.70\frac{kJ}{kg}$$

$$h_3 = h_{g@P_3} = 2698.82\frac{kJ}{kg}, s_3 = s_{g@P_3} = 7.18\frac{kJ}{kg.K}$$

$$h_4 = h_{g@T_4} = 2714.63\frac{kJ}{kg}, s_4 = s_{g@T_4} = 7.07\frac{kJ}{kg.K}$$

$$h_5 = h_{f@T_5} = 529.64\frac{kJ}{kg}, s_5 = s_{f@T_5} = 1.59\frac{kJ}{kg.K}$$

$$h_6 = \left(\frac{C_{P@T_6} + C_{P@T_{ref}}}{2}\right)T_6 + v_6\left(P_6 - P_0\right)$$

$$= 3.8554(74) + 0.00097329(400 - 101.325) = 285.59\frac{kJ}{kg}$$

$$h_7 = \left(\frac{C_{P@T_7} + C_{P@T_{ref}}}{2}\right)T_7 + v_7\left(P_7 - P_0\right)$$

$$= 3.8573(79) + 0.00097599(350 - 101.325) = 304.97\frac{kJ}{kg}$$

$$h_8 = \left(\frac{C_{P@T_8} + C_{P@T_{ref}}}{2}\right)T_8 + v_8\left(P_8 - P_0\right)$$

$$= 3.9149(62) + 0.00097671(600 - 101.325) = 243.03\frac{kJ}{kg}$$

$$h_9 = \left(\frac{C_{P@T_9} + C_{P@T_{ref}}}{2}\right)T_9 + v_9\left(P_9 - P_0\right)$$

$$= 3.9166(67) + 0.00097901(550 - 101.325) = 262.66\frac{kJ}{kg}$$

Mass and material balance for steady state:

$$\sum \dot{m}_i = \sum \dot{m}_o \to \begin{cases} \dot{m}_4 = \dot{m}_5 \\ \dot{m}_6 = \dot{m}_7 \\ \dot{m}_8 = \dot{m}_9 \\ TS_1\dot{m}_1 = TS_2\dot{m}_2 \\ \dot{m}_1 = \dot{m}_2 + \dot{m}_3 \end{cases} \to \begin{cases} \dot{m}_4 = \dot{m}_5 \\ \dot{m}_6 = \dot{m}_7 \\ \dot{m}_8 = \dot{m}_9 \\ 0.1051(3.5) = 0.1404\dot{m}_2 \\ 3.5 = \dot{m}_2 + \dot{m}_3 \end{cases} \to \begin{cases} \dot{m}_2 = 2.62\,{}^{kg}\!/_{s} \\ \dot{m}_3 = 0.88\,{}^{kg}\!/_{s} \\ \dot{m}_7 = 3.5\,{}^{kg}\!/_{s} \\ \dot{m}_9 = 3.5\,{}^{kg}\!/_{s} \end{cases}$$

Energy balance for steady state:

$$\sum \dot{E}_i = \sum \dot{E}_o$$

$$\to \dot{m}_1 h_1 + \dot{m}_4 h_4 + \dot{m}_6 h_6 + \dot{m}_8 h_8 + \dot{W} = \dot{m}_2 h_2 + \dot{m}_3 h_3 + \dot{m}_5 h_5 + \dot{m}_7 h_7 + \dot{m}_9 h_9$$

$$\to 3.5(341.43) + \dot{m}_4(2714.63) + 3.5(285.59) + 3.5(243.03) + 1$$
$$= 2.62(443.70) + 0.88(2698.82) + \dot{m}_5(529.64)$$
$$+ 3.5(304.97) + 3.5(262.66)$$

$$\to \dot{m}_4 = \dot{m}_5 = 1.13\frac{kg}{s}$$

Equations 6.8 and 5.10 were used to calculate the exergy of juice and water flows, respectively:

$$ex_1 = \left(\frac{C_{p@T_1} + C_{p@T_0}}{2}\right)\left(T_1 - T_0 - T_0 \ln\left(\frac{T_1}{T_0}\right)\right) + v_{@T_1}(P_1 - P_0)$$

$$= 3.92\left[360.15 - 298.15 - 298.15\ln\left(\frac{360.15}{298.15}\right)\right]$$

$$+(360 - 101.325) = 22.52\frac{\text{kJ}}{\text{kg}}$$

$$ex_2 = \left(\frac{C_{p@T_2} + C_{p@T_0}}{2}\right)\left(T_2 - T_0 - T_0 \ln\left(\frac{T_2}{T_0}\right)\right) + v_{@T_2}(P_2 - P_0)$$

$$= 3.85\left[388.55 - 298.15 - 298.15\ln\left(\frac{388.55}{298.15}\right)\right]$$

$$+(170 - 101.325) = 44.10\frac{\text{kJ}}{\text{kg}}$$

$$ex_3 = (h_3 - h_0) - T_0(s_3 - s_0)$$

$$= (2698.82 - 104.92) - 298.15(7.18 - 0.37) = 562.30\frac{\text{kJ}}{\text{kg}}$$

$$ex_4 = (h_4 - h_0) - T_0(s_4 - s_0)$$

$$= (2714.63 - 104.92) - 298.15(7.07 - 0.37) = 612.43\frac{\text{kJ}}{\text{kg}}$$

$$ex_5 = (h_5 - h_0) - T_0(s_5 - s_0)$$

$$= (529.64 - 104.92) - 298.15(1.59 - 0.37) = 59.25\frac{\text{kJ}}{\text{kg}}$$

$$ex_6 = \left(\frac{C_{p@T_6} + C_{p@T_0}}{2}\right)\left(T_6 - T_0 - T_0 \ln\left(\frac{T_6}{T_0}\right)\right) + v_{@T_6}(P_6 - P_0)$$

$$= 3.86\left[347.15 - 298.15 - 298.15\ln\left(\frac{347.15}{298.15}\right)\right]$$

$$+(400 - 101.325) = 14.31\frac{\text{kJ}}{\text{kg}}$$

$$ex_7 = \left(\frac{C_{p@T_7} + C_{p@T_0}}{2}\right)\left(T_7 - T_0 - T_0 \ln\left(\frac{T_7}{T_0}\right)\right) + v_{@T_7}(P_7 - P_0)$$

$$= 3.86\left[352.15 - 298.15 - 298.15\ln\left(\frac{352.15}{298.15}\right)\right]$$

$$+(350 - 101.325) = 17.11\frac{\text{kJ}}{\text{kg}}$$

$$ex_8 = \left(\frac{C_{p@T_8} + C_{p@T_0}}{2}\right)\left(T_8 - T_0 - T_0 \ln\left(\frac{T_8}{T_0}\right)\right) + v_{@T_8}(P_8 - P_0)$$

$$= 3.91\left[335.15 - 298.15 - 298.15 \ln\left(\frac{335.15}{298.15}\right)\right]$$

$$+(600 - 101.325) = 8.79 \frac{\text{kJ}}{\text{kg}}$$

$$ex_9 = \left(\frac{C_{p@T_9} + C_{p@T_0}}{2}\right)\left(T_9 - T_0 - T_0 \ln\left(\frac{T_9}{T_0}\right)\right) + v_{@T_9}(P_9 - P_0)$$

$$= 3.92\left[340.15 - 298.15 - 298.15 \ln\left(\frac{340.15}{298.15}\right)\right]$$

$$+(550 - 101.325) = 11.04 \frac{\text{kJ}}{\text{kg}}$$

Calculation of exergy of fuel, product, and loss:

$$\dot{E}x_F = \left(\dot{E}x_4 - \dot{E}x_5\right) + \dot{E}x_1 + \dot{W} = \left(\dot{m}_4 ex_4 - \dot{m}_5 ex_5\right) + \dot{m}_1 ex_1 + \dot{W}$$

$$= 1.13(612.43 - 59.25) + 3.50(22.52) + 1 = 707.16\text{kW}$$

$$\dot{E}x_p = \dot{E}x_2 + \dot{E}x_3 + (\dot{E}x_7 - \dot{E}x6) + (\dot{E}x_9 + \dot{E}x_8)$$

$$= \dot{m}_2 ex_2 + \dot{m}_3 ex_3 + (\dot{m}_7 ex_7 - \dot{m}_6 ex_6) + (\dot{m}_9 ex_9 - \dot{m}_8 ex_8)$$

$$= 2.62(44.10) + 0.88(562.30) + 3.5(17.11 - 14.31)$$

$$+3.5(11.04 - 8.79) = 628.03\text{kw}$$

Exergy destruction rate:

$$\dot{E}x_{des} = \dot{E}x_F - \dot{E}x_P = 707.16 - 628.03 = 79.13\text{kW}$$

Exergy efficiency:

$$\eta_{ex} = \frac{\dot{E}x_P}{\dot{E}x_F} \times 100 = \frac{628.03}{707.16} \times 100 = 88.81\%$$

6.10.2 Exergy Analysis of Syrup Preheating Line

Example 6.22: In sugar factories, the temperature of the diluted syrup entering the evaporation has great importance. If the syrup temperature is low, part of the enthalpy of steam entering the line is used to raise the syrup temperature to boiling, and the rest is used for evaporation. Therefore, in these factories, the syrup entering the evaporation line is preheated by the vapor of the evaporators (Piri et al., 2022). The figure shows the syrup

preheating line in a sugar factory, which uses three shell-tube exchangers to heat the syrup with Brix 14.2%. Table 6.33 presents information about the flows of this heat exchanger. The surface temperature of the equipment for heat exchangers 1, 2, and 3 has been measured as 45°C, 49°C, and 56°C, respectively. Calculate the exergy destruction rate, exergy loss rate, and exergy efficiency for each of the heat exchangers.

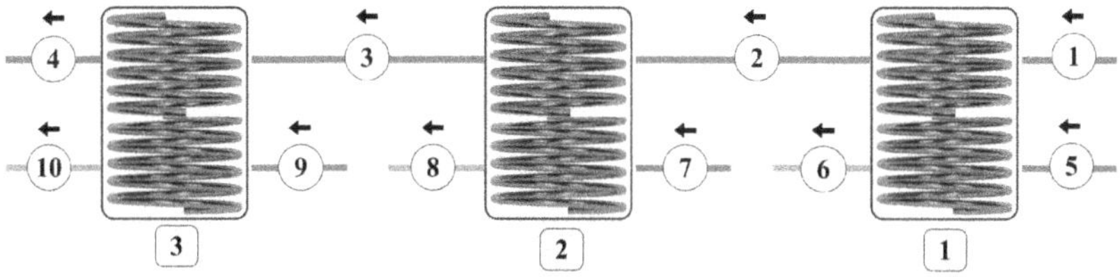

FIGURE 6.49 Schematic of the syrup preheating line related to Example 6.22.

TABLE 6.33
Information Related to the Flows of Example 6.22

State No.	Fluid Type	T (°C)	P (kPa)	Phase Description	$\dot{m}\left(\frac{kg}{s}\right)$
1	Syrup	90.10	551.33	—	48.60
2	Syrup	108.00	491.33	—	—
3	Syrup	118.00	441.33	—	—
4	Syrup	124.00	391.33	—	—
5	Steam	—	171.33	Saturated steam	1.54
6	Water	—	141.33	Saturated water	—
7	Steam	—	241.33	Saturated steam	0.88
8	Water	—	201.33	Saturated water	—
9	Steam	—	369.45	Saturated steam	0.55
10	Water	—	340.33	Saturated water	

✍Solution

Assumptions:

1. The process is considered steady state.
2. Kinetic and potential energy changes are ignored.
3. The dead state temperature and pressure are 25°C and 101.325 kPa.
4. The reference temperature and pressure are 0°C and 101.325 kPa.
5. Condensate water coming out of heat exchangers is a saturated liquid.
6. Syrup ingredients are only water and sucrose (carbohydrates).

To calculate the specific heat and density of syrup compounds, the equations in Table 6.1 were used, and the results are presented in Table 6.34.

TABLE 6.34
Specific Heat and Density of Syrup Compounds for Example 6.22

	Carbohydrate		Water	
T **(°C)**	C_P	ρ	C_P	ρ
$T_{ref} = 0$	1.55	—	4.18	—
$T_0 = 25$	1.59	—	4.18	—
$T_1 = 90.10$	1.68	1571.13	4.21	966.96
$T_2 = 108$	1.69	1565.57	4.23	953.69
$T_3 = 118$	1.70	1562.47	4.24	945.23
$T_4 = 124$	1.70	1560.60	4.25	939.80

Specific heat and specific volume of syrup were determined using Equations 6.4 and 6.7, respectively, and the results are presented in Table 6.35.

TABLE 6.35
Specific Heat and Density of Syrup Flows for Example 6.22

T **(°C)**	$C_P\left(\frac{\text{kJ}}{\text{kg.K}}\right)$	$v\left(\frac{\text{m}^3}{\text{kg}}\right)$
$T_{ref} = 0$	3.80	—
$T_0 = 25$	3.81	—
$T_1 = 90.10$	3.85	0.000978
$T_2 = 108$	3.87	0.000990
$T_3 = 118$	3.88	0.000999
$T_4 = 124$	3.89	0.001004

Equation 6.6 was used to determine the enthalpy of syrup flows. The enthalpy and entropy of water flows were determined from the table of the thermodynamic properties of water:

$$h_1 = \left(\frac{C_{P@T_1} + C_{P@T_{ref}}}{2}\right)T_1 + v_1(P_1 - P_0)$$

$$= 3.83(90) + 0.000978(551.33 - 101.33) = 345.32\frac{\text{kJ}}{\text{kg}}$$

$$h_2 = \left(\frac{C_{P@T_2} + C_{P@T_{ref}}}{2}\right)T_2 + v_2(P_2 - P_0)$$

$$= 3.84(108) + 0.000990(491.33 - 101.33) = 414.72\frac{\text{kJ}}{\text{kg}}$$

$$h_3 = \left(\frac{C_{P@T_3} + C_{P@T_{ref}}}{2}\right)T_3 + v_3\left(P_3 - P_0\right)$$

$$= 3.84(118) + 0.000999(441.33 - 101.33) = 453.67\frac{\text{kJ}}{\text{kg}}$$

$$h_4 = \left(\frac{C_{P@T_4} + C_{P@T_{ref}}}{2}\right)T_4 + v_4\left(P_4 - P_0\right)$$

$$= 3.85(124) + 0.001004(391.33 - 101.33) = 477.09\frac{\text{kJ}}{\text{kg}}$$

$$@25°\text{C}, 101.325\text{kPa}\begin{cases} h_0 = 104.92\dfrac{\text{kJ}}{\text{kg}} \\ s_0 = 0.37\dfrac{\text{kJ}}{\text{kg}}.\text{K} \end{cases}$$

$$@171.33\text{kPa}\begin{cases} h_5 = h_g = 2699.17\dfrac{\text{kJ}}{\text{kg}} \\ s_5 = s_g = 7.18\dfrac{\text{kJ}}{\text{kg}}.\text{K} \end{cases}$$

$$@141.33\text{kPa}\begin{cases} h_6 = h_f = 459.60\dfrac{\text{kJ}}{\text{kg}} \\ s_6 = s_f = 1.41\dfrac{\text{kJ}}{\text{kg}}.\text{K} \end{cases}$$

$$@241.33\text{kPa}\begin{cases} h_7 = h_g = 2714.89\dfrac{\text{kJ}}{\text{kg}} \\ s_7 = s_g = 7.06\dfrac{\text{kJ}}{\text{kg}}.\text{K} \end{cases}$$

$$@201.33\text{kPa}\begin{cases} h_8 = h_f = 505.59\dfrac{\text{kJ}}{\text{kg}} \\ s_8 = s_f = 1.53\dfrac{\text{kJ}}{\text{kg}}.\text{K} \end{cases}$$

$$@369.45\text{kPa}\begin{cases} h_9 = h_g = 2734.45\dfrac{\text{kJ}}{\text{kg}} \\ s_9 = s_g = 6.92\dfrac{\text{kJ}}{\text{kg}}.\text{K} \end{cases}$$

$$@340.33\text{kPa}\begin{cases} h_{10} = h_f = 580.05\dfrac{\text{kJ}}{\text{kg}} \\ s_{10} = s_f = 1.72\dfrac{\text{kJ}}{\text{kg}}.\text{K} \end{cases}$$

Equations 6.8 and 5.10 were used to calculate the exergy of syrup and water flows, respectively:

$$ex_1 = \left(\frac{C_{p@T_1} + C_{p@T_0}}{2}\right)\left(T_1 - T_0 - T_0 \ln\left(\frac{T_1}{T_0}\right)\right) + v_{@T_1}(P_1 - P_0)$$

$$= 3.83\left[363.25 - 298.15 - 298.15 \ln\left(\frac{363.25}{298.15}\right)\right]$$

$$+0.000978(551.33 - 101.325) = 24.26\frac{\text{kJ}}{\text{kg}}$$

$$ex_2 = \left(\frac{C_{p@T_2} + C_{p@T_0}}{2}\right)\left(T_2 - T_0 - T_0 \ln\left(\frac{T_2}{T_0}\right)\right) + v_{@T_2}(P_2 - P_0)$$

$$= 3.84\left[381.15 - 298.15 - 298.15 \ln\left(\frac{381.15}{298.15}\right)\right]$$

$$+0.000990(491.33 - 101.325) = 37.93\frac{\text{kJ}}{\text{kg}}$$

$$ex_3 = \left(\frac{C_{p@T_3} + C_{p@T_0}}{2}\right)\left(T_3 - T_0 - T_0 \ln\left(\frac{T_3}{T_0}\right)\right) + v_{@T_3}(P_3 - P_0)$$

$$= 3.85\left[391.15 - 298.15 - 298.15 \ln\left(\frac{391.15}{298.15}\right)\right]$$

$$+0.000999(441.33 - 101.325) = 46.69\frac{\text{kJ}}{\text{kg}}$$

$$ex_4 = \left(\frac{C_{p@T_4} + C_{p@T_0}}{2}\right)\left(T_4 - T_0 - T_0 \ln\left(\frac{T_4}{T_0}\right)\right) + v_{@T_4}(P_4 - P_0)$$

$$= 3.85\left[397.15 - 298.15 - 298.15 \ln\left(\frac{397.15}{298.15}\right)\right]$$

$$+0.001004(391.33 - 101.325) = 52.31\frac{\text{kJ}}{\text{kg}}$$

$$ex_5 = (h_5 - h_0) - T_0(s_5 - s_0)$$

$$= (2699.17 - 104.92) - 298.15(7.18 - 0.37) = 563.43\frac{\text{kJ}}{\text{kg}}$$

$$ex_6 = (h_6 - h_0) - T_0(s_6 - s_0)$$

$$= (459.60 - 104.92) - 298.15(1.41 - 0.37) = 42.55\frac{\text{kJ}}{\text{kg}}$$

$$ex_7 = (h_7 - h_0) - T_0(s_7 - s_0)$$

$$= (2714.89 - 104.92) - 298.15(7.06 - 0.37) = 613.23\frac{\text{kJ}}{\text{kg}}$$

$$ex_8 = (h_8 - h_0) - T_0(s_8 - s_0)$$
$$= (505.59 - 104.92) - 298.15(1.53 - 0.37) = 53.26\frac{kJ}{kg}$$

$$ex_9 = (h_9 - h_0) - T_0(s_9 - s_0)$$
$$= (2734.45 - 104.92) - 298.15(6.92 - 0.37) = 675.18\frac{kJ}{kg}.$$

$$ex_{10} = (h_{10} - h_0) - T_0(s_{10} - s_0)$$
$$= (580.05 - 104.92) - 298.15(1.72 - 0.37) = 72.64\frac{kJ}{kg}$$

Mass and energy balance for heat exchanger 1:

$$\dot{m}_2 = \dot{m}_1 = 48.60\frac{kg}{s}$$

$$\dot{m}_6 = \dot{m}_5 = 1.54\frac{kg}{s}$$

$$\dot{m}_1h_1 + \dot{m}_5h_5 = \dot{m}_2h_2 + \dot{m}_6h_6 + \dot{Q}_{L,1} \rightarrow \dot{Q}_{L,1} = \dot{m}_5(h_5 - h_6) - \dot{m}_1(h_2 - h_1)$$
$$= 1.54(2699.17 - 459.60) - 48.60(414.72 - 345.32)$$
$$= 76.33\text{kW}$$

Mass and energy balance for heat exchanger 2:

$$\dot{m}_3 = \dot{m}_2 = 48.60\frac{kg}{s}$$

$$\dot{m}_8 = \dot{m}_7 = 0.88\frac{kg}{s}$$

$$\dot{m}_2h_2 + \dot{m}_7h_7 = \dot{m}_3h_3 + \dot{m}_8h_8 + \dot{Q}_{L,2} \rightarrow \dot{Q}_{L,2} = \dot{m}_7(h_7 - h_8) - \dot{m}_2(h_3 - h_2)$$
$$= 0.88(2714.89 - 505.59) - 48.60(453.67 - 414.72)$$
$$= 51.23\text{kW}$$

Mass and energy balance for heat exchanger 3:

$$\dot{m}_4 = \dot{m}_3 = 48.60\frac{kg}{s}$$

$$\dot{m}_{10} = \dot{m}_9 = 0.55\frac{kg}{s}$$

$$\dot{m}_3h_3 + \dot{m}_9h_9 = \dot{m}_4h_4 + \dot{m}_{10}h_{10} + \dot{Q}_{L,3} \rightarrow \dot{Q}_{L,3}$$
$$= \dot{m}_9(h_9 - h_{10}) - \dot{m}_3(h_4 - h_3)$$
$$= 0.55(2734.45 - 580.05) - 48.60(477.09 - 453.67)$$
$$= 46.52\text{kW}$$

Exergy balance for heat exchanger 1:

$$\dot{E}x_{F,1} = \dot{E}x_5 - \dot{E}x_6 = \dot{m}_5\left(ex_5 - ex_6\right) = 1.54\left(563.43 - 42.55\right)$$
$$= 802.16\text{kW}$$

$$\dot{E}x_{P,1} = \dot{E}x_2 - \dot{E}x_1 = \dot{m}_1\left(ex_2 - ex_1\right) = 48.60\left(37.93 - 24.26\right)$$
$$= 664.12\text{kW}$$

$$\dot{E}x_{L,1} = \dot{Q}_{L,1}\left(1 - \frac{T_0}{T_{b,1}}\right) = 76.33\left(1 - \frac{25 + 273.15}{45 + 273.15}\right) = 4.80\text{kW}$$

$$\dot{E}x_{des,1} = \dot{E}x_{F,1} - \dot{E}x_{P,1} - \dot{E}x_{L,1} = 802.16 - 664.12 - 4.80 = 133.24\text{kW}$$

$$\eta_{ex,1} = \frac{\dot{E}x_{P,1}}{\dot{E}x_{F,1}} = \frac{664.12}{802.16} = 82.79$$

Exergy balance for heat exchanger 2:

$$\dot{E}x_{F,2} = \dot{E}x_7 - \dot{E}x_8 = \dot{m}_7\left(ex_7 - ex_8\right) = 0.88\left(613.23 - 53.26\right)$$
$$= 492.77\text{kW}$$

$$\dot{E}x_{P,2} = \dot{E}x_3 - \dot{E}x_2 = \dot{m}_2\left(ex_3 - ex_2\right) = 48.60\left(46.69 - 37.93\right)$$
$$= 426.10\text{kW}$$

$$\dot{E}x_{L,2} = \dot{Q}_{L,2}\left(1 - \frac{T_0}{T_{b,2}}\right) = 51.23\left(1 - \frac{25 + 273.15}{49 + 273.15}\right) = 3.82\text{kW}$$

$$\dot{E}x_{des,2} = \dot{E}x_{F,2} - \dot{E}x_{P,2} - \dot{E}x_{L,2} = 492.77 - 426.10 - 3.82 = 62.86\text{kW}$$

$$\eta_{ex,2} = \frac{\dot{E}x_{P,2}}{\dot{E}x_{F,2}} = \frac{426.10}{492.77} = 86.47$$

Exergy balance for heat exchanger 3:

$$\dot{E}x_{F,3} = \dot{E}x_9 - \dot{E}x_{10} = \dot{m}_9\left(ex_9 - ex_{10}\right) = 0.55\left(675.18 - 72.64\right)$$
$$= 331.40\text{kW}$$

$$\dot{E}x_{P,3} = \dot{E}x_4 - \dot{E}x_3 = \dot{m}_3\left(ex_4 - ex_3\right) = 48.60\left(52.31 - 46.69\right)$$
$$= 272.98\text{kW}$$

$$\dot{E}x_{L,3} = \dot{Q}_{L,3}\left(1 - \frac{T_0}{T_{b,3}}\right) = 46.52\left(1 - \frac{25 + 273.15}{56 + 273.15}\right) = 4.38\text{kW}$$

$$\dot{E}x_{des,3} = \dot{E}x_{F,3} - \dot{E}x_{P,3} - \dot{E}x_{L,3} = 331.40 - 272.98 - 4.38 = 54.04\text{kW}$$

$$\eta_{ex,3} = \frac{\dot{E}x_{P,3}}{\dot{E}x_{F,3}} = \frac{272.98}{331.40} = 82.37$$

6.10.3 Exergy Analysis of Honey Pasteurization

Example 6.23: A shell-tube heat exchanger is used for pasteurization of honey in a honey packaging factory according to the figure. A total of 120 kg of honey with a temperature of 30°C and Brix of 80% is poured into a tank and circulated by a pump for 15 minutes in a heat exchanger until its temperature reaches 74°C. The steam is mixed with water in a mixing chamber and enters the heat exchanger, and the resulting mixture is circulated by a power pump in the heat exchanger and performs the heating process. The surface temperature of the heat exchanger is 60°C, and the power consumption of the pumps is 1 kW. Table 6.36 presents information about the steam flow and heat exchanger condensate. Calculate the exergy destruction and exergy loss in the process.

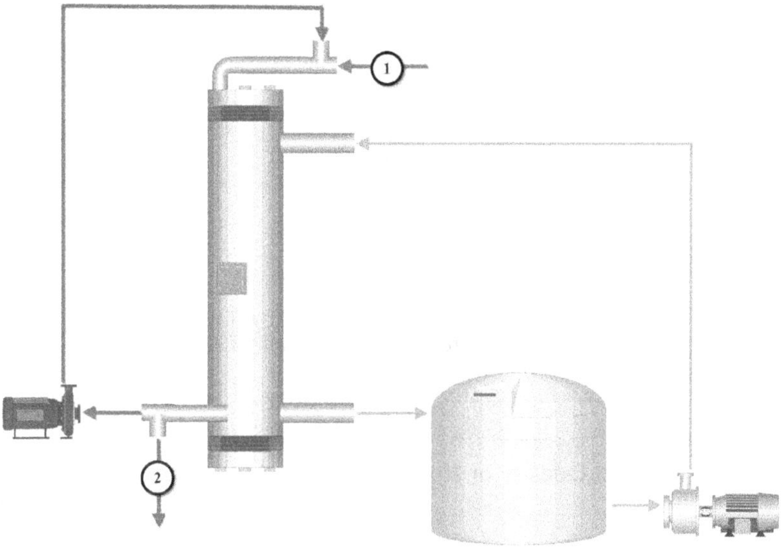

FIGURE 6.50 Schematic of the heat pump related to Example 6.23.

TABLE 6.36
Information Related to the Flows of Example 6.23

State No.	Fluid Type	T (°C)	P (kPa)	$\dot{m}\left(\frac{kg}{s}\right)$
1	Steam	147.73	151.33	0.01
2	Water	84.30	121.33	—

✍Solution

Assumptions:

1. Kinetic and potential energy changes are ignored.
2. The dead state temperature and pressure are 25°C and 101.325 kPa.
3. The reference temperature and pressure are 0°C and 101.325 kPa.
4. The components of honey are only water and carbohydrates.

Characteristics of streams:

To calculate the specific heat and density of honey compounds, the equations in Table 6.1 were used, and the results are presented in Table 6.37.

TABLE 6.37
Specific Heat and Density of Honey Compounds for Example 6.23

	Carbohydrate		Water	
T (°C)	C_P	ρ	C_P	ρ
$T_{ref}=0$	1.55	—	4.18	—
$T_0=25$	1.59	1599.10	4.18	997.18
$T_i=30$	1.60		4.18	993.89
$T_f=74$	1.66		4.20	976.84

The specific heat and volume of honey were determined using Equations 6.4 and 6.7, respectively, and the results are presented in Table 6.38.

TABLE 6.38
Specific Heat and Density of Honey for Example 6.23

T (°C)	$C_P\left(\frac{\text{kJ}}{\text{kg.}\mathbf{K}}\right)$	$V\left(\frac{\mathbf{m}^3}{\text{kg}}\right)$
$T_{ref}=0$	2.07	—
$T_0=25$	2.11	0.000701
$T_i=30$	2.12	0.000704
$T_f=74$	2.17	0.000712

Equation 3.26 was used to determine the internal energy of honey. The enthalpy and entropy of water flows were determined from the table of the thermodynamic properties of water:

$$u_i=\left(\frac{C_{P@T_i}+C_{P@T_{ref}}}{2}\right)T_i=2.10(30)=62.88\frac{\text{kJ}}{\text{kg}}$$

$$u_f = \left(\frac{C_{P@T_i} + C_{P@T_{ref}}}{2}\right) T_f = 2.12(74) = 157.00 \frac{\text{kJ}}{\text{kg}}$$

$$@25°\text{C},101.325\text{kPa} \begin{cases} h_0 = 104.92 \dfrac{\text{kJ}}{\text{kg}} \\ s_0 = 0.37 \dfrac{\text{kJ}}{\text{kg}}.\text{K} \end{cases}$$

$$@147.73°\text{C},151.325\text{kPa} \begin{cases} h_1 = 2768.20 \dfrac{\text{kJ}}{\text{kg}} \\ s_1 = 7.41 \dfrac{\text{kJ}}{\text{kg}}.\text{K} \end{cases}$$

$$@84.3°\text{C},121.325\text{kPa} \begin{cases} h_2 = 353.12 \dfrac{\text{kJ}}{\text{kg}} \\ s_2 = 1.13 \dfrac{\text{kJ}}{\text{kg}}.\text{K} \end{cases}$$

$$ex_i = \left(\frac{C_{p@T_i} + C_{p@T_0}}{2}\right)\left(T_i - T_0 - T_0 i\left(\frac{T_i}{T_0}\right)\right) + P_0\left(v_i - v_0\right)$$

$$= 2.11\left[303.15 - 298.15 - 298.15 \ln\left(\frac{303.15}{298.15}\right)\right]$$

$$+101.325(0.000704 - 0.000701) = 0.09 \frac{\text{kJ}}{\text{kg}}$$

$$ex_f = \left(\frac{C_{p@T_f} + C_{p@T_0}}{2}\right)\left(T_f - T_0 - T_0 \ln\left(\frac{T_f}{T_0}\right)\right) + P_0\left(v_f - v_0\right)$$

$$= 2.14\left[374.15 - 298.15 - 298.15 \ln\left(\frac{374.15}{298.15}\right)\right]$$

$$+101.325(0.000712 - 0.000701) = 7.78 \frac{\text{kJ}}{\text{kg}}$$

$$ex_1 = (h_1 - h_0) - T_0(s_1 - s_0)$$

$$= (2768.20 - 104.92) - 298.15(7.41 - 0.37) = 564.76 \frac{\text{kJ}}{\text{kg}}$$

$$ex_2 = (h_2 - h_0) - T_0(s_2 - s_0)$$

$$= (353.12 - 104.92) - 298.15(1.13 - 0.37) = 21.87 \frac{\text{kJ}}{\text{kg}}$$

Mass balance:

$$m_{inlet} = m_{outlet} \rightarrow \dot{m}_1 = \dot{m}_2 = \dot{m}_s = 0.01 \frac{\text{kg}}{\text{s}}$$

$$\frac{dm_{honey}}{dt} = 0 \rightarrow \dot{m}_{i,\text{honey}} = \dot{m}_{f,\text{honey}} = \dot{m}_{\text{honey}} = 120\text{kg}$$

Energy balance:

$$E_{inlet} - E_{outlet} = \Delta E_{system}$$

$$\rightarrow m_1 h_1 + W_i - m_2 h_2 - Q_o = \Delta E_{system}$$

$$\rightarrow \left(\dot{m}_1 h_1 + \dot{W} - \dot{m}_2 h_2\right)\Delta t - Q_o = \Delta E_{system}$$

$$\rightarrow \left[\dot{m}_s \left(h_1 - h_2\right) + \dot{W}\right]\Delta t - Q_o = m_{honey}\left(u_f - u_i\right)$$

$$\rightarrow Q_o = \left[\dot{m}_s \left(h_1 - h_2\right) + \dot{W}\right]\Delta t - m_{honey}\left(u_f - u_i\right)$$

$$\rightarrow Q_o = \left[0.01(2768.20 - 353.12) + 1\right](15 \times 60) - 120(157 - 62.88)$$
$$= 11340.40\text{kJ}$$

Exergy balance:

$$Ex_{inlet} - Ex_{outlet} - Ex_{des} = \Delta Ex_{system}$$

$$m_1 ex_1 + W_i - m_2 ex_2 - Ex_{Q_o} - Ex_{des} = \Delta Ex_{system}$$

$$\rightarrow \left(\dot{m}_1 ex_1 + \dot{W} - \dot{m}_2 ex_2\right)\Delta t - Ex_{Q_o} - Ex_{des} = \Delta E_{system}$$

$$\rightarrow \left[\dot{m}_s \left(ex_1 - ex_2\right) + \dot{W}\right]\Delta t - Ex_{Q_o} - Ex_{des} = m_{honey}\left(ex_f - ex_i\right)$$

$$\rightarrow \left[\dot{m}_s \left(ex_1 - ex_2\right) + \dot{W}\right]\Delta t - Ex_{Q_o} - Ex_{des} = m_{honey}\left(ex_f - ex_i\right)$$

$$\rightarrow Ex_{des} = \left[\dot{m}_s \left(\text{ex}_1 - \text{ex}_2\right) + \dot{W}\right]\Delta t - \text{Ex}_{Q_o} - m_{honey}\left(\text{ex}_f - \text{ex}_i\right)$$

$$\dot{E}x_{Q_o} = Q_o\left(1 - \frac{T_0}{T_b}\right) = 11340.40\left(1 - \frac{25 + 273.15}{60 + 273.15}\right) = 1191.40\text{kJ}$$

$$\rightarrow Ex_{des} = \left[0.01(564.76 - 21.87) + 1\right](15 \times 60) - 1191.40 - 120(7.78 - 0.09)$$
$$= 3672.05\text{kJ}$$

7 Psychrometrics and Drying

7.1 MIXTURE IN THERMODYNAMICS

From thermodynamics, we know that a gas mixture is a combination of several different components with pressure, temperature, mass, and volume attributes that can be experimentally measured, but they alone are not sufficient to determine the thermodynamic state of the mixture. In the process engineering, we face several types of mixtures or gas mixtures that play important roles in energy transfer, reactions, and mass exchange. Drying is the most famous process, mostly known as an intensive energy-consuming one. Ranging from the wood industry to food processing, drying is frequently used for preservation, size reduction, waste recovery, and quality improvement objectives. Moist air is the energy/mass-carrying element of most typical drying techniques. Thermodynamically speaking, it is a gas mixture. Here we bring in some basic concepts from thermodynamics through which the thermodynamic state of a mixture and the amount of each component in the mixture are specified. These would be preliminary steps for more advanced applications. However, owing to psychrometric charts developed by ASHRAE where air is assumed to be an ideal gas (with 1% error of assumption), engineers are more comfortable finding physical and thermal properties of air such as absolute humidity, relative humidity, dry/wet bulb temperature, specific volume, and enthalpy. The following equations are presented for a quick review. For deeper insight, thermodynamic textbooks are recommended.

7.1.1 Mole Fraction

The mole fraction is defined as follows:

$$y_i = \frac{n_i}{n} \tag{7.1}$$

y_i is the mole fraction of the i-th component, n_i is the number of moles of component i, and n is the total number of moles of the mixture:

$$n = \sum_i n_i \tag{7.2}$$

By combining these equations, we can write:

$$\sum_i y_i = 1 \tag{7.3}$$

DOI: 10.1201/9781003424680-7

7.1.2 Mass Fraction

Mass fraction is defined as follows:

$$x_i = \frac{m_i}{m} \tag{7.4}$$

x_i is the mass fraction of the *i*-th components, m_i is the mass of component *i*, and *m* is the total mass of the mixture:

$$m = \sum_i m_i \tag{7.5}$$

By combining these equations, we can write:

$$\sum_i x_i = 1 \tag{7.6}$$

7.2 GAS MIXTURES

In the thermodynamic analysis of gas mixtures, it is assumed that the mixture operates under ideal conditions and adheres to the ideal gas equation. Two prominent models, those of Dalton and Amagat, are employed for the thermodynamic analysis of gas mixtures.

7.2.1 Dalton Model

In the Dalton model, each component is considered alone in the total volume of the mixture and the temperature of the mixture, and the total pressure of the mixture is equal to the sum of the partial pressures of all the components of the mixture. When dealing with a mixture comprising two distinct gas types, the aggregate number of moles in the mixture equals the summation of the moles of each individual component.

$$n_{total} = n_{mix} = n_A + n_B \tag{7.7}$$

By using the equation of the ideal gas state and simplifying the equations, we can write:

$$\frac{PV}{\mathcal{R}T} = \frac{P_A V}{\mathcal{R}T} + \frac{P_B V}{\mathcal{R}T} \Rightarrow P_{mix} = P_A + P_B \tag{7.8}$$

Therefore, according to Dalton's model, in ideal gas mixtures, the total pressure is equal to the sum of the partial pressures of the components of the mixture:

$$P_{mix} = \sum_{i=1}^{k} P_i \tag{7.9}$$

According to this model, in an ideal gas mixture, the pressure of each component is equal to the product of the corresponding mole fraction in the pressure of the mixture:

$$P_i = y_i P_{mix} \tag{7.10}$$

7.2.2 Amagat Model

This model operates under the assumption that the overall volume of the mixture matches the combined sum of its individual partial volumes, with all components experiencing the pressure and temperature of the mixture. Similar to the analysis presented for the Dalton model, it can be affirmed that, in accordance with the Amagat model, the total volume within an ideal gas mixture is equivalent to the cumulative sum of the partial volumes associated with its constituent components.

$$V_{mix} = V_A + V_B \tag{7.11}$$

Therefore

$$\mathrm{V}_{\mathrm{mix}} = \sum_{\mathrm{i=1}}^{\mathrm{k}} \mathrm{V}_{\mathrm{i}} \tag{7.12}$$

and

$$V_i = y_i V_{mix} \tag{7.13}$$

✐Example 7.1: A rigid tank contains 2 kg of nitrogen gas (N_2) and 4 kg of carbon dioxide gas (CO_2) at a temperature of 25°C and a pressure of 2 MPa. Calculate the partial pressure of both gases and the gas constant of the mixture.

✍Solution

To calculate the partial pressure of gases, it is necessary to determine the number of moles:

$$\begin{cases} n_{N_2} = \dfrac{m_{N_2}}{M_{N_2}} = \dfrac{2}{28} = 0.0714\text{mol} \\ n_{co_2} = \dfrac{m_{co_2}}{M_{co_2}} = \dfrac{4}{44} = 0.0909\text{mol} \end{cases} \rightarrow n = 0.0714 + 0.0909 = 0.1623\text{mol}$$

Determining the mole fraction:

$$y_{N_2} = \frac{n_{N_2}}{n} = \frac{0.0714}{0.1623} = 0.44$$

$$y_{co_2} = \frac{n_{co_2}}{n} = \frac{0.0909}{0.1623} = 0.56$$

Partial pressure:

$$P_{N_2} = y_{N_2}P = (0.44)(2) = 0.88\text{MPa}$$
$$P_{CO_2} = y_{CO_2}P = (0.56)(2) = 1.12\text{MPa}$$

Molar mass of mixture:

$$M = \sum(y_iM_i) = (0.44)(28) + (0.56)(44) = 36.96\,{}^{kg}\!/_{kmol}$$

The gas constant is equal to:

$$R = \frac{\mathcal{R}}{M} = \frac{8.314}{36.96} = 0.225\frac{\text{kJ}}{\text{kg.K}}$$

✐Example 7.2: A container contains 2 kg of nitrogen gas (N_2), 4 kg of oxygen gas (O_2), and 6 kg of carbon dioxide gas (CO_2). Calculate the mass fraction, mole fraction, molecular mass of the mixture, and gas constant of the mixture.

✍Solution

Total mass of mixture:

$$m = m_{N_2} + m_{O_2} + m_{CO_2} = 2 + 4 + 6 = 12\text{kg}$$

Mass fraction:

$$c_{N_2} = \frac{m_{N_2}}{m} = \frac{2}{12} = 0.1667$$
$$c_{N_2} = \frac{m_{O_2}}{m} = \frac{4}{12} = 0.3333$$
$$c_{CO_2} = \frac{m_{CO_2}}{m} = \frac{6}{12} = 0.5$$

Number of moles:

$$\begin{cases} n_{N_2} = \dfrac{m_{N_2}}{M_{N_2}} = \dfrac{2}{28} = 0.0714\text{mol} \\ n_{O_2} = \dfrac{m_{O_2}}{M_{O_2}} = \dfrac{4}{32} = 0.125\text{mol} \\ n_{CO_2} = \dfrac{m_{CO_2}}{M_{CO_2}} = \dfrac{6}{44} = 0.1364\text{mol} \end{cases}$$

$$\rightarrow n = 0.0714 + 0.125 + 0.1364 = 0.3328\text{mol}$$

Determining the mole fraction:

$$y_{N_2} = \frac{n_{N_2}}{n} = \frac{0.0714}{0.3328} = 0.215$$

$$y_{O_2} = \frac{n_{O_2}}{n} = \frac{0.125}{0.3328} = 0.376$$

$$y_{CO_2} = \frac{n_{CO_2}}{n} = \frac{0.1364}{0.3328} = 0.41$$

Molar mass of mixture:

$$M = \sum (y_i M_i) = (0.215)(28) + (0.376)(32) + (0.41)(44) = 36.1 \,{}^{kg}/_{kmol}$$

Determining the gas constant is equal to:

$$R = \frac{\mathcal{R}}{M} = \frac{8.314}{36.1} = 0.23 \frac{kJ}{kg.K}$$

7.3 PROPERTIES OF A MIXTURE OF IDEAL GASES

The extensive coordinates of a mixture, such as enthalpy (H), internal energy (U), and entropy (S), can be calculated simply by adding up the contribution of each component. For example, the enthalpy of a mixture is:

$$H = \sum H_i = H_1 + H_2 + H_3 + \ldots \tag{7.14}$$

And according to specific enthalpy

$$H = mh = \sum m_i h_i \tag{7.15}$$

By dividing the previous equation by m, we can write:

$$h = \sum x_i h_i \tag{7.16}$$

The enthalpy changes of each component are equal to:

$$\Delta h_i = C_{p,i} \Delta T \tag{7.17}$$

Therefore

$$\Delta h = C_p \Delta T = \sum x_i \left(C_{p,i} \Delta T \right) \tag{7.18}$$

By dividing both sides of the equation by ΔT, the result is:

$$C_p = \sum x_i C_{p,i} \tag{7.19}$$

In the same way, it can be written:

$$C_v = \sum x_i C_{v,i} \tag{7.20}$$

Example 7.3: A mixture of an ideal gas consists of 2 kmol of methane gas (CH_4), 1 kmol of nitrogen gas (N_2), and 1 kmol of carbon dioxide gas (CO_2) at a temperature of 20°C and a pressure of 20 kPa. This mixture is heated to reach 400°C under constant pressure. Calculate the amount of heat transfer, amount of work, and entropy changes. Table 7.1 presents the specific heats of each component.

TABLE 7.1
Information Related to Example 7.3

	C_p (kJ/kg.K)	C_v (kJ/kg.K)
CH_4	2.254	1.735
N_2	1.042	0.745
CO_2	0.842	0.653

Solution

Assumptions:

1. Kinetic and potential energy changes are ignored.
2. The mixture is an ideal gas.
3. Specific heats are constant.

Determining the mass of the mixture:

$$m = nM \rightarrow \begin{cases} m_{CH_4} = (2)(16) = 32\text{kg} \\ m_{N_2} = (1)(14) = 14\text{kg} \\ m_{CO_2} = (1)(44) = 44\text{kg} \end{cases} \rightarrow m = 32 + 14 + 44 = 90\text{kg}$$

Mass fraction of each component:

$$c_{CH_4} = \frac{m_{CH_4}}{m} = \frac{32}{90} = 0.356$$

$$c_{N_2} = \frac{m_{N_2}}{m} = \frac{14}{90} = 0.1556$$

$$c_{CO_2} = \frac{m_{CO_2}}{m} = \frac{44}{90} = 0.498$$

Determination of specific heat:

$$C_p = \sum c_i C_{p,i} = (0.356)(2.254) + (0.1556)(1.042) + (0.489)(0.842) = 1.376 \, \text{kJ}/\text{kg.K}$$

$$C_v = \sum c_i C_{v,i} = (0.356)(1.735) + (0.1556)(0.745) + (0.489)(0.653) = 1.053 \, \text{kJ}/\text{kg.K}$$

Ignoring the changes in kinetic energy and potential, the energy balance for the gas mixture is as follows:

$$Q_i - W_o = \Delta U \rightarrow W_o = Q_i - \Delta U$$

In the constant pressure process, the amount of heat exchanged is equal to the enthalpy changes, so according to the ideal gas properties, the energy balance can be written as follows:

$$W_o = mC_p\Delta T - mC_v\Delta T = m(T_2 - T_1)(C_p - C_v)$$
$$= 90(400 - 20)(1.376 - 1.053) = 11046.6\text{kJ}$$

The process is constant pressure, so the entropy change of the ideal gas can be calculated as follows:

$$\Delta S = m\left(C_p \ln\frac{T_2}{T_1} - R\ln\frac{P_2}{P_1}\right) = 90\left[(1.376)\ln\frac{273.15+400}{273.15+20}\right] = 103\frac{\text{kJ}}{\text{kg.K}}$$

✐Example 7.4: Table 7.2 presents the information about the inlet and outlet flows of the steam boiler.

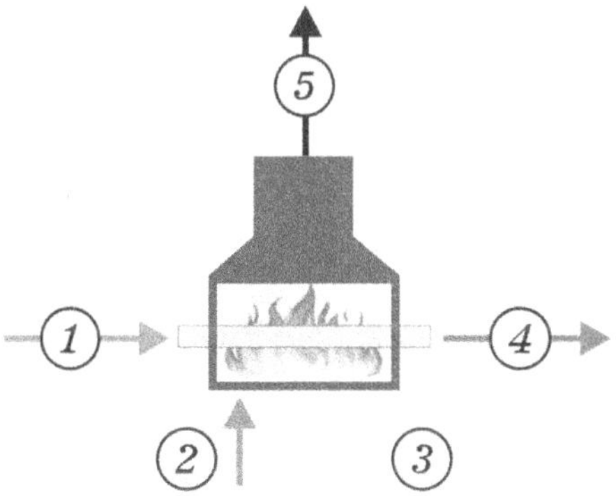

FIGURE 7.1 Schematic view of flows in Example 7.4.

TABLE 7.2
Information about the Flows of Example 7.4

State No.	Fluid Type	P (kPa)	T (°C)	$\dot{m}\left(\frac{\text{kg}}{\text{s}}\right)$
1	Water	950.00	70.00	3.24
2	Fuel	100.00	25.00	0.20
3	Air	100.00	25.00	3.50
4	Steam	800.00	180.00	3.24
5	Exhaust gases	190.00	130.00	3.70

Using the following information, calculate the heat loss from the surface of the boiler:

1. The LHV of fuel is 44,661 kJ/kg.
2. The specific heat of air is 1.005 kJ/kg°C.
3. The power consumption of the boiler is 1 kW.
4. The mass fraction and specific heat of outlet steam components are according to Table 7.3.

TABLE 7.3

Information about the Components of Exhaust Gases of the Boiler, Example 7.4

Gas	Mass Fraction	Special heat $\left(\frac{kJ}{kg.K}\right)$
O_2	0.027577	0.946103
N_2	0.706995	1.044997
H_2O	0.123572	1.915275
CO_2	0.129677	0.942402
Ar	0.012164	0.52
He	0.000001	5.193
Ne	0.000012	1.03
Kr	0.000002	0.274

Solution

Assumptions:

1. The process is considered steady state.
2. Kinetic and potential energy changes are neglected.
3. Inlet air and output exhaust gases are assumed to be ideal gases.
4. The reference temperature is 0°C.

Calculation of the specific heat of outlet steam:

$$\begin{aligned} C_p = \sum c_i C_{p,i} &= (0.027577)(0.946103) + (0.706995)(1.044997) \\ &+ (0.123572)(1.915275) + (0.129677)(0.942402) \\ &+ (0.012164)(0.52) + (0.000001)(5.193) + (0.000012)(1.03) \\ &+ (0.000002)(0.274) = 1.13\, \frac{kJ}{kg.K} \end{aligned}$$

Determination of flow characteristics:

$$h_1 = h_{f\,@70^\circ C} = 293.07 \frac{kJ}{kg}$$

$$h_3 = C_p T_3 = 1.005(25) = 25.13 \frac{\text{kJ}}{\text{kg}}$$

$$h_4 = h_{@180°\text{C},800\text{kPa}} = 2791.47 \frac{\text{kJ}}{\text{kg}}$$

$$h_5 = C_p T_5 = 1.13(130) = 146.9 \frac{\text{kJ}}{\text{kg}}$$

The energy balance for the boiler is as follows:

$$\dot{m}_1 h_1 + \dot{m}_2 LHV_{fuel} + \dot{m}_3 h_3 + \dot{W} = \dot{m}_4 h_4 + \dot{m}_5 h_5 + \dot{Q}_{loss}$$

$$\rightarrow 3.24(293.07) + 0.2(44661) + 3.5(25.126) + 1$$
$$= 3.24(2791.47) + 3.7(146.9) + \dot{Q}_{loss}$$

$$\rightarrow \dot{Q}_{loss} = 382.87 kW$$

7.4 ANALYSIS OF HUMID AIR

The Dalton model is used for air analysis, in which both dry air and water vapor can be assumed to be ideal gases.

$$P = P_a + P_v \quad (7.21)$$

$$V_a = V_v = V \quad (7.22)$$

$$m = m_a + m_v \quad (7.23)$$

$$P_a V = m_a R_a T \quad (7.24)$$

$$P_v V = m_v R_v T \quad (7.25)$$

Usually, the partial pressure of water vapor is called vapor pressure (P_v), and it is the pressure that water vapor alone has at the temperature and volume of atmospheric air. In the previous equations, the properties related to dry air are specified with index a, and the properties related to vapor are specified with index v.

7.4.1 Dew Point Temperature

The dew point temperature is the temperature at which distillation begins when the air is cooled to that temperature at constant pressure; in other words, T_{dp} is the saturation temperature corresponding to the vapor pressure.

$$T_{dp} = T_{sat @ P_v} \quad (7.26)$$

7.4.2 Humidity Ratio, Specific Humidity, or Absolute Humidity

The ratio of the mass of water vapor in the air to the mass of dry air is called the humidity ratio, specific humidity, or absolute humidity.

$$\omega = \frac{m_v}{m_a} = \frac{\frac{P_v V}{R_v T}}{\frac{P_a V}{R_a T}} = \frac{\frac{P_v}{R_v}}{\frac{P_a}{R_a}} = 0.622\frac{P_v}{P_a} = \frac{0.622P_v}{P - P_v}\left(\frac{\text{kg}_{\text{water}}}{\text{kg}_{\text{dry air}}}\right) \tag{7.27}$$

7.4.3 Relative Humidity

The ratio of the amount of humidity in the air (m_v) to the maximum humidity that the air can hold at the same temperature and pressure (m_g) is called relative humidity:

$$\varnothing = \frac{m_v}{m_g} = \frac{\frac{P_v V}{R_v T}}{\frac{P_g V}{R_v T}} = \frac{P_v}{P_g} \tag{7.28}$$

such that

$$P_g = P_{sat@T} \tag{7.29}$$

By combining the equations of humidity ratio and relative humidity, we can write:

$$\varnothing = \frac{\omega P}{(0.622 + \omega)P_g} \tag{7.30}$$

$$\omega = \frac{0.622\varnothing P_g}{P - \varnothing P_g} \tag{7.31}$$

7.4.4 Enthalpy of Humid Air

The total enthalpy of the atmospheric air is equal to the sum of the enthalpy of dry air and water vapor:

$$H = H_a + H_v = m_a h_a + m_v h_v \tag{7.32}$$

By dividing by m_a:

$$h = \frac{H}{m_a} = h_a + \frac{m_v}{m_a} h_v = h_a + h_v \omega \left(\frac{\text{kJ}}{\text{kg}_{\text{dry air}}}\right) \tag{7.33}$$

The enthalpy of dry air relative to 0°C is determined as follows:

$$h = C_{p,air} T + h_v \omega \left(\frac{\text{kJ}}{\text{kg}_{\text{dry air}}}\right) \tag{7.34}$$

7.4.5 Special Heat of Moist Air

The specific heat of humid air can be obtained from the following equation:

$$C_p = C_{p,a} + C_{p,v}\omega \tag{7.35}$$

The following equations are used to calculate the specific heat of water vapor and dry air (Cengel & Boles, 2011):

$$C_{p,a} = \frac{28.11 + 0.1967\times10^{-2}T + 0.4802\times10^{-5}T^2 - 1.966\times10^{-9}T^3}{28.97} \tag{7.36}$$

$$C_{p,v} = \frac{32.24 + 0.1923\times10^{-2}T + 1.055\times10^{-5}T^2 - 3.595\times10^{-9}T^3}{18} \tag{7.37}$$

In these equations, the temperature is in Kelvin and specific heat is in kJ/kg.K.

✐Example 7.5: The air inside a room with a volume of 150 m^3, a temperature of 25°C, and a pressure of 100 kPa has a relative humidity of 60%. Calculate the humidity ratio, dew point temperature, mass of water vapor in the air, and mole fraction of water vapor.

✍Solution

By calculating the partial pressure of air and water vapor, the ratio of humidity can be calculated:

$$\varnothing = \frac{P_v}{P_g} \rightarrow P_v = \varnothing P_{g@25°C} = (0.6)(3.1698) = 1.9019\text{kPa}$$

$$P = P_a + P_v \rightarrow P_a = P - P_v = 100 - 1.9019 = 98.0981\text{kPa}$$

$$\omega = 0.622\frac{P_v}{P_a} = 0.622\left(\frac{1.9019}{98.0981}\right) = 0.012059\frac{\text{kg}_{\text{water}}}{\text{kg}_{\text{dry air}}}$$

The dew point temperature is equal to the saturation temperature at vapor pressure:

$$T_{dp} = T_{sat@P_v} = 16.7°C$$

The mass of water vapor can be calculated using the ideal gas equation:

$$\omega = \frac{m_v}{m_a} \rightarrow m_v = \omega m_a = \omega\left(\frac{P_aV_a}{R_aT}\right)$$

$$= (0.012059)\left[\frac{(98.0981)(150)}{(0.287)(273.15+25)}\right] = 2.07\text{kg}$$

The molar masses of water and air are 18 and 28.97 kg/mol, respectively. By calculating the number of moles of water and air, the mole fraction of water vapor can be determined:

$$\begin{cases} n_v = \dfrac{m_v}{M_v} = \dfrac{2.07}{18} = 0.1152\text{mol} \\ n_a = \dfrac{m_a}{M_a} = \dfrac{\left[\dfrac{(98.0981)(150)}{(0.287)(298.15)}\right]}{28.97} = 5.94\text{mol} \end{cases}$$

So, the mole fraction of water vapor is equal to:

$$y_v = \frac{0.1152}{0.1152 + 5.94} = 0.019$$

7.5 ADIABATIC SATURATION PROCESS

Relative humidity and humidity ratio are widely used in engineering and meteorological sciences, so it is better to relate them to simple measurable quantities, such as temperature and pressure. One way to calculate relative humidity is to calculate the dew point temperature of the air. By knowing the dew point temperature, the vapor pressure can be determined; then the relative humidity can be calculated. Although this method is simple, it is not completely practical. Another method of calculating humidity ratio or relative humidity is related to an adiabatic process.

During this process, an unsaturated air-vapor mixture with the specific humidity ω_1 and temperature T_1 is placed in contact with water in a completely insulated channel. As the air passes over the water, part of the water evaporates and mixes with the air flow; the amount of air humidity increases, and the temperature of the air-vapor mixture decreases, since a part of the latent heat of evaporation is supplied through the air. If the channel is long enough, the air flow is saturated (relative humidity 100%) and exits with a temperature of T_2, which is called the adiabatic saturation temperature. For the process to be a steady state, compensating water at the adiabatic saturation temperature (T_2) is added to the system in the same amount as the flow that evaporates, and the pressure is considered constant.

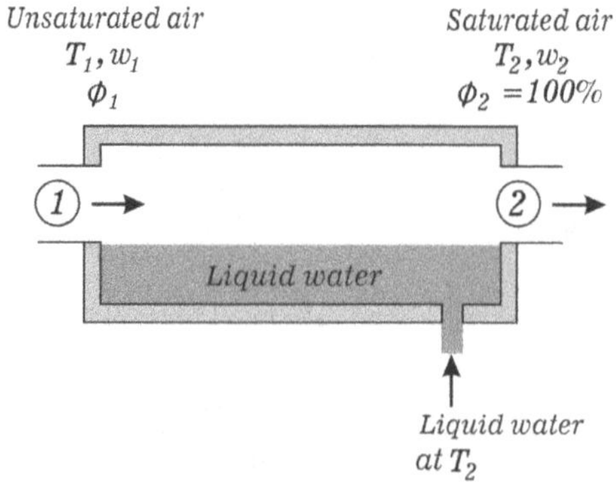

FIGURE 7.2 Adiabatic saturation process.

7.5.1 Conservation of Mass for Adiabatic Saturation Process

For dry air:

$$\dot{m}_{a1} = \dot{m}_{a2} = \dot{m}_a \tag{7.38}$$

For water:

$$\dot{m}_{w1} + \dot{m}_f = \dot{m}_{w2} \tag{7.39}$$

Or:

$$\dot{m}_{a1}\omega_1 + \dot{m}_f = \dot{m}_{a2}\omega_2 \tag{7.40}$$

m_f is the mass flow rate of liquid water injected; therefore, we can write:

$$\dot{m}_f = \dot{m}_a\left(\omega_2 - \omega_1\right) \tag{7.41}$$

7.5.2 Conservation of Energy for Adiabatic Saturation Process

Neglecting changes in kinetic and potential energy:

$$\dot{m}_a h_1 + \dot{m}_f h_{f@T_2} = \dot{m}_a h_2 \tag{7.42}$$

By dividing both sides of the equation by $\dot{m}_a$:

$$h_1 + \left(\omega_2 - \omega_1\right)h_{f@T_2} = h_2 \tag{7.43}$$

Or

$$\left(c_p T_1 + h_{g@T_1}\omega_1\right) + \left(\omega_2 - \omega_1\right)h_{f@T_2} = \left(c_p T_2 + h_{g@T_2}\omega_2\right) \tag{7.44}$$

Which results in:

$$\omega_1 = \frac{c_p\left(T_2 - T_1\right) + \omega_2 h_{fg@T_2}}{h_{g@T_1} - h_{f@T_2}} \tag{7.45}$$

By inserting the previous into the equation:

$$\omega = \frac{0.622\varnothing P_g}{P - \varnothing P_g}, \varnothing = 1 \Rightarrow \omega_2 = \frac{0.622 P_{g2}}{P_2 - P_{g2}} = \frac{0.622 P_{sat@T_2}}{P_2 - P_{sat@T_2}} \tag{7.46}$$

Adiabatic saturation process makes it possible to calculate the relative humidity or humidity ratio, but this requires a long channel to achieve saturation conditions at the output. A more practical method is to use a thermometer whose reservoir is covered with a cotton wick soaked in water, where air flow is maintained over the wick (Figure 7.3). The measured temperature in this case is called wet bulb temperature (wb).

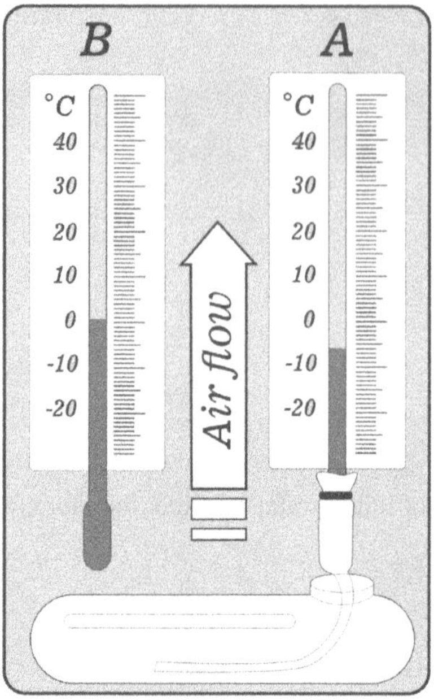

FIGURE 7.3 Diagram of thermometer. A) Wet bulb temperature. B) Dry bulb temperature (db: dry bulb).

The basic principle used is like that used in the adiabatic saturation process. When unsaturated air passes over the wick, some of the water on the wick evaporates. As a result, the water temperature decreases, and a temperature difference occurs between the air and water due to heat transfer. After some time, the heat that the water loses due to evaporation equals the heat that the air takes, and the air temperature stabilizes. The temperature that the thermometer shows at this moment is called the wet bulb temperature.

A device that works on such a principle is called a suspended hygrometer. Usually, the adiabatic saturation temperature and the wet bulb temperature are not the same; However, for air–water–vapor mixtures at atmospheric pressure, the wet bulb temperature is approximately equal to the adiabatic saturation temperature; therefore, to determine the air humidity ratio, the wet bulb temperature T_{wb} can be used instead of T_2.

Example 7.6: The dry and wet bulb temperatures of atmospheric air at 95 kPa pressure are 25°C and 17°C, respectively. It is desirable to calculate the humidity ratio, relative humidity, and air enthalpy.

Solution

Assumptions:

1. The specific heat of air is 1.005 kJ/kg.K.

A)

$$\omega_2 = \frac{0.622P_{sat@T_2}}{P_2 - P_{sat@T_2}} = \frac{0.622(1.9591)}{95-1.9591} = 0.01301\frac{kg_{water}}{kg_{dry\ air}}$$

$$\omega_1 = \frac{C_p(T_2 - T_1) + \omega_2 h_{fg@T_2}}{h_{g@T_1} - h_{f@T_2}} = \frac{1.005(17-25)+0.01301(2460.64)}{2546.5-71.3552}$$

$$= 0.00968\frac{kg_{water}}{kg_{dry\ air}}$$

B)

$$\varnothing_1 = \frac{\omega_1 P_1}{(0.622+\omega_1)P_{g1}} = \frac{\omega_1 P_1}{(0.622+\omega_1)P_{sat@25°C}}$$

$$= \frac{0.00968(95)}{(0.622+0.00968)(3.1698)} = 0.459 \text{ or } 45.9\%$$

C)

$$h_1 = h_{a1} + h_{v1}\omega_1 \cong C_{p,air}T_1 + h_{g@T_1}\omega_1$$

$$= \left(1.005\frac{kJ}{kg°C}\right)(25°C) + (0.00968)\left(2546.5\frac{kJ}{kg}\right)$$

$$= 49.78\frac{kJ}{kg_{dry\ air}}$$

7.6 PSYCHROMETRY CHART

The state of atmospheric air at a certain pressure can be completely determined by using two independent and concentrated properties. The remaining properties can then be computed accordingly. Psychrometric charts are a graphical representation of the physical (vapor pressure, relative humidity, humidity ratio, and specific volume) and thermal (dry-bulb temperature, wet-bulb temperature, enthalpy, and dew point temperature) properties of atmospheric air. The general form of a psychrometric chart is shown in Figure 7.4. Dry-bulb temperatures are displayed on the horizontal axis and the ratio of humidity on the vertical axis.

Instead of a straight line on the left side of the graph, there is a curve (called the saturation line). All states of saturated air are placed on this curve. Other constant relative humidity curves have the same general shape as this curve. Lines of constant wet-bulb temperature appear to slope to the right. The lines of constant specific volume also have the same shape, with the difference that they have a steeper slope. The lines of constant enthalpy lie almost parallel to the lines of constant wet bulb temperature. The values read from the psychrometric chart inevitably have some errors; today, the use of psychometric software has mitigated such a problem.

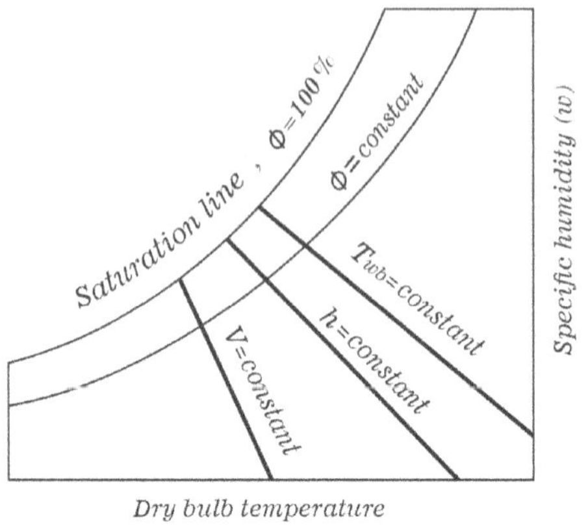

FIGURE 7.4 Outline of psychometric chart.

7.7 THERMODYNAMIC PROCESSES OF HUMID AIR

In operations such as drying and cooling, the conditions of the air must be changed. For example, to dry a product, the ambient air is heated so that it heats up and its ability to absorb humidity increases. Changing the state of air from one state to another is accompanied by a change in energy or air temperature, which is called the thermodynamic process of air. In this part, three widely used thermodynamic processes of air change are examined.

7.7.1 Sensible Heating

Sensible heating of the air occurs when the air temperature increases without changing the absolute humidity of the air. During this process, relative humidity of the air decreases. The reason for this is that the relative humidity is equal to the ratio of the amount of humidity to the humidity capacity of the air at the same temperature, and the humidity capacity increases with temperature.

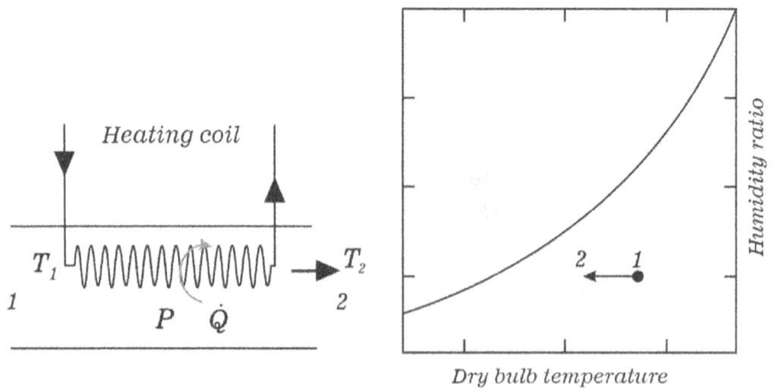

FIGURE 7.5 During the simple heating process, the amount of specific humidity remains constant.

Mass balance:

$$\dot{m}_{a1} = \dot{m}_{a2} = \dot{m}_a \quad (7.47)$$

$$\omega_1 = \omega_2 \quad (7.48)$$

Energy balance:

$$\dot{Q} = \dot{m}_a \left(h_2 - h_1\right) \quad (7.49)$$

7.7.2 Sensible Cooling

Sensible cooling is lowering the air temperature without changing its absolute humidity. With sensible cooling of the air, the dry-bulb temperature and the enthalpy of the air decrease, but its relative humidity increases.

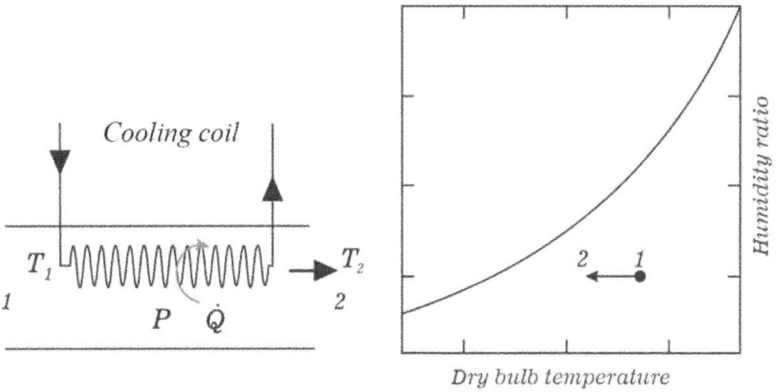

FIGURE 7.6 During the simple cooling process, the amount of specific humidity remains constant.

Mass balance:

$$\dot{m}_{a1} = \dot{m}_{a2} = \dot{m}_a \quad (7.50)$$

$$\omega_1 = \omega_2 \quad (7.51)$$

Energy balance:

$$\dot{Q} = \dot{m}_a \left(h_1 - h_2\right) \quad (7.52)$$

✐Example 7.7: Air enters a heating chamber with a flow rate of 6 m³/min at a temperature of 20°C, a pressure of 95 kPa, and a relative humidity of 30% and leaves it at a temperature of 25°C. Calculate

the relative humidity of the outlet air and the heat transfer rate in the heating chamber.

✍Solution

Assumptions:

1. The process is considered steady state.
2. Dry air and water vapor are ideal gases.
3. Kinetic and potential energy changes are negligible.
4. The specific heat of air is 1.005 kJ/kg.K.

For air with a temperature of 20°C and a relative humidity of 30%:

$$P_{g@20^\circ C} = 2.3392\text{kPa}$$

$$\varnothing = \frac{P_v}{P_g} \rightarrow P_{v1} = \varnothing_1 P_{g@20^\circ C} = 0.3(2.3392) = 0.70176\text{kPa}$$

Since the mass of water vapor and the mass of dry air are constant in the simple heating process, then the vapor pressure at the end of the process is equal to its pressure at the beginning of the process:

$$P_{v2} = P_{v1} = 0.70176\text{kPa}$$

So, the relative humidity can be calculated as follows:

$$\varnothing_2 = \frac{P_{v2}}{P_{g@25^\circ C}} = \frac{0.70176}{3.1698} = 0.2214 \text{ or } 22.14\%$$

For air at the beginning of the process:

$$\omega_1 = \frac{0.622P_{v1}}{P - P_{v1}} = \frac{0.622(0.70176)}{95 - 0.70176} = 0.00463 \frac{\text{kg}_{\text{water}}}{\text{kg}_{\text{dry air}}}$$

In simple heating, the value of the humidity ratio remains constant, so:

$$\omega_2 = \omega_1 = 0.00463 \frac{\text{kg}_{\text{water}}}{\text{kg}_{\text{dry air}}}$$

Calculation of enthalpy:

$$\begin{aligned} h_1 &= h_{a1} + h_{v1}\omega_1 \cong C_{p,air}T_1 + h_{g@T_1}\omega_1 \\ &= \left(1.005\frac{\text{kJ}}{\text{kg}^\circ\text{C}}\right)(20^\circ\text{C}) + (0.00463)\left(2537.4\frac{\text{kJ}}{\text{kg}}\right) \\ &= 31.85\frac{\text{kJ}}{\text{kg}_{\text{dry air}}} \end{aligned}$$

$$h_2 = h_{a2} + h_{v2}\omega_2 \cong C_{p,air}T_2 + h_{g@T_2}\omega_2$$
$$= \left(1.005\frac{\text{kJ}}{\text{kg°C}}\right)(25°\text{C}) + (0.00463)\left(2546.5\frac{\text{kJ}}{\text{kg}}\right)$$
$$= 36.92\frac{\text{kJ}}{\text{kg}_{\text{dry air}}}$$

Mass balance for dry air:

$$\dot{m}_{a1} = \dot{m}_{a2} = \dot{m}_a$$

Calculation of dry air flow rate:

$$\dot{m}_a = \frac{P_{a1}\dot{V}_a}{R_aT_1} = \frac{(P - P_{v1})\dot{V}_a}{R_aT_1} = \frac{(95 - 0.70176)(6)}{(0.287)(20 + 273.15)} = 6.72\frac{\text{kg}}{\text{min}}$$

Calculation of heat transfer rate:

$$\dot{Q} = \dot{m}_a(h_2 - h_1) = (6.72)(36.92 - 31.85) = 34.07\frac{\text{kg}}{\text{min}}$$

7.7.3 Adiabatic Mixing

Occasionally, two distinct states of air combine, giving rise to fresh air conditions. This phenomenon is referred to as air mixing. If no energy enters the air in this mixing (adiabatic mixing), the new air obtained has characteristics between the two primary conditions of air.

7.7.3.1 Mass and Energy Balance

Mass balance of dry air:

$$\dot{m}_{a1} + \dot{m}_{a2} = \dot{m}_{a3} \tag{7.53}$$

Water mass balance:

$$\dot{m}_{a1}\omega_1 + \dot{m}_{a2}\omega_2 = \dot{m}_{a3}\omega_3 \tag{7.54}$$

Energy balance:

$$\dot{m}_{a1}h_1 + \dot{m}_{a2}h_2 = \dot{m}_{a3}h_3 \tag{7.55}$$

Using the mass balance, we can write:

$$\omega_3 = \frac{\dot{m}_{a1}\omega_1 + \dot{m}_{a2}\omega_2}{\dot{m}_{a1} + \dot{m}_{a2}} \tag{7.56}$$

Using the energy balance, we can write:

$$h_3 = \frac{\dot{m}_{a1}h_1 + \dot{m}_{a2}h_2}{\dot{m}_{a1} + \dot{m}_{a2}} \tag{7.57}$$

By removing $\dot{m}_{a3}$ from the mass and energy balance:

$$\frac{\dot{m}_{a1}}{\dot{m}_{a2}} = \frac{\omega_2 - \omega_3}{\omega_3 - \omega_1} = \frac{h_2 - h_3}{h_3 - h_1} \tag{7.58}$$

When two air flows with two different states (states 1 and 2) are mixed adiabatically, the final mixed state (state 3) is located on the interface line of states 1 and 2, and the ratio of the distances 2–3 to 1–3 is equal to the ratio of $\dot{m}_{a1}$ and $\dot{m}_{a2}$.

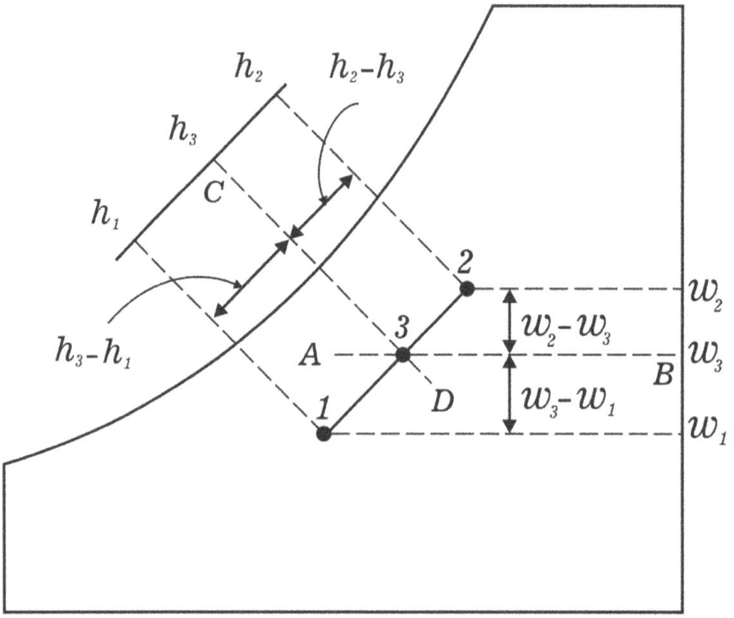

FIGURE 7.7 Psychrometric diagram of mixing two air flows.

It's important to understand that when one of the two primary airs has a higher percentage, the mixed air will exhibit characteristics more closely aligned with the primary air featuring the higher percentage.

Example 7.8: Two air streams, whose characteristics are given in Table 7.4, are mixed during an adiabatic process. Calculate the volumetric air flow rate of stream 1. The process is carried out at a pressure of 100 kPa.

TABLE 7.4
Information Related to the flows of Example 7.8

State No.	T (°C)	$\varnothing$ (%)	$\dot{V}\left(\frac{m^3}{min}\right)$
1	10	30	—
2	30	60	50
3	22	—	—

✍Solution

Assumptions:

1. The process is considered steady state.
2. Dry air and water vapor are ideal gases.
3. Changes in kinetic and potential energy are negligible.
4. There is no heat transfer from the system to the environment, and vice versa.
5. The specific heat of air is 1.005 kJ/kg.K.

Calculate the partial pressure of vapor and dry air:

$$\varnothing = \frac{P_v}{P_g} \rightarrow \begin{cases} P_{v,1} = \varnothing_1 P_{g@10°C} = 0.3(1.2281) = 0.3684\text{kPa} \\ P_{v,2} = \varnothing_2 P_{g@30°C} = 0.6(4.2469) = 2.5481\text{kPa} \end{cases}$$

$$P_a = P - P_v \rightarrow \begin{cases} P_{a,1} = 100 - 0.3684 = 99.6316\text{kPa} \\ P_{a,2} = 100 - 2.5481 = 97.4519\text{kPa} \end{cases}$$

Calculate the density of dry air:

$$\rho = \frac{P}{RT} \rightarrow \begin{cases} \rho_{a,1} = \frac{P_{a,1}}{R_a T_1} = \frac{99.6316}{(0.287)(10+273.15)} = 1.226\frac{\text{kg}}{\text{m}^3} \\ \rho_{a,2} = \frac{P_{a,2}}{R_a T_2} = \frac{97.4519}{(0.287)(30+273.15)} = 1.120\frac{\text{kg}}{\text{m}^3} \end{cases}$$

Calculate the humidity ratio:

$$\omega = \frac{0.622P_v}{P_a} \rightarrow \begin{cases} \omega_1 = \frac{0.622P_{v,1}}{P_{a,1}} = \frac{0.622(0.3684)}{99.6316} = 0.0023\frac{\text{kg}_{\text{water}}}{\text{kg}_{\text{dry air}}} \\ \omega_2 = \frac{0.622P_{v,2}}{P_{a,2}} = \frac{0.622(2.5481)}{97.4519} = 0.0163\frac{\text{kg}_{\text{water}}}{\text{kg}_{\text{dry air}}} \end{cases}$$

Calculation of enthalpy of flows:

$$h_1 = h_{a1} + h_{v1}\omega_1 \cong C_{p,air}T_1 + h_{g@T_1}\omega_1$$
$$= \left(1.005\frac{\text{kJ}}{\text{kg°C}}\right)(10°\text{C}) + (0.0023)\left(2519.2\frac{\text{kJ}}{\text{kg}}\right)$$
$$= 15.844\frac{\text{kJ}}{\text{kg}_{\text{dry air}}}$$

$$h_2 = h_{a2} + h_{v2}\omega_2 \cong C_{p,air}T_2 + h_{g@T_2}\omega_2$$
$$= \left(1.005\frac{\text{kJ}}{\text{kg°C}}\right)(30°\text{C}) + (0.0163)\left(2555.6\frac{\text{kJ}}{\text{kg}}\right)$$
$$= 71.806\frac{\text{kJ}}{\text{kg}_{\text{dry air}}}$$

$$h_3 = h_{a3} + h_{v3}\omega_3 \cong C_{p,air}T_3 + h_{g@T_3}\omega_3$$
$$= \left(1.005\frac{\text{kJ}}{\text{kg°C}}\right)(22°\text{C}) + (\omega_3)\left(2541.0\frac{\text{kJ}}{\text{kg}}\right)$$
$$= (22.11 + 2541\omega_3)\frac{\text{kJ}}{\text{kg}_{\text{dry air}}}$$

Mass and energy balance:

$$\begin{cases} \dot{m}_{a1}h_1 + \dot{m}_{a2}h_2 = \dot{m}_{a3}h_3 \\ \dot{m}_{a1} + \dot{m}_{a2} = \dot{m}_{a3} \end{cases} \rightarrow \dot{m}_{a1}h_1 + \dot{m}_{a2}h_2 = (\dot{m}_{a1} + \dot{m}_{a2})h_3$$

$\dot{m}$ is equal to $\rho\dot{V}$:

$$\rightarrow (\rho_{a,1}\dot{V}_1)h_1 + (\rho_{a,2}\dot{V}_2)h_2 = \left[\rho_{a,1}\dot{V}_1 + \rho_{a,2}\dot{V}_2\right]h_3$$
$$\rightarrow (1.226\dot{V}_1)(15.844) + (1.120)(50)(71.806)$$
$$= \left[1.226\dot{V}_1 + (1.120)(50)\right](22.11 + 2541\omega_3)$$

Mass balance for water:

$$\dot{m}_{a1}\omega_1 + \dot{m}_{a2}\omega_2 = \dot{m}_{a3}\omega_3$$
$$\rightarrow (1.226\dot{V}_1)(0.0023) + (1.120)(50)(0.0163)$$
$$= \left[1.226\dot{V}_1 + (1.120)(50)\right]\omega_3$$

By simultaneously solving the energy balance and the mass balance for water, the volumetric flow rate of flow 3 is calculated:

$$\rightarrow \dot{V}_1 = 31.22\frac{\text{m}^3}{\text{min}}$$

$$\omega_3 = 0.01062\frac{\text{kg}_{\text{water}}}{\text{kg}_{\text{dry air}}}$$

7.8 COOLING TOWER

The function of a cooling tower is to absorb heat from a process and expel it to the atmosphere, which is done by evaporation. Therefore, cooling towers are commonly used to remove heat from refrigeration systems, air conditioning, and industrial processes. The cooling tower works in such a way that water enters the spray section from a series of branches and falls on surfaces called packing media. At this stage, at the same time as the water passes through the surface of the packing media, it comes into contact with the air, and a part of it evaporates and enters the air. The air coming out of the device is close to saturated air (100% relative humidity). The water coming out of the packing media, whose temperature has decreased, is collected in a part called the pan and is used to send to the desired process (Figure 7.8).

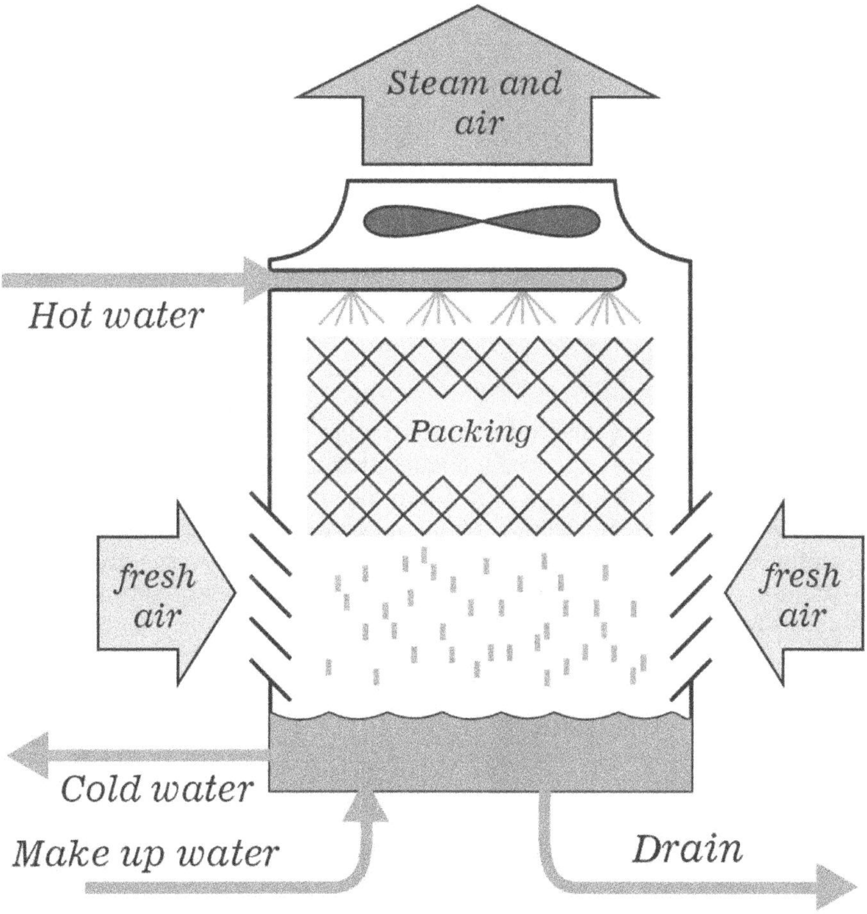

FIGURE 7.8 The schematic view of performance of the cooling tower.

7.8.1 Mass and Energy Balances

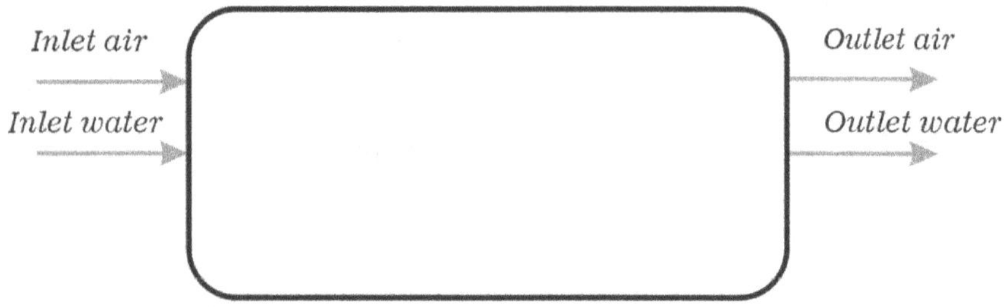

FIGURE 7.9 Flow to the cooling tower.

Mass balance of dry air:

$$\dot{m}_{a1} = \dot{m}_{a3} = \dot{m}_a \tag{7.59}$$

Water mass balance:

$$\dot{m}_{a1}\omega_1 + \dot{m}_2 = \dot{m}_{a3}\omega_3 + \dot{m}_4 \tag{7.60}$$

Energy balance:

$$\dot{m}_{a1}h_1 + \dot{m}_2h_2 + \dot{W}_{fan} = \dot{m}_{a3}h_3 + \dot{m}_4h_4 \tag{7.61}$$

✐Example 7.9: A cooling tower is designed to cool water with a mass flow rate of 5.5 kg/s and a temperature of 44°C. The power consumption of the fan is 4.75 kW, and the volumetric flow rate of the air flow created by it is 9 m³/s. The temperature and relative humidity of the air entering the cooling tower are 18°C and 60%, respectively, and the temperature of the outlet air is 26°C. If the pressure inside the tower is assumed to be constant and 101.325 kPa, calculate: (a) the mass flow rate of the compensation water and (b) the temperature of the water leaving the cooling tower.

✍Solution

Assumptions:

1. The process is considered steady state.
2. Dry air and water vapor are ideal gases.
3. Changes in kinetic and potential energy are negligible.

4. The process is adiabatic.
5. The air leaving the cooling tower is saturated.
6. The specific heat of air is 1.005 kJ/kg.K.

a) Determining the partial pressure of vapor:

$$\varnothing = \frac{P_v}{P_g} \rightarrow \begin{cases} P_{v,1} = \varnothing_1 P_{g@18°C} = 0.6(2.0858) = 1.2515\text{kPa} \\ P_{v,3} = \varnothing_3 P_{g@26°C} = 1(3.3852) = 3.3852\text{kPa} \end{cases}$$

Determining the partial pressure of dry air:

$$P_a = P - P_v \rightarrow \begin{cases} P_{a1} = 101.325 - 1.2515 = 100.0735\text{kPa} \\ P_{a3} = 101.325 - 3.3852 = 97.9398\text{kPa} \end{cases}$$

Humidity ratio of inlet air:

$$\omega = 0.622\frac{P_v}{P_a} \rightarrow \begin{cases} \omega_1 = \dfrac{0.622(1.2515)}{100.0735} = 0.00779\dfrac{\text{kg}_{\text{water}}}{\text{kg}_{\text{dry air}}} \\ \omega_3 = \dfrac{0.622(3.3852)}{97.9398} = 0.0215\dfrac{\text{kg}_{\text{water}}}{\text{kg}_{\text{dry air}}} \end{cases}$$

Determining the flow rate of dry air:

$$\dot{m}_{a1} = \frac{P_{a1}\dot{V}}{R_a T_1} = \frac{(100.0735)(9)}{(0.287)(273.15+18)} = 10.78\frac{\text{kg}}{\text{s}}$$

Compensatory water discharge:

$$\dot{m}_{make\ up} = \dot{m}_a(\omega_3 - \omega_1) = 10.78(0.0215 - 0.00779) = 0.148\frac{\text{kg}}{\text{s}}$$

b) The outlet water temperature can be calculated from the energy balance:

$$\dot{m}_{a1}h_1 + \dot{m}_2 h_2 + \dot{W}_{fan} = \dot{m}_{a3}h_3 + \dot{m}_4 h_4$$

Calculation of enthalpy of flows:

$$\begin{aligned} h_1 &= h_{a1} + h_{v1}\omega_1 \cong C_{p,air}T_1 + h_{g@T_1}\omega_1 \\ &= \left(1.005\frac{\text{kJ}}{\text{kg°C}}\right)(18°\text{C}) + (0.00779)\left(2533.8\frac{\text{kJ}}{\text{kg}}\right) \\ &= 37.83\frac{\text{kJ}}{\text{kg}_{\text{dry air}}} \end{aligned}$$

$$h_2 \cong h_{f@44°C} = 184.26\frac{\text{kJ}}{\text{kg}}$$

$$h_3 = h_{a3} + h_{v3}\omega_3 \cong C_{p,air}T_3 + h_{g@T_3}\omega_3$$
$$= \left(1.005\frac{\text{kJ}}{\text{kg°C}}\right)(26°\text{C}) + (0.0215)\left(2548.3\frac{\text{kJ}}{\text{kg}}\right)$$
$$= 80.92\frac{\text{kJ}}{\text{kg}_{\text{dry air}}}$$

Calculation of mass flow rate of outlet water:

$$\dot{m}_4 = \dot{m}_2 - \dot{m}_{Make\ up} = 5.5 - 0.148 = 5.352\frac{\text{kg}}{\text{s}}$$

Calculation of the enthalpy of outlet water from the energy balance:

$$h_4 = \frac{\dot{m}_{a1}h_1 + \dot{m}_2h_2 + \dot{W}_{fan} - \dot{m}_{a3}h_3}{\dot{m}_4}$$
$$= \frac{(10.78)(37.83) + (5.5)(184.26) + 4.75 - (10.78)(80.92)}{5.352}.$$
$$= 103.45\frac{\text{kJ}}{\text{kg}}$$

Determining the outlet water temperature:

$$h_3 \cong h_{f@T_4} = 103.45\frac{\text{kJ}}{\text{kg}} \rightarrow T_4 = 24.19°\text{C}$$

7.9 EXERGY OF MOIST AIR

The physical exergy of humid air can be obtained from the following equation:

$$ex = \left(C_{p,a} + C_{p,v}\omega\right)(T - T_0) - T_0\left(\left(C_{p,a} + C_{p,v}\omega\right)\ln\frac{T}{T_0} - \left(\frac{1}{M_a} + \frac{\omega}{M_w}\right)R\ln\frac{P}{P_0}\right) \quad (7.62)$$

where M_a and M_w are the molar mass of dry air and water, respectively.

7.9.1 Exergy Analysis of Simple Heating and Cooling Processes

The exergy balance of a simple heating process is to be calculated. Figure 7.10 illustrates a simple heating process.

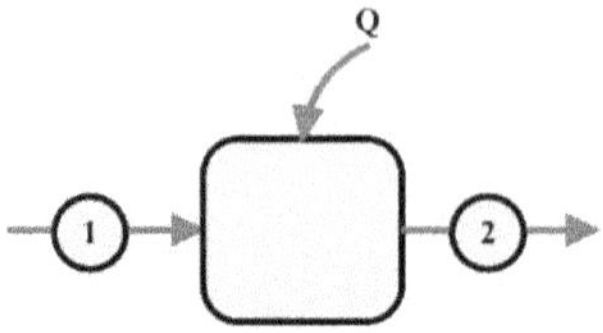

FIGURE 7.10 Simple heating.

$$\dot{E}x_F = \left(1 - \frac{T_0}{T_b}\right)\dot{Q} \tag{7.63}$$

$$\dot{E}x_P = \dot{E}x_2 - \dot{E}x_1 \tag{7.64}$$

$$\dot{E}x_{des} = \left(1 - \frac{T_0}{T_b}\right)\dot{Q} - \left(\dot{E}x_2 - \dot{E}x_1\right) \tag{7.65}$$

Exergy efficiency of the heating process:

$$\eta_{ex} = \frac{\dot{E}x_2 - \dot{E}x_1}{\left(1 - \frac{T_0}{T_b}\right)\dot{Q}} \tag{7.66}$$

The exergy balance of the cooling process can be similarly calculated. A simple schematic of the cooling process is shown in Figure 7.11.

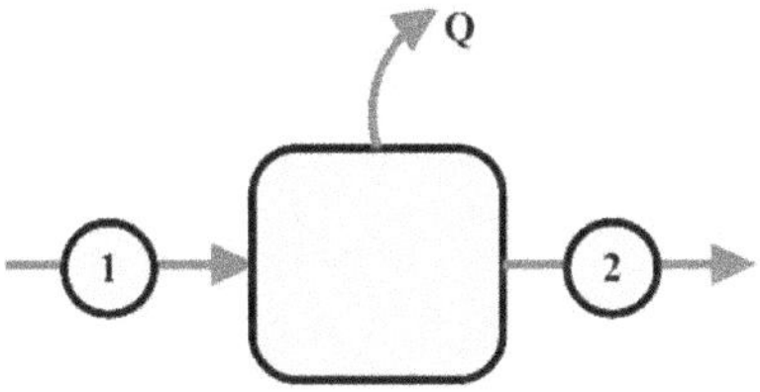

FIGURE 7.11 Simple cooling.

$$\dot{E}x_F = \dot{E}x_1 - \dot{E}x_2 \tag{7.67}$$

$$\dot{E}x_P = \left(1 - \frac{T_0}{T_b}\right)\dot{Q} \tag{7.68}$$

$$\dot{E}x_{des} = \left(\dot{E}x_1 - \dot{E}x_2\right) - \left(1 - \frac{T_0}{T_b}\right)\dot{Q} \tag{7.69}$$

Exergy efficiency of the cooling process:

$$\eta_{ex} = \frac{\left(1 - \frac{T_0}{T_b}\right)\dot{Q}}{\dot{E}x_1 - \dot{E}x_2} \tag{7.70}$$

✐Example 7.10: Air with a volumetric flow rate of 1 m³/s, which is at an initial temperature of 15°C, a pressure of 101.325, and a relative

humidity of 66%, is heated to 50°C at the same pressure for use in a dryer. Calculate the rate of destruction and exergy efficiency in this process. The temperature of the heater is 200°C.

✍Solution

Assumptions:

1. Kinetic and potential energy changes are ignored.
2. The dead state temperature and pressure are 25°C and 101.325 kPa.
3. The process is considered steady state.
4. Dry air and water vapor are ideal gases.
5. The specific heat of air is 1.005 kJ/kg.K.

Determining the partial pressure of vapor:

$$\varnothing = \frac{P_v}{P_g} \rightarrow P_{v,1} = \varnothing_1 P_{g@15°C} = 0.66(1.7057) = 1.1258\text{kPa}$$

Determining the partial pressure of dry air:

$$P_a = P - P_v \rightarrow P_{a1} = 101.325 - 1.1258 = 100.1992\text{kPa}$$

Humidity ratio of inlet air:

$$\omega = 0.622\frac{P_v}{P_a} \rightarrow \omega_1 = \frac{0.622(1.1258)}{100.1992} = 0.007\frac{\text{kg}_{\text{water}}}{\text{kg}_{\text{dry air}}}$$

Determining the flow rate of dry air:

$$\dot{m}_{a1} = \frac{P_{a1}\dot{V}}{R_a T_1} = \frac{(100.1992)(1)}{(0.287)(273.15+15)} = 1.21\frac{\text{kg}}{\text{s}}$$

In the simple heating, the value of the humidity ratio remains constant, so:

$$\omega_2 = \omega_1 = 0.007\frac{\text{kg}_{\text{water}}}{\text{kg}_{\text{dry air}}}$$

Calculation of enthalpy:

$$\begin{aligned} h_1 &= h_{a1} + h_{v1}\omega_1 \cong C_{p,air}T_1 + h_{g@T_1}\omega_1 \\ &= \left(1.005\frac{\text{kJ}}{\text{kg°C}}\right)(15°\text{C}) + (0.007)\left(2528.3\frac{\text{kJ}}{\text{kg}}\right) \\ &= 32.77\frac{\text{kJ}}{\text{kg}_{\text{dry air}}} \end{aligned}$$

$$h_2 = h_{a2} + h_{v2}\omega_2 \cong C_{p,air}T_2 + h_{g@T_2}\omega_2$$

$$= \left(1.005\frac{kJ}{kg°C}\right)(50°C) + (0.007)\left(2591.3\frac{kJ}{kg}\right)$$

$$= 68.39\frac{kJ}{kg_{dry\ air}}$$

Calculation of heat transfer rate:

$$\dot{Q} = \dot{m}_a(h_2 - h_1) = (1.21)(68.39 - 32.77) = 43.10kW$$

Calculation of exergy specific for flows:

$$ex = C_p(T - T_0) - T_0\left(C_p ln\frac{T}{T_0} - \left(\frac{1}{M_a} + \frac{\omega}{M_w}\right)R\ln\frac{P}{P_0}\right)$$

$$\rightarrow \begin{cases} ex_1 = 0.1724\frac{kJ}{kg} \\ ex_2 = 0.998\frac{kJ}{kg} \end{cases}$$

Exergy destruction rate:

$$\dot{E}x_{des} = \left(1 - \frac{T_0}{T_b}\right)\dot{Q} - \left(\dot{E}x_2 - \dot{E}x_1\right) = \left(1 - \frac{T_0}{T_b}\right)\dot{Q} - \dot{m}_a(ex_2 - ex_1)$$

$$= \left[1 - \frac{(25 + 273.15)}{(200 + 273.15)}\right](43.10)$$

$$-1.21(0.998 - 0.1724) = 14.94kW$$

Exergy efficiency:

$$\eta_{ex} = \frac{\dot{E}x_2 - \dot{E}x_1}{\left(1 - \frac{T_0}{T_b}\right)\dot{Q}} = \frac{\dot{m}_a(ex_2 - ex_1)}{\left(1 - \frac{T_0}{T_b}\right)\dot{Q}} = \frac{1.21(0.998 - 0.1724)}{\left[1 - \frac{(25 + 273.15)}{(200 + 273.15)}\right](43.10)}$$

$$= \frac{1.00}{15.94} = 0.0627 \text{ or } 6.27\%$$

7.9.2 Exergy Analysis of Adiabatic Mixing Process

Exergy balance:

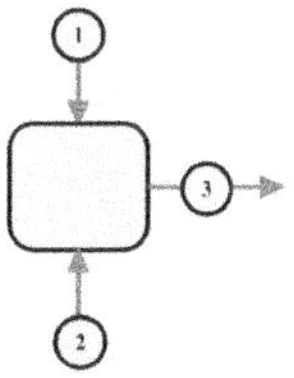

FIGURE 7.12 Mixing.

$$\dot{E}x_F = \dot{E}x_1 + \dot{E}x_2 \tag{7.71}$$

$$\dot{E}x_P = \dot{E}x_3 \tag{7.72}$$

$$\dot{E}x_{des} = \left(\dot{E}x_1 + \dot{E}x_2\right) - \dot{E}x_3 \tag{7.73}$$

Exergy efficiency of adiabatic mixing:

$$\eta_{ex} = \frac{\dot{E}x_3}{\dot{E}x_1 + \dot{E}x_2} \tag{7.74}$$

✐Example 7.11: Two air flows are mixed together in an adiabatic enclosure. The first flow, with a temperature of 35°C, relative humidity of 35%, and flow rate of 15 m³/min, and the second flow, with a temperature of 12°C, relative humidity of 90%, and flow rate of 25 m³/min, enter the mixing chamber. The temperature and relative humidity of the outlet flow are 20.19°C and 60%, respectively. Calculate the rate of destruction and exergy efficiency in this process.

✍Solution

Assumptions:

1. Kinetic and potential energy changes are ignored.
2. The dead state temperature and pressure are 25°C and 101.325 kPa.
3. The process is considered steady state.
4. Dry air and water vapor are ideal gases.
5. The specific heat of air is 1.005 kJ/kg.K.
6. The process is adiabatic.
7. The process is carried out at a constant pressure of 101.325 kPa.

Determining the partial pressure of vapor:

$$\varnothing = \frac{P_v}{P_g} \rightarrow \begin{cases} P_{v,1} = \varnothing_1 P_{g@35^\circ C} = 0.35(4.2469) = 1.4865\text{kPa} \\ P_{v,2} = \varnothing_2 P_{g@12^\circ C} = 0.9(1.4191) = 1.2772\text{kPa} \\ P_{v,3} = \varnothing_3 P_{g@20.19^\circ C} = 0.6(2.3708) = 1.4225\text{kPa} \end{cases}$$

Determining the partial pressure of dry air:

$$P_a = P - P_v \rightarrow \begin{cases} P_{a1} = 101.325 - 1.4865 = 99.8385\text{kPa} \\ P_{a2} = 101.325 - 1.2772 = 100.0478\text{kPa} \\ P_{a3} = 101.325 - 1.4225 = 99.9025\text{kPa} \end{cases}$$

Humidity ratio of inlet air:

$$\omega = 0.622\frac{P_v}{P_a} \rightarrow \begin{cases} \omega_1 = \dfrac{0.622(1.4865)}{99.8385} = 0.00926\dfrac{kg_{water}}{kg_{dry\ air}} \\ \omega_2 = \dfrac{0.622(1.2772)}{100.0478} = 0.00794\dfrac{kg_{water}}{kg_{dry\ air}} \\ \omega_3 = \dfrac{0.622(1.4225)}{99.9025} = 0.00886\dfrac{kg_{water}}{kg_{dry\ air}} \end{cases}$$

Determining the flow rate of dry air:

$$\dot{m}_a = \frac{P_a\dot{V}}{R_aT} \rightarrow \begin{cases} \dot{m}_{a1} = \dfrac{P_{a1}\dot{V}}{R_aT_1} = \dfrac{(99.8385)(15)}{(0.287)(273.15+35)} = 18.11\dfrac{kg}{min} \\ \dot{m}_{a2} = \dfrac{P_{a1}\dot{V}}{R_aT_1} = \dfrac{(100.0478)(25)}{(0.287)(273.15+12)} = 30.56\dfrac{kg}{min} \end{cases}$$

$$\dot{m}_{a3} = \dot{m}_{a1} + \dot{m}_{a2} = 48.67\frac{kg}{min}$$

Calculation of exergy specific for flows:

$$ex = C_p(T - T_0) - T_0\left(C_p\, ln\frac{T}{T_0} - \left(\frac{1}{M_a} + \frac{\omega}{M_w}\right)R\ln\frac{P}{P_0}\right)$$

$$\rightarrow \begin{cases} ex_1 = 0.1649\dfrac{kJ}{kg} \\ ex_2 = 0.2934\dfrac{kJ}{kg} \\ ex_3 = 0.0394\dfrac{kJ}{kg} \end{cases}$$

Exergy destruction rate:

$$\begin{aligned} \dot{E}x_{des} &= \left(\dot{E}x_1 + \dot{E}x_2\right) - \dot{E}x_3 = \left(\dot{m}_{a1}ex_1 + \dot{m}_{a1}ex_1\right) - \dot{m}_{a3}ex_3 \\ &= \left[(18.11)(0.1649) + (30.56)(0.2934)\right] \\ &\quad -(48.67)(0.0394) = 10.04\frac{kJ}{min} \end{aligned}$$

Exergy efficiency:

$$\begin{aligned} \eta_{ex} &= \frac{\dot{E}x_3}{\dot{E}x_1 + \dot{E}x_2} = \frac{\dot{m}_{a3}ex_3}{\dot{m}_{a1}ex_1 + \dot{m}_{a1}ex_1} \\ &= \frac{(48.67)(0.0394)}{(18.11)(0.1649) + (30.56)(0.2934)} \\ &= 0.1604 \text{ or } 16.04\% \end{aligned}$$

7.9.3 Exergy Analysis of Cooling Tower

According to Figure 7.9, exergy balance can be written:

$$\dot{E}x_F = \dot{E}x_1 + \dot{E}x_2 + \dot{W}_{fan} \tag{7.75}$$

$$\dot{E}x_P = \dot{E}x_3 + \dot{E}x_4 \tag{7.76}$$

$$\dot{E}x_{des} = \left(\dot{E}x_1 + \dot{E}x_2 + \dot{W}_{fan}\right) - \left(\dot{E}x_3 + \dot{E}x_4\right) \tag{7.77}$$

Exergy efficiency:

$$\eta_{ex} = \frac{\dot{E}x_3 + \dot{E}x_4}{\dot{E}x_1 + \dot{E}x_2 + \dot{W}_{fan}} \tag{7.78}$$

✐**Example 7.12: The cooling tower shown in the figure is used to cool water in a concentrate production line. If the power consumption of the fan is 22 kW, calculate the destruction rate and exergy efficiency using the information in Table 7.5.**

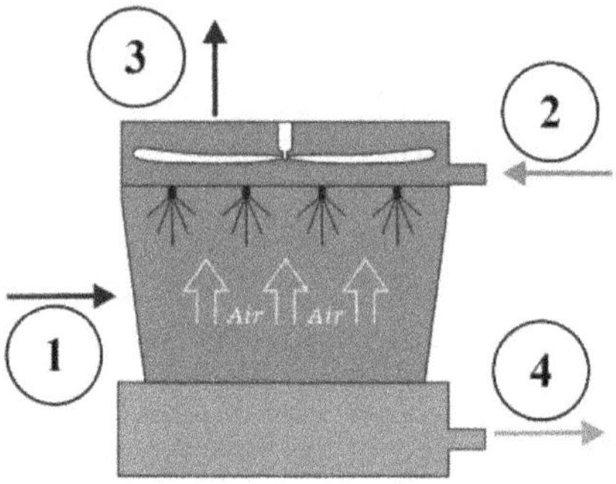

FIGURE 7.13 Diagram of Example 7.12.

TABLE 7.5

Information Related to the Flows in Example 7.12

State No.	*T* (°C)	*P* (kPa)	Rate	Relative Humidity (%)
1	25	101.325	19.63 m3/s	35
2	40	101.325	29.40 kg/s	—
3	35	101.325	—	100
4	22	101.325	—	—

✍Solution

Assumptions:

1. Kinetic and potential energy changes are ignored.
2. The dead state temperature and pressure are 25°C and 101.325 kPa.
3. The process is considered steady state.
4. Dry air and water vapor are ideal gases.
5. The specific heat of air is 1.005 kJ/kg.K.
6. The process is adiabatic.

Determining the partial pressure of vapor:

$$\varnothing = \frac{P_v}{P_g} \rightarrow \begin{cases} P_{v,1} = \varnothing_1 P_{g@25°C} = 0.35(3.1698) = 1.1094\,\text{kPa} \\ P_{v,3} = \varnothing_3 P_{g@35°C} = 1(5.6291) = 5.6291\,\text{kPa} \end{cases}$$

Determining the partial pressure of dry air:

$$P_a = P - P_v \rightarrow \begin{cases} P_{a1} = 101.325 - 1.1094 = 100.2156\,\text{kPa} \\ P_{a3} = 101.325 - 5.6291 = 95.6959\,\text{kPa} \end{cases}$$

Humidity ratio of inlet air:

$$\omega = 0.622\frac{P_v}{P_a} \rightarrow \begin{cases} \omega_1 = \dfrac{0.622(1.1094)}{100.2156} = 0.00689\dfrac{\text{kg}_{\text{water}}}{\text{kg}_{\text{dry air}}} \\ \omega_3 = \dfrac{0.622(5.6291)}{95.6959} = 0.03659\dfrac{\text{kg}_{\text{water}}}{\text{kg}_{\text{dry air}}} \end{cases}$$

Determining the flow rate of dry air:

$$\dot{m}_{a1} = \frac{P_{a1}\dot{V}}{R_a T_1} = \frac{(100.2156)(19.63)}{(0.287)(273.15+25)} = 22.99\frac{\text{kg}}{\text{s}}$$

$$\dot{m}_{a2} = \dot{m}_{a1} = 22.99\frac{\text{kg}}{\text{s}} = \dot{m}_a$$

Determining the output water flow rate:

$$\dot{m}_{a1}\omega_1 + \dot{m}_2 = \dot{m}_{a3}\omega_3 + \dot{m}_4 \rightarrow \dot{m}_4 = \dot{m}_2 + \dot{m}_a(\omega_1 - \omega_3)$$

$$= 29.40 + 22.99(0.00689 - 0.03659) = 28.72\frac{\text{kg}}{\text{s}}$$

Exergy calculation for air flows:

$$ex = C_p(T - T_0) - T_0\left(C_p ln\frac{T}{T_0} - \left(\frac{1}{M_a} + \frac{\omega}{M_w}\right)R\ln\frac{P}{P_0}\right)$$

$$\rightarrow \begin{cases} ex_1 = 0.1724\dfrac{\text{kJ}}{\text{kg}} \\ ex_3 = 0.998\dfrac{\text{kJ}}{\text{kg}} \end{cases}$$

Exergy calculation for water flows:

$$@25°C, 101.325 kPa \begin{cases} h_0 = 104.92 \dfrac{kJ}{kg} \\ s_0 = 0.37 \dfrac{kJ}{kg}.K \end{cases}$$

$$h_2 \cong h_{f@40°C} = 167.53 \frac{kJ}{kg}$$

$$h_4 \cong h_{f@22°C} = 92.28 \frac{kJ}{kg}$$

$$s_2 \cong s_{f@40°C} = 0.5724 \frac{kJ}{kg.K}$$

$$s_4 \cong s_{f@22°C} = 0.3248 \frac{kJ}{kg.K}$$

$$ex = (h - h_0) - T_0(s - s_0) \rightarrow \begin{cases} ex_2 = 2.26 \dfrac{kJ}{kg} \\ ex_4 = 0.84 \dfrac{kJ}{kg} \end{cases}$$

Exergy destruction rate:

$$\begin{aligned} \dot{E}x_{des} &= \left(\dot{E}x_1 + \dot{E}x_2 + \dot{W}_{fan}\right) - \left(\dot{E}x_3 + \dot{E}x_4\right) \\ &= \left(\dot{m}_{a1}ex_1 + \dot{m}_2 ex_2 + \dot{W}_{fan}\right) - \left(\dot{m}_{a3}ex_3 + \dot{m}_4 ex_4\right) \\ &= \left[(22.99)(0.1724) + (29.40)(2.26) + 22\right] - \left[(22.99)(0.998) + (28.72)(0.84)\right] \\ &= 45.34 kW \end{aligned}$$

Exergy efficiency:

$$\begin{aligned} \eta_{ex} &= \frac{\dot{E}x_3 + \dot{E}x_4}{\dot{E}x_1 + \dot{E}x_2 + \dot{W}_{fan}} = \frac{\dot{m}_{a3}ex_3 + \dot{m}_4 ex_4}{\dot{m}_{a1}ex_1 + \dot{m}_2 ex_2 + \dot{W}_{fan}} \\ &= \frac{(22.99)(0.998) + (28.72)(0.84)}{(22.99)(0.1724) + (29.40)(2.26) + 22} \\ &= 0.5093 \text{ or } 50.93\% \end{aligned}$$

7.10 DRYING

Drying can be defined as removing part of the solvent or water in a solid substance. The food industry might be a leading industry utilizing this process, but practically, drying is widely used in a wide variety of operations to:

- increase storage and diminish the possibility of microbial spoilage,
- reduce the volume and weight of material,

- make the transportation and handling easier and cheaper,
- diversify the production of new products, and
- change the functionality of materials.

Drying is generally a time- and energy-consuming process due to the high latent heat of water. Effective thermodynamic analysis is the first vital step in identifying the potential of reducing energy consumption and thus reducing environmental effects and making the process more economical.

7.10.1 Drying Process

Drying is a simultaneous unit operation of mass and heat transfer. As a result of the evaporation of moisture from the product, a region of low-pressure water vapor is created on its surface in which, due to the pressure difference between the inside of the product and the low-pressure region on its surface, a pressure gradient from the inside to the environment is created. This gradient, along with the forces caused by the difference in concentration and temperature and the capillary force inside the product, provides the driving force necessary to repel water, and the water in the product, as a combination of moisture diffusion in the form of liquid and vapor and hydraulic flow, moves towards the surface. The moisture that has reached the surface of the product evaporates, and the driving force is strengthened again. This process continues until the end of the drying process.

7.10.2 Drying Process by Convection Method

Most of the processes of drying agricultural products are done using airflow. More than 85% of industrial dryers are convection dryers. Air enters the dryer with a certain temperature and humidity, and after entering the dryer, it absorbs moisture from the surroundings and the surface of the product. If the moisture on the surface of the product is in the form of vapor, this vapor enters the airflow and increases its relative humidity, and the air temperature remains constant (state 1 in Figure 7.14).

If the moisture on the surface of the product is in the form of water, some of the energy of the air is used for water evaporation and the water vapor enters the air; in this case, the air enthalpy is constant, the air temperature decreases, and the relative humidity increases (state 2 in Figure 7.14).

As the process progresses, the surface humidity of the product decreases due to drying, and the humidity must move from inside the product and reach its surface. In this stage, some of the energy of the incoming air is used to increase the temperature of the product, and as a result, the temperature and enthalpy of the air leaving the dryer are lower than the air entering, but its relative humidity increases due to the absorption of humidity (state 3 in Figure 7.14).

When the product reaches its equilibrium humidity with the inlet air conditions, the drying process should be stopped. If the airflow continues after this stage, only the temperature of the product will increase, and the temperature of the outlet air will decrease, but its absolute humidity will be constant (state 4 in Figure 14.7).

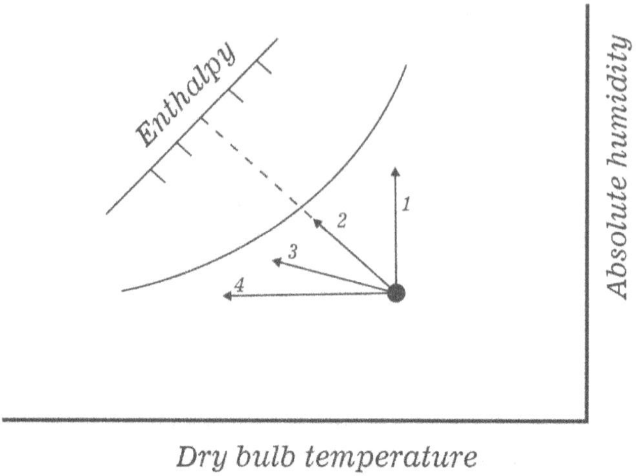

FIGURE 7.14 Representation of different states of drying with hot air on the psychrometric chart.

7.10.3 Types of Hot Air Dryers

7.10.3.1 Cabinet Dryers

Cabinet tray dryers are one of the most versatile and common types of dryers in food processing. This dryer is simple in terms of construction, is flexible in terms of use, and has the lowest initial investment. This dryer can be seen in Figure 7.15, which includes a dryer compartment that is heated by steam, electric coil, or heat sources. The drying air enters the heater by a fan and is heated in contact with the heating elements.

7.10.3.2 Tunnel Dryers

This dryer is a developed form of a cabinet dryer with trays in which the drying operation is done continuously. Tunnel dryers are relatively long (20 to 200 m), where trays containing wet products are placed on wagons and passed through the tunnel. These dryers usually heat the air in a heat exchanger with steam. The stay time of the wagons in the tunnel is adjusted so that the humidity of the product reaches the desired value.

The direction or pattern of flow in these dryers is crucial, and depending on the direction of the air flow and the movement of the food, they are made in aligned and non-aligned types. In the aligned system (Figure 7.16), the direction of air flow and food movement is the same. In other words, in this system, the hottest air contacts the coldest product. As a result, the rate of water evaporation in the wet opening (that is, the place where the fresh product enters) is high.

In non-aligned systems (Figure 7.17), the directions of movement of air and food are opposite each other; that is, the hottest air upon entering the dryer contacts the driest material leaving the system, so the initial speed of drying is relatively low and

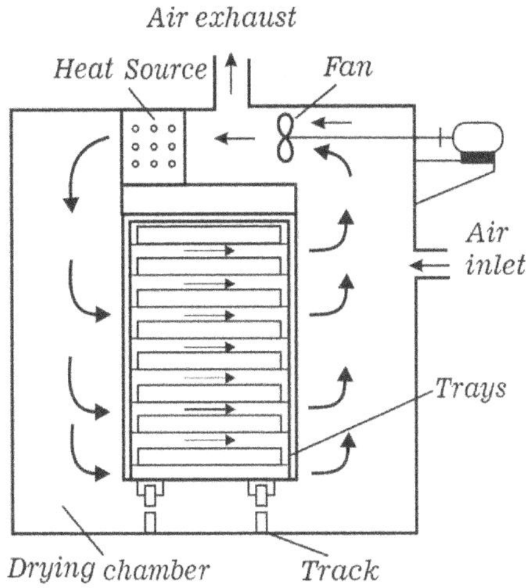

FIGURE 7.15 Cabinet dryer.

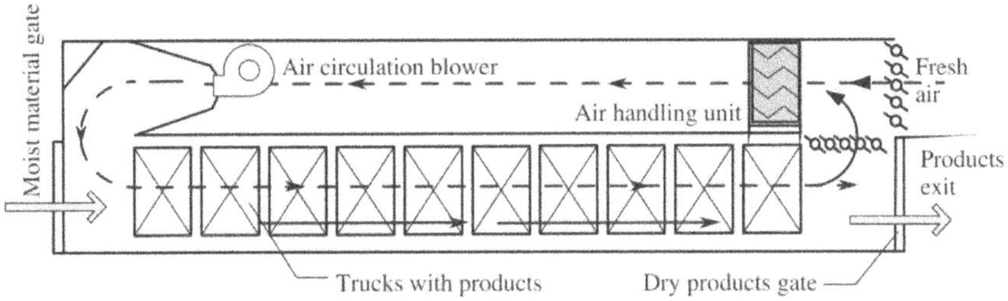

FIGURE 7.16 Parallel flow tunnel dryer (Dinçer & Zamfirescu, 2016).

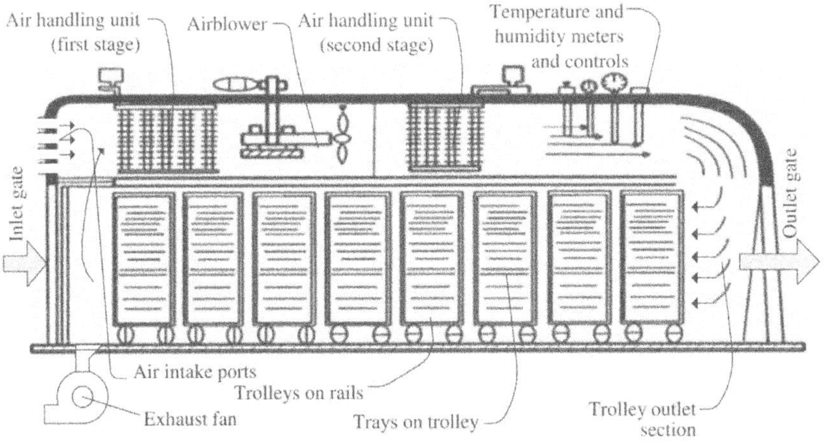

FIGURE 7.17 Unaligned flow tunnel dryer (Dinçer & Zamfirescu, 2016).

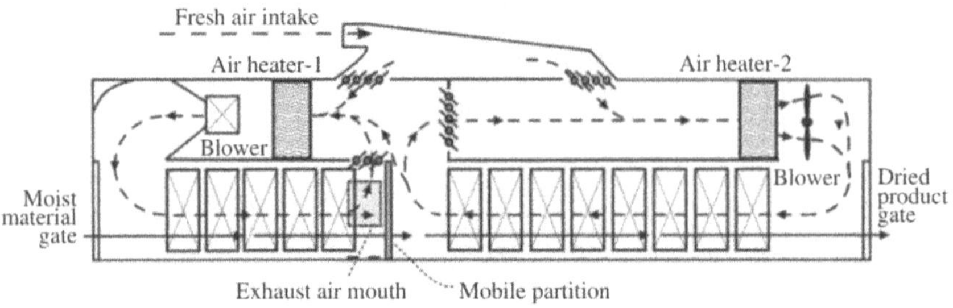

FIGURE 7.18 Combined tunnel dryer (Dinçer & Zamfirescu, 2016).

causes the drying time to be prolonged; as a result, the phenomenon of surface hardening does not occur, and because the product contacts hot air at the outlet, it is possible to produce a product with low humidity. But in this system, there is a possibility of surface burn of the output product. It should be noted that the energy efficiency in non-aligned dryers is far higher than in the aligned type.

Sometimes both systems are used together, that is, a preliminary tunnel in an aligned and shorter length, and then a longer secondary tunnel that works in a non-aligned way (Figure 7.18). In this system, a fan is located in the middle part, and the air is sucked into the system from both sides by the fan; as a result, we see both aligned and non-aligned types together.

7.10.3.3 Belt Dryers (Strip or Conveyor Belt)

In these dryers, the wet product is poured on a mesh conveyor belt, it enters the dryer chamber at a certain rate, and after being dried by the cold air flow, it leaves the dryer. (Figure 7.19). These types of dryers offer the following advantages:

- full continuity of the production line,
- very little need for manpower, and
- proper uniformity of the quality of the final product.

7.10.3.4 Fluidized Bed Dryers

The fluidized bed dryer (FBD) is used to dry all kinds of granular materials and wet powders in the food, pharmaceutical, and chemical industries. In this system, the food substance, after being placed on a mesh screen, is placed in contact with hot air that has a relatively high flow rate and floats at a short distance from the mesh screen. The intense mixing created causes uniform temperature and humidity conditions during drying, so the drying speed is the same in different parts and layers of the product. The height of the column should be great enough to prevent particles from leaving the system. There is also a cyclone in the exit part (Figure 7.20). In this dryer, the power of the heater and the amount of ventilation of the blower can be changed.

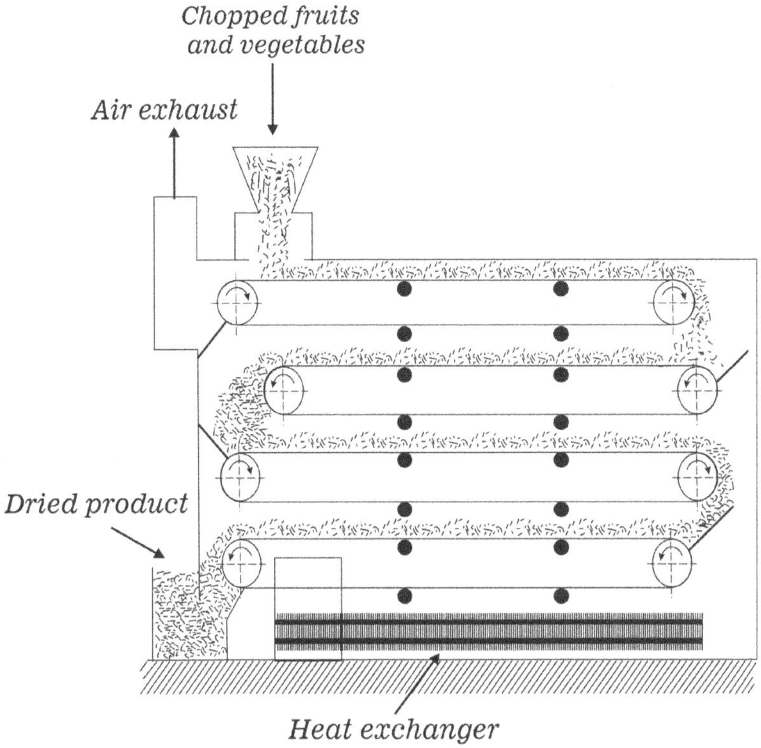

FIGURE 7.19 Strip dryer.

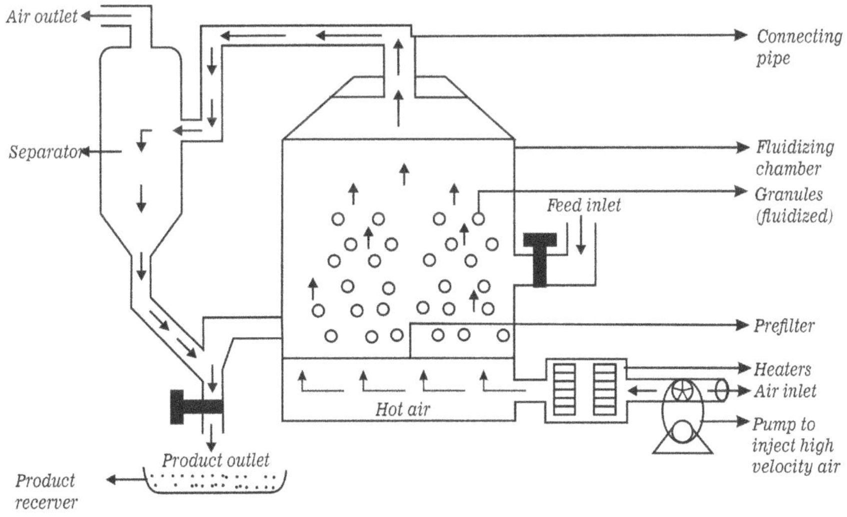

FIGURE 7.20 Schematic of fluidized bed dryer.

7.10.3.5 Rotary Dryers

Rotating dryers are often used to dry wet powder and granular materials. In these dryers, the wet product enters a cylinder with a gentle slope from one side, and while moving forward, it turns upside down, dries in contact with hot air, and exits from the other side. Hot air can flow parallel or non-aligned, and the source of hot air production can be direct or indirect.

7.10.3.6 Spray Dryers

Spray drying is the most suitable method for removing moisture from dilute liquids, such as milk, and producing powder. In this method, like the fluid bed method, the food substance is suspended in contact with hot air. The basis of the device's work is that the food substance is made into very fine droplets by an atomizer and is sprayed inside the dryer compartment where hot air flows. These tiny droplets with a large surface area dry in a few seconds due to contact with hot air, and the food substance is transported out of the dryer chamber along with the air. Usually, before entering the dryer, the liquid food is condensed in one step by the evaporator.

7.10.4 Drying by Conduction

The second largest category of dryers is dryers based on conduction heat transfer. Heat in these types of dryers is generally transferred to the product by water or hot steam condensation through conduction through the metal wall. The disadvantage of conductive drying is the relatively high temperature of the product, which is equal to the boiling temperature and can negatively affect the quality of the product (Qiu, 2019).

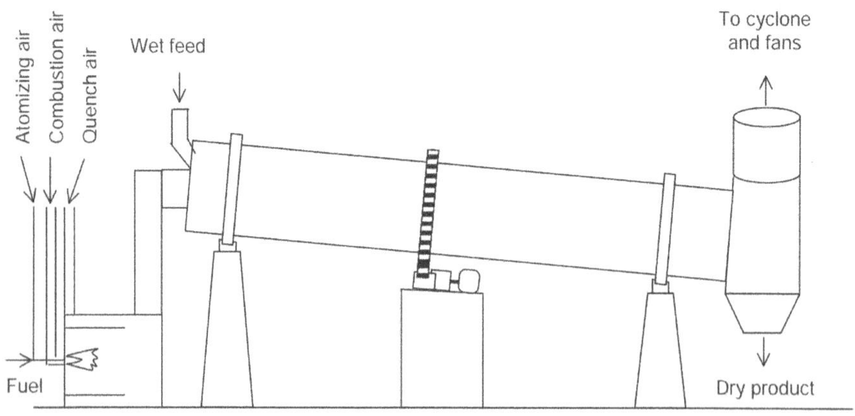

FIGURE 7.21 Schematic of a rotary dryer.

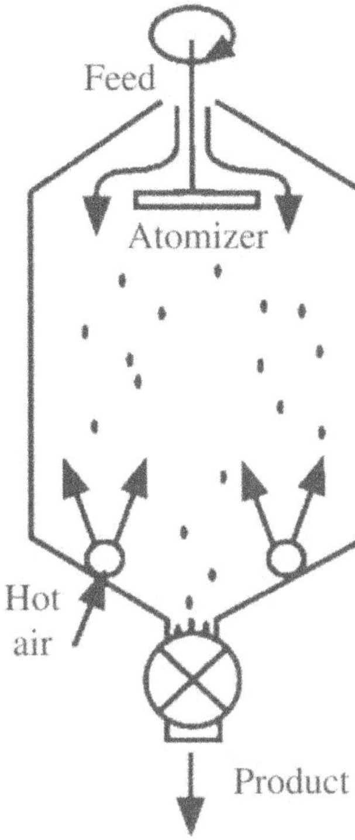

FIGURE 7.22 Principles of spray dryer operation.

7.10.4.1 Roller Dryers

The roller dryer is made of one or two hollow cylinders made of stainless steel. This cylinder is heated from the inside with steam until the surface temperature of the roller reaches about 150°C. In the single-roller type, a thin, uniform layer of food is spread over the surface due to contact with the surface of the roller. Before the cylinder makes a full revolution (between 20 seconds and 3 minutes), the dried food is scraped by a special blade that is uniformly in contact with the surface of the cylinder along its length. In the two-roller type, the liquid is poured between two rollers that are a short distance from each other. These two rollers rotate against each other. The thinner the liquid, the smaller the distance between the two rollers.

In roller dryers, the energy efficiency and drying speed are high. These types of dryers are more suitable for drying liquids that contain pasty or semi-liquid materials than spray dryers, but due to the damage caused to heat-sensitive foods, the tendency toward spray dryers is increasing in the case of high-volume foods. These dryers are used to dry curd, puree, and animal feed. Drying rollers can be placed in vacuum tanks to dry food at a lower temperature, but the heavy cost of such a system limits its use for valuable and heat-sensitive foods.

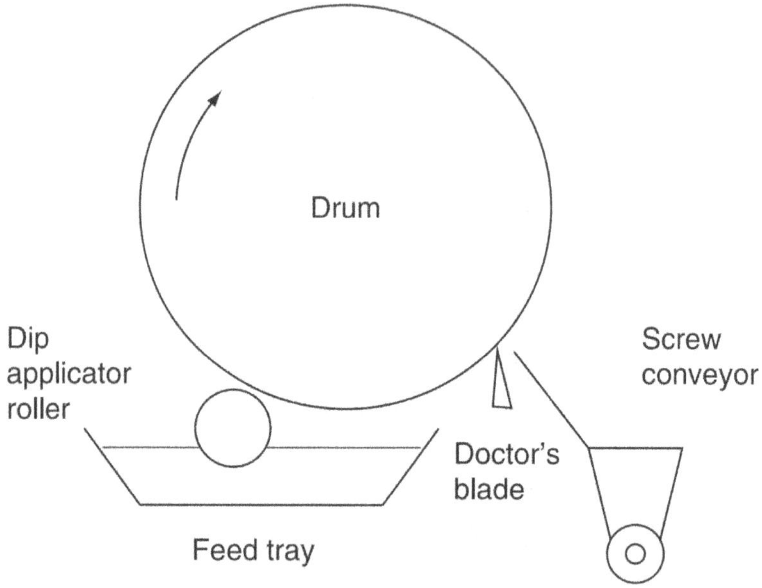

FIGURE 7.23 Single-roller dryer (Mujumdar, 2006).

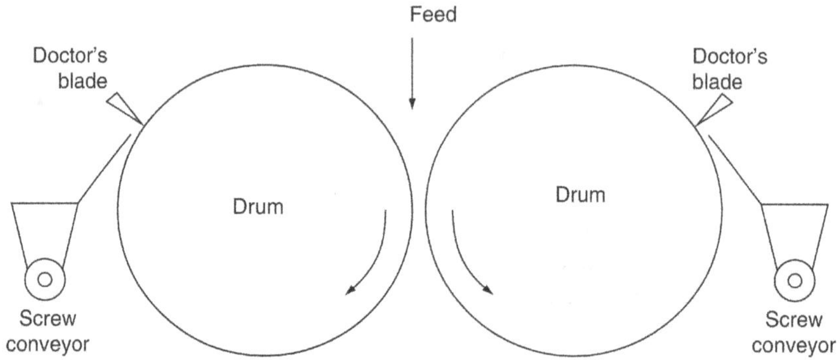

FIGURE 7.24 Double-roller dryer (Mujumdar, 2006).

7.10.4.2 Rotary Dryers with Conduction Heating

Figure 7.25 shows the scheme of a rotary conveying dryer. In this system, heat is transferred to the product through the walls; the chamber has two walls, and steam or hot combustion gases pass through the annular space and transfer heat to the process.

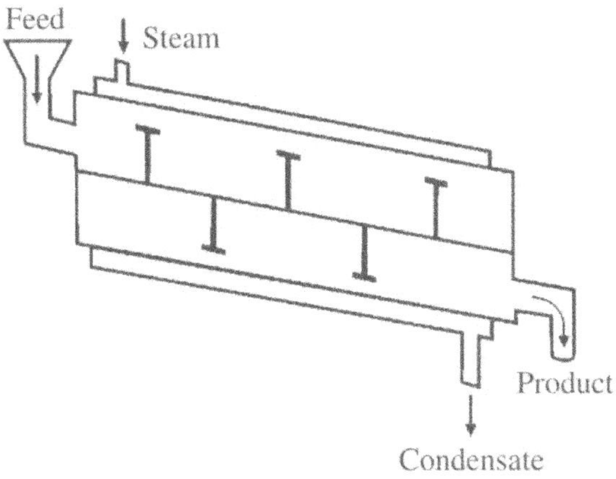

FIGURE 7.25 Schematic of rotary dryer with conduction heating.

7.10.5 Vacuum Drying

In this method, the product is dried under vacuum. In drying heat-sensitive materials, the pressure should be lowered to lower the boiling point temperature. This method is called evaporation under vacuum. If the pressure is lower than the pressure of the triple point of water, there will be no liquid phase, and by giving heat, all the moisture will be converted directly from the solid phase to the vapor phase (sublimation process). The presence of a vacuum in the drying process causes the temperature of the boiling point of water to decrease and the vapor pressure gradient between the inner layers of the product and its surface to rise and the drying rate of the product to increase. In this way, the color, texture, and smell of the dried product are preserved. Therefore, this method is very suitable for drying products sensitive to high temperature, such as fruits with high sugar content and protein materials. Figure 7.26 represents a typical vacuum dryer.

7.10.6 Infrared Drying

The absorption of infrared waves by water molecules causes the molecules to vibrate at a frequency of 60,000 to 150,000 MHz, and as a result, heat is produced (Aboud et al., 2019). The depth of infrared penetration depends on the thickness of the product, water activity, product composition, and the wavelength of infrared radiation.

Absorption of radiation by food mainly depends on the amount of water, thickness, and physicochemical nature of the product. Results have shown that most food products have high wave transmission capabilities at wavelengths less than 2.5 mm. Also, with the increase in the thickness of the food material, a simultaneous decrease occurs in the ability to transmit waves and an increase in the absorption ability, and products with a low thickness are recommended for processing with infrared rays. Proper adjustment of the radiation wavelength prevents the formation of the surface

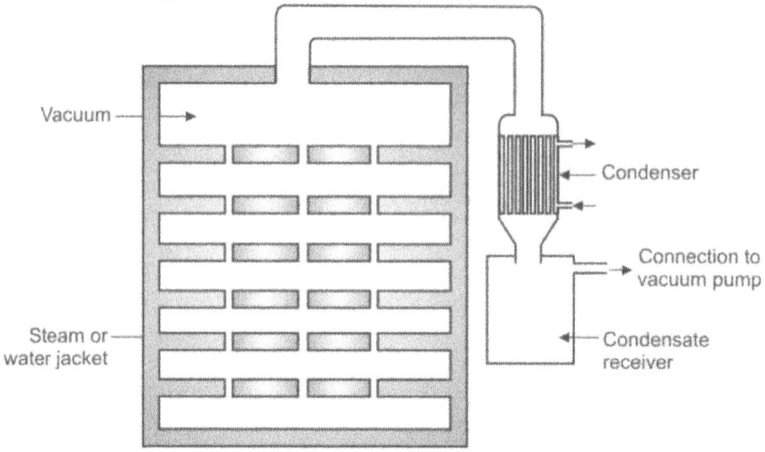

FIGURE 7.26 Vacuum dryer.

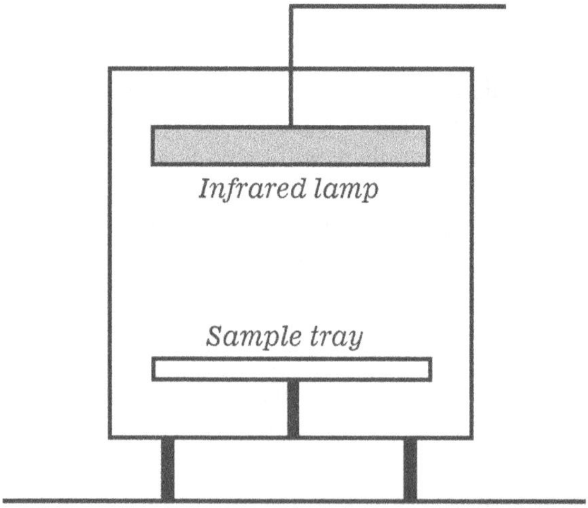

FIGURE 7.27 Infrared drying.

hardening phenomenon during drying, which occurs naturally in the hot air drying method. Failure to properly adjust the radiation intensity (product temperature regulator) causes damage or destruction of heat-sensitive food and agricultural products.

7.10.6.1 Limitation of Using Infrared for Drying

Infrared is used to dry food with a thin layer by the thin-layer method, because with an increase in the thickness of the food, part of the heat transfer into the material is done by the conduction method, which increases the duration of the process.

7.10.7 Combined Methods

Various methods have been used to dry agricultural products (fruits and vegetables). Compared to conventional drying methods, the infrared radiation method has advantages, such as faster heat transfer, shorter time, and lower energy consumption, and as a result, the quality of the product increases. Also, to increase the drying speed, hot air can be used in combination with infrared. The results reported by researchers indicate the high quality of the dried product, brighter color compared to other methods, less wrinkling, higher water absorption and porosity, shorter process time, and low energy consumption of the combined infrared hot air drying method. Another advantage of these types of dryers is the ability to be used separately as a hot air, infrared, or combined infrared–hot air dryer.

Creating a vacuum in an infrared dryer can improve the quality of the dried product and reduce the drying time. According to the researchers' reports, color changes

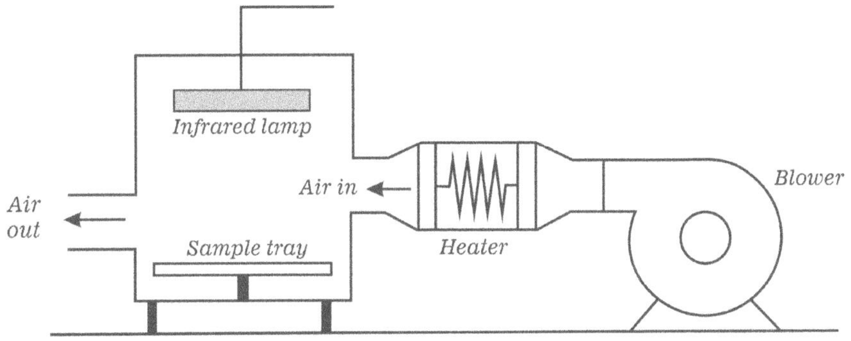

FIGURE 7.28 Combined infrared–hot air dryer.

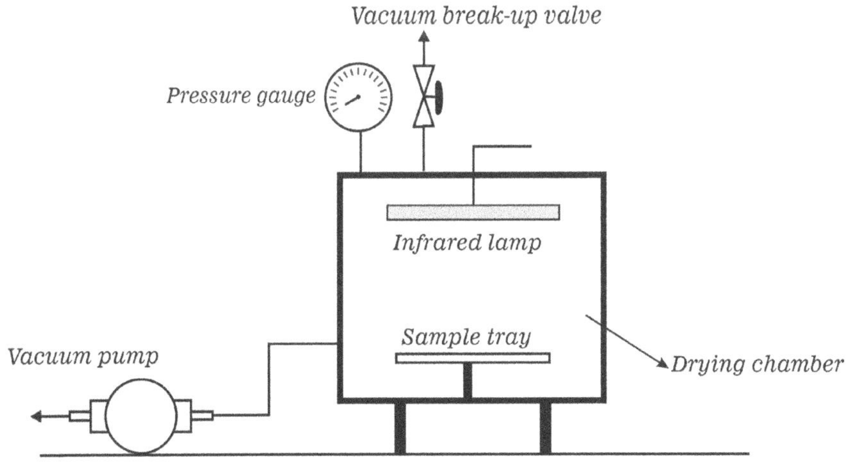

FIGURE 7.29 Combined infrared-vacuum dryer.

in infrared conditions under vacuum are less compared to other drying methods, and also the amount of porosity and water reabsorption increases.

7.10.8 Freeze Drying

In this method, the product is first frozen, and then a high vacuum is applied to it; moisture is converted from solid phase to vapor phase by sublimation and leaves the dryer. Freeze drying is a slow method for drying heat-sensitive products, such as some pharmaceutical and biological products. This method is the best drying method in terms of maintaining product quality, because the temperature of the process is low and oxygen is not available during the process. But it is time-consuming and expensive due to the use of refrigeration and vacuum systems. Food products that are sensitive to heat are often dried through sublimation of humidity. In a freeze dryer or sublimation dryer, the desired product is first frozen. Then the solvent (which is usually water) is frozen and removed from the food material through sublimation under vacuum. Sublimated ice is sucked from inside the drying chamber by the vacuum pump. The heat required for sublimation is provided through conduction or radiation (Figure 7.31). Frozen water sublimates at temperatures below 0°C or less, under a pressure of 610 Pa or less (Figure 7.30).

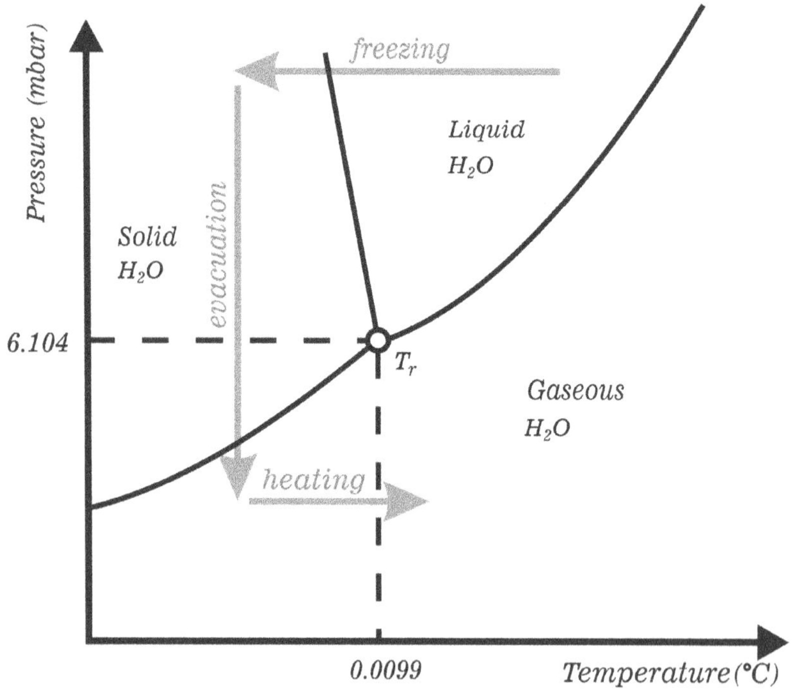

FIGURE 7.30 Freeze drying process on P-T phase diagram (Klepzig et al., 2020).

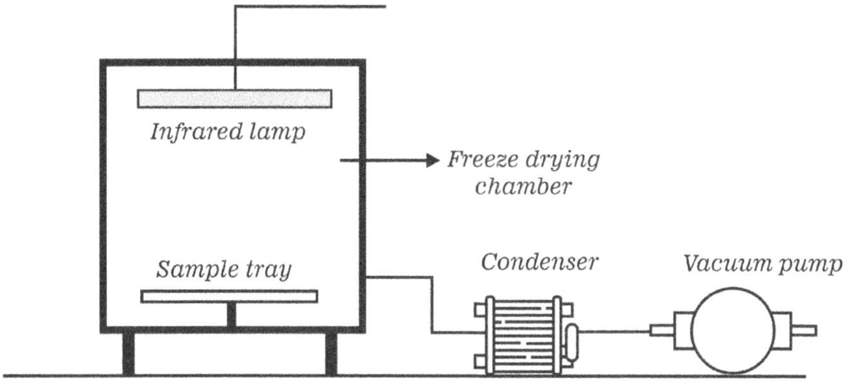

FIGURE 7.31 Freeze dryer with infrared heating.

Sublimation-dried products usually have high quality. This is because the structure of the food material, when the water is separated from it through sublimation, is not a porous and wrinkled structure. A product that is dried in this way is easily re-moistened. During drying, the taste and smell of the food will not decrease, or the decrease will be slight, because the low temperature of drying prevents the undesirable processes of destruction and spoilage, and the dried product will have high quality. According to researchers' reports, a freeze dryer is a suitable method to produce a product with a porous structure and high quality, but it is expensive and cannot be used for every product.

7.10.9 Equilibrium Equations for Continuous Dryers

The drying chamber can be modeled as a control volume, as shown in Figure 7.32, and the equilibrium equations can be written for it:

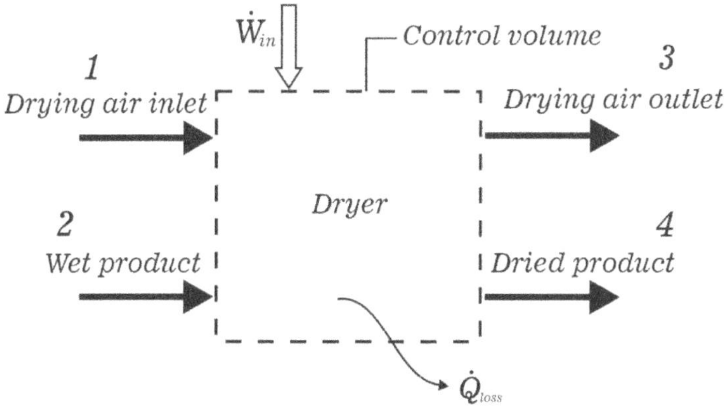

FIGURE 7.32 Schematic of the thermodynamic model of the drying process.

7.10.9.1 Mass Balance

Mass balance equation for dry air:

$$\left(\dot{m}_a\right)_1 = \left(\dot{m}_a\right)_3 = \dot{m}_a \tag{7.79}$$

Mass balance equation for dry product:

$$\left(1 - x_{water,2}\right)\dot{m}_2 = \left(1 - x_{water,4}\right)\dot{m}_4 \tag{7.80}$$

Mass balance equation for moisture content:

$$\omega_1\left(\dot{m}_a\right)_1 + x_{water,2}\dot{m}_2 = \omega_3\left(\dot{m}_a\right)_3 + x_{water,4}\dot{m}_4 \tag{7.81}$$

General mass balance equation:

$$\dot{m}_1 + \dot{m}_2 = \dot{m}_3 + \dot{m}_4 \tag{7.82}$$

7.10.9.2 Energy Balance

The energy balance equation, ignoring kinetic and potential energy, can be written as follows:

$$\left(\dot{m}_a\right)_1 h_1 + \dot{m}_2 h_2 + \dot{W}_{in} = \left(\dot{m}_a\right)_3 h_3 + \dot{m}_4 h_4 + \dot{Q}_{loss} \tag{7.83}$$

$$h_1 = \left(\frac{C_{p,air@T_1} + C_{p,air@0°C}}{2}\right)T_1 + h_{g@T_1}\omega_1 \tag{7.84}$$

$$h_2 = \left(\frac{C_{p@T_2} + C_{p@0°C}}{2}\right)T_2 + v_{@T_2}\left(P_2 - P_0\right) \tag{7.85}$$

$$h_3 = \left(\frac{C_{p,air@T_3} + C_{p,air@0°C}}{2}\right)T_3 + h_{g@T_3}\omega_3 \tag{7.86}$$

$$h_4 = \left(\frac{C_{p@T_4} + C_{p@0°C}}{2}\right)T_4 + v_{@T_4}\left(P_4 - P_0\right) \tag{7.87}$$

7.10.9.3 Exergy Balance

Exergy of flows:

$$\dot{E}x_1 = \left(\dot{m}_a\right)_1 ex_1 \tag{7.88}$$

$$\dot{E}x_2 = \dot{m}_2 ex_2 \tag{7.89}$$

$$\dot{E}x_3 = \left(\dot{m}_a\right)_3 ex_3 \tag{7.90}$$

$$\dot{E}x_4 = \dot{m}_4 ex_4 \tag{7.91}$$

$$ex_1 = \left(C_{p,a1} + C_{p,v1}\omega_1\right)\left(T_1 - T_0\right) - T_0\left[\begin{array}{l}\left(C_{p,a1} + C_{p,v1}\omega_1\right)\ln\dfrac{T_1}{T_0}\\ -\left(\dfrac{1}{M_a} + \dfrac{\omega_1}{M_w}\right)R\ln\dfrac{P_1}{P_0}\end{array}\right] \tag{7.92}$$

$$C_{p,a1} = \frac{C_{p,air@T_1} + C_{p,air@T_0}}{2} \tag{7.93}$$

$$C_{p,v1} = \frac{C_{p,v@T_1} + C_{p,v@T_0}}{2} \tag{7.94}$$

$$ex_2 = \left(\frac{C_{p@T_2} + C_{p@T_0}}{2}\right)\left(T_2 - T_0 - T_0\, ln\left(\frac{T_2}{T_0}\right)\right) + v_{@T_2}\left(P_2 - P_0\right) \tag{7.95}$$

$$ex_3 = \left(C_{p,a3} + C_{p,v3}\omega_3\right)\left(T_3 - T_0\right) - T_0\left[\begin{array}{l}\left(C_{p,a3} + C_{p,v3}\omega_3\right)\ln\dfrac{T_3}{T_0}\\ -\left(\dfrac{1}{M_a} + \dfrac{\omega_3}{M_w}\right)R\ln\dfrac{P_3}{P_0}\end{array}\right] \tag{7.96}$$

$$C_{p,a3} = \frac{C_{p,air@T_3} + C_{p,air@T_0}}{2} \tag{7.97}$$

$$C_{p,v3} = \frac{C_{p,v@T_3} + C_{p,v@T_0}}{2} \tag{7.98}$$

$$ex_4 = \left(\frac{C_{p@T_4} + C_{p@T_0}}{2}\right)\left(T_4 - T_0 - T_0\ln\left(\frac{T_4}{T_0}\right)\right) + v_{@T_4}\left(P_4 - P_0\right) \tag{7.99}$$

Exergy balance:

$$\dot{E}x_F = \dot{E}x_1 + \dot{E}x_2 + \dot{W}_{in} \tag{7.100}$$

$$\dot{E}x_P = \dot{E}x_3 + \dot{E}x_4 \tag{7.101}$$

$$\dot{E}x_L = \left(1 - \frac{T_0}{T_b}\right)\dot{Q} \tag{7.102}$$

$$\dot{E}x_{des} = \dot{E}x_F - \dot{E}x_P - \dot{E}x_L \tag{7.103}$$

Exergy efficiency:

$$\eta_{ex} = \frac{\dot{E}x_3 + \dot{E}x_4}{\dot{E}x_1 + \dot{E}x_2 + \dot{W}_{in}} \tag{7.104}$$

7.10.9.4 Exergy efficiency of the drying process

The exergy efficiency of the drying process can be defined as the ratio of the exergy used to evaporate the moisture of the product to the exergy sent to the drying system. In exergy hot air dryers, the drying process is defined as follows:

$$\eta_{ex} = \frac{exergy\ used\ for\ evaporation}{exergy\ supplied\ to\ the\ system} = \frac{\dot{E}x_{evaporation}}{\dot{E}x_{supply}} \tag{7.105}$$

$$\dot{E}x_{evaporation} = \left(1 - \frac{T_0}{T_p}\right)\dot{Q}_{evaporation} \tag{7.106}$$

$$\dot{Q}_{evaporation} = \dot{m}_w h_{fg} \tag{7.107}$$

$$\dot{m}_w = \dot{m}_2 - \dot{m}_4 \tag{7.108}$$

t_p is the drying temperature of the product and m_w is the rate of evaporation of water from the product.

7.10.9.5 Examples

✐Example 7.13: Carrot slices containing 12% solid matter undergo drying in a continuous tunnel dryer until achieving a final moisture content of 4%. The dryer is supplied with air at a rate of 10 m³/s, featuring a specific humidity of 0.01 kg_{water}/kg_{dry} air and a specific volume of 1.02 $m^3/kg_{dry\ air}$. The exiting air carries a specific humidity of 0.02 $kg_{water}/kg_{dry\ air}$. Determine the mass flow rates of the incoming and outgoing carrot slices.

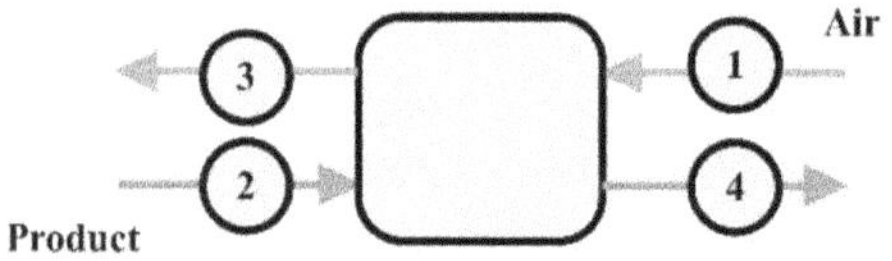

FIGURE 7.33 Schematic of Example 7.13.

✍solution

Assumptions:

Mass balance for dry air:

$$(\dot{m}_a)_1 = (\dot{m}_a)_3 = \dot{m}_a = \frac{\dot{V}}{v_a} = \frac{10}{1.02} = 9.8\frac{kg_{dry\ air}}{s}$$

Simultaneous solution of mass balance for dry matter and moisture content:

$$\begin{cases} (1-x_{water,2})\dot{m}_2 = (1-x_{water,4})\dot{m}_4 \\ \omega_1(\dot{m}_a)_1 + x_{water,2}\dot{m}_2 = \omega_3(\dot{m}_a)_3 + x_{water,4}\dot{m}_4 \end{cases}$$

$$\rightarrow \begin{cases} (1-0.12)\dot{m}_2 = (1-0.04)\dot{m}_4 \\ 0.01(9.8)+0.12\dot{m}_2 = 0.02(9.8)+0.04\dot{m}_4 \end{cases} \rightarrow \begin{cases} \dot{m}_2 = 1.176\frac{kg}{s} \\ \dot{m}_4 = 1.078\frac{kg}{s} \end{cases}$$

✐**Example 7.14: In a honey packaging factory, honey jars are washed before filling and then completely dried in a continuous tunnel dryer. Table 7.6 presents the information related to the flows of this dryer. The volumetric flow rate of the incoming air to the dryer is 2.56 m³/s, the relative humidity of the incoming air is 23%, and the outgoing air is 10%. The electric power input to the dryer is 2.5 kW, and the surface temperature of the dryer is 40°C. Calculate the rate of heat loss from the surface of the dryer, the exergy destruction rate, the exergy efficiency of the dryer, and the exergy efficiency of the drying process. Assume the air temperature provided for drying is 70°C**

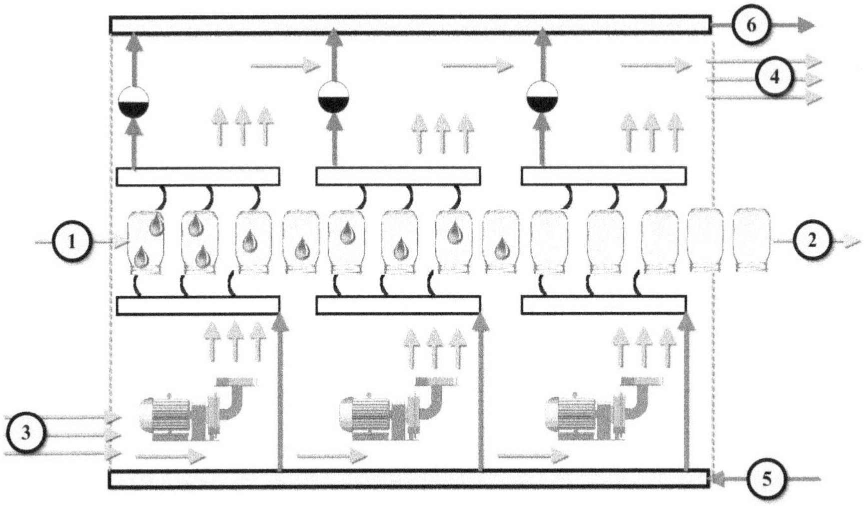

FIGURE 7.34 Schematic of dryer related to Example 7.14.

TABLE 7.6
Information Related to the Flows in Example 7.14

State No.	Fluid Type	T (°C)	P (kPa)	$\dot{m}\left(\frac{kg}{s}\right)$
1	Jar	16.00	101.33	—
2	Jar	45.00	101.33	0.07
3	Air	25.00	101.33	—
4	Air	47.00	101.33	—
5	Steam	150.00	200.00	0.10
6	Water	95.00	120.00	—

✍Solution

Assumptions:

1. Kinetic and potential energy changes are ignored.
2. The dead state temperature and pressure are 25°C and 101.325 kPa.
3. The reference temperature and pressure are 0°C and 101.325 kPa.
4. The process is considered steady state.
5. Dry air and water vapor are ideal gases.
6. The specific heat of air is 1.005 kJ/kg.k.
7. The specific heat of glass is 0.8 kJ/kg.k.
8. The specific heat of water is 4.4 kJ/kg.k.
9. The humidity of the glass leaving the dryer is zero.

Calculation of the partial pressure of vapor and dry air at the inlet and outlet:

$$\varnothing = \frac{P_v}{P_g} \rightarrow \begin{cases} P_{v,3} = \varnothing_3 P_{g@25°C} = 0.23(3.1698) = 0.7291\text{kPa} \\ P_{v,4} = \varnothing_4 P_{g@47°C} = 0.1(10.6980) = 1.0698\text{kPa} \end{cases}$$

$$P_a = P - P_v \rightarrow \begin{cases} P_{a,3} = 101.325 - 0.7291 = 100.5959\text{kPa} \\ P_{a,4} = 101.325 - 1.0698 = 100.2552\text{kPa} \end{cases}$$

Calculation of humidity ratio of inlet and outlet air:

$$\omega = \frac{0.622P_v}{P_a} \rightarrow \begin{cases} \omega_3 = \dfrac{0.622P_{v,3}}{P_{a,3}} = \dfrac{0.622(0.7291)}{100.5959} = 0.00451\dfrac{\text{kg}_{\text{water}}}{\text{kg}_{\text{dry air}}} \\ \omega_4 = \dfrac{0.622P_{v,4}}{P_{a,4}} = \dfrac{0.622(1.0698)}{100.2552} = 0.00664\dfrac{\text{kg}_{\text{water}}}{\text{kg}_{\text{dry air}}} \end{cases}$$

Calculation of enthalpy of inlet and outlet air flows:

$$h_3 = h_{a3} + h_{v3}\omega_3 \cong c_{p,air}T_3 + h_{g@t_3}\omega_3 = \left(1.005\frac{\text{kJ}}{\text{kg°C}}\right)(25°\text{C}) + (0.00451)\left(2546.5\frac{\text{kJ}}{\text{kg}}\right)$$

$$= 36.61\frac{\text{kJ}}{\text{kg}_{\text{dry air}}}$$

$$h_4 = h_{a4} + h_{v4}\omega_4 \cong c_{p,air}T_4 + h_{g@T_4}\omega_4 = \left(1.005\frac{\text{kJ}}{\text{kg°C}}\right)(47°\text{C}) + (0.00664)\left(2586.0\frac{\text{kJ}}{\text{kg}}\right)$$

$$= 64.41\frac{\text{kJ}}{\text{kg}_{\text{dry air}}}$$

Calculate the density of dry air:

$$\rho = \frac{P}{RT} \rightarrow \rho_{a,3} = \frac{P_{a,3}}{R_a T_3} = \frac{100.5959}{(0.287)(25+273.15)} = 1.1756\frac{\text{kg}}{\text{m}^3}$$

Determining the flow rate of dry air:

$$\dot{m}_{a3} = \frac{P_{a,3}\dot{V}}{R_a T_3} = \frac{(100.5959)(2.56)}{(0.287)(273.15+25)} = 3.01\frac{\text{kg}}{\text{s}}$$

$$\dot{m}_{a4} = \dot{m}_{a3} = 3.01\frac{\text{kg}}{\text{s}} = \dot{m}_a$$

Calculation of the evaporated water flow rate during the drying process:

$$\dot{m}_w = \dot{m}_a(\omega_4 - \omega_3) = 3.01(0.00664 - 0.00451) = 0.0064\frac{\text{kg}}{\text{s}}$$

Calculation of the flow rate of the input glass to the dryer:

$$\dot{m}_1 = \dot{m}_2 + \dot{m}_w = 0.07 + 0.0064 = 0.0764\frac{\text{kg}}{\text{s}}$$

Calculation of the enthalpy of the input glass to the dryer:

$$c_{p,1} = \left(\frac{0.07}{0.0764}\right)c_{p,jar} + \left(\frac{0.0064}{0.0764}\right)c_{p,ave,water} = 1.1\frac{\text{kJ}}{\text{kg°C}}$$

$$h_1 = c_{p,1}T_1 = 1.1(16) = 17.6\frac{\text{kJ}}{\text{kg}}$$

Calculation of the enthalpy of the glass leaving the dryer:

$$h_2 = c_{p,jar}T_2 = 0.8(45) = 36\frac{\text{kJ}}{\text{kg}}$$

Properties of water vapor (flow 5) and condensation (flow 6):

$$@25°\text{C}, 101.325\text{kPa}\begin{cases} h_0 = 104.92\dfrac{\text{kJ}}{\text{kg}} \\ s_0 = 0.37\dfrac{\text{kJ}}{\text{kg}}.\text{K} \end{cases}$$

$$@150°C, 200kpa \begin{cases} h_5 = 2769.1 \dfrac{kJ}{kg} \\ s_5 = 7.2810 \dfrac{kJ}{kg}.K \end{cases}$$

$$@95°C, 120kpa \begin{cases} h_6 \cong h_{f@95°C} = 398.09 \dfrac{kJ}{kg} \\ s_6 \cong s_{f@95°C} = 1.2504 \dfrac{kJ}{kg}.K \end{cases}$$

Mass and energy balance for dryer:

$$\dot{m}_5 = \dot{m}_6$$

$$\dot{m}_{a4} = \dot{m}_{a3} = \dot{m}_a$$

$$\dot{m}_1 h_1 + (\dot{m}_a)_3 h_3 + \dot{m}_5 h_5 + \dot{W}_{in} = \dot{m}_2 h_2 + (\dot{m}_a)_4 h_4 + \dot{m}_6 h_6 + \dot{Q}_{loss}$$

$$\begin{aligned} \rightarrow \dot{W}_{loss} &= \dot{m}_1 h_1 - \dot{m}_2 h_2 + \dot{m}_a (h_3 - h_4) + \dot{m}_5 (h_5 - h_6) + \dot{W}_{in} \\ &= 0.0764(17.6) - 0.07(36) + 3.01(36.61 - 64.41) \\ &\quad + 0.1(2769.1 - 398.09) + 2.5 = 154.75 kW \end{aligned}$$

Calculation of exergy of flows:

$$ex_1 = c_{p,1}\left(T_1 - T_0 - T_0 \ln\left(\frac{T_1}{T_0}\right)\right) = 1.1\left[289.15 - 298.15 - 298.15 \ln\left(\frac{289.15}{298.15}\right)\right] = 0.1525 \frac{kJ}{kg}$$

$$ex_2 = c_{p,2}\left(T_2 - T_0 - T_0 \ln\left(\frac{T_2}{T_0}\right)\right) = 0.8\left[318.15 - 298.15 - 298.15 \ln\left(\frac{318.15}{298.15}\right)\right] = 0.5138 \frac{kJ}{kg}$$

$$ex_3 = c_{p,3}\left(T_3 - T_0 - T_0 \ln\left(\frac{T_3}{T_0}\right)\right) = 1.005\left[298.15 - 298.15 - 298.15 \ln\left(\frac{298.15}{298.15}\right)\right] = 0$$

$$ex_4 = c_{p,4}\left(T_4 - T_0 - T_0 \ln\left(\frac{T_4}{T_0}\right)\right) = 1.005\left[320.15 - 298.15 - 298.15 \ln\left(\frac{320.15}{298.15}\right)\right] = 0.7777 \frac{kJ}{kg}$$

$$ex_5 = (h_5 - h_0) - T_0 (s_5 - s_0) = (2769.1 - 104.92) - 298.15(7.2810 - 0.37) = 603.67 \frac{kJ}{kg}$$

$$ex_6 = (h_6 - h_0) - T_0 (s_6 - s_0) = (398.09 - 104.92) - 298.15(1.2504 - 0.37) = 30.68 \frac{kJ}{kg}$$

Calculation of exergy of fuel, product, and waste:

$$\begin{aligned} \dot{e}x_f &= (\dot{e}x_5 - \dot{e}x_6) + \dot{e}x_1 + \dot{e}x_3 + \dot{W}_{in} \\ &= (\dot{m}_5 ex_5 - \dot{m}_6 ex_6) + \dot{m}_1 ex_1 + \dot{m}_3 ex_3 \\ &= \dot{m}_5 (ex_5 - ex_6) + \dot{m}_1 ex_1 + \dot{m}_3 ex_3 \\ &= 0.1(603.67 - 30.68) + 0.0764(0.1525) + 3.01(0) + 2.5 \\ &= 59.81 kw \end{aligned}$$

$$\dot{E}x_p = \dot{E}x_2 + \dot{E}x_4 = \dot{m}_2 ex_2 + \dot{m}_4 ex_4 = 0.07(0.5138) + 3.01(0.7777) = 2.38\text{kW}$$

$$\dot{E}x_l = \left(1 - \frac{T_0}{T_b}\right)\dot{Q} = \left(1 - \frac{298.15}{273.15 + 40}\right)154.75 = 7.41\text{kW}$$

Exergy destruction rate:

$$\dot{E}x_{des} = \dot{E}x_f - \dot{E}x_p - \dot{E}x_l = 59.81 - 2.38 - 7.41 = 50.02\text{kW}$$

Dryer exergy efficiency:

$$\eta_{ex} = \frac{\dot{E}x_p}{\dot{E}x_f} \times 100 = \frac{2.38}{59.81} \times 100 = 3.98\%$$

Exergy efficiency of the drying process:

$$\dot{Q}_{evaporation} = \dot{m}_w h_{fg@70^\circ C} = 0.0064(2333) = 14.93\text{kW}$$

$$\dot{E}x_{evaporation} = \left(1 - \frac{T_0}{T_p}\right)\dot{Q}_{evaporation} = \left(1 - \frac{298.15}{273.15 + 70}\right)14.93 = 1.96\text{kW}$$

$$\eta_{ex} = \frac{\dot{E}x_{evaporation}}{\dot{E}x_5 - \dot{E}x_6} = \frac{1.96}{57.30} = 0.0342 \text{ or } 3.42\%$$

8 Exergy-Economic Analysis

8.1 CONCEPTS OF EXERGY-ECONOMIC ANALYSIS

Exergy-economic analysis provides a valuable tool for conducting economic evaluations of energy systems. Exergy-economy is a branch of engineering sciences that combines exergy analysis and economic principles and provides information that cannot be obtained by energy analysis and economic evaluation. It is clear that exergy analysis can determine the location and actual amount of exergy losses and destruction; exergy-economic analysis enables engineers to calculate the cost of exergy loss and destruction in a system.

8.1.1 Exergy Costing

For the exergy-economics analysis, various perspectives have been proposed, the most common of which is the specific exergy costing (SPECO) perspective, which includes three stages:

1. identification of exergy flows
2. definition of fuel and product for each component
3. allocation of cost equations

Steps 1 and 2 are discussed in detail in Chapter 6. In exergy costing, a cost is determined for each of the exergy flows. The cost rate related to the *j*-th flow is as follows:

$$\dot{C}_j = c_j \dot{E}x_j = \dot{m}_j c_j ex_j \tag{8.1}$$

$\dot{C}$ is the cost rate of flow (USD/s), and c is the unit cost of exergy of flow (USD/kJ). The cost rate of heat transfer and work is also determined as follows:

$$\dot{C}_Q = c_Q \dot{E}x_Q \tag{8.2}$$

$$\dot{C}_W = c_W \dot{W} \tag{8.3}$$

$\dot{C}_Q$ is the heat cost rate (USD/s), c_Q is the heat exergy unit cost (USD/kJ), $\dot{C}_W$ is the work cost rate (USD/s), and c_W is the work exergy unit cost (USD/kJ).

Exergy costing includes cost equations that are usually formulated separately for each system. The cost equation applied to a system indicates that the final cost of each of the existing flows is equal to the final cost of the inlet flows including the equipment cost. The cost balance equation for a system in general is written as follows:

DOI: 10.1201/9781003424680-8

$$\sum\left(c_i\dot{E}x_i\right)+\left(c_W\dot{W}\right)_i+\left(c_Q\dot{E}x_Q\right)_i+\dot{Z}=\sum\left(c_o\dot{E}x_o\right)+\left(c_W\dot{W}\right)_o+\left(c_Q\dot{E}x_Q\right)_o \quad (8.4)$$

8.1.1.1 Economic Analysis

The cost rate of each equipment for a system is due to the initial investment and maintenance costs.

$$\dot{Z}=\dot{Z}^{CI}+\dot{Z}^{OM} \quad (8.5)$$

$\dot{Z}_k$ is the equipment cost rate (USD/s), $\dot{Z}^{CI}$ is the investment cost rate (USD/s), and $\dot{Z}^{OM}$ is the maintenance cost rate (USD/s). The equipment cost rate can be calculated from the following equation:

$$\dot{Z}=\frac{Z\times CRF\times\varphi}{3600\times N} \quad (8.6)$$

Z is the initial purchase cost of the component, ϕ is the coefficient related to the maintenance cost, N is the number of hours of annual operation of the system, and CRF is the return-on-investment coefficient and is determined from the following equation:

$$CRF=\frac{i\left(1+i\right)^n}{\left(1+i\right)^n-1} \quad (8.7)$$

where i is the capital profit rate, and n is the number of years of system operation.

8.1.2 Exergy-Economic Analysis Parameters

8.1.2.1 Cost of Fuel and Product

The definition of fuel and product for a system leads to the introduction of cost rates related to fuel $\dot{C}_F$ and product $\dot{C}_P$ in that system. $\dot{C}_F$ indicates the flow rate cost at which fuel exergy is created for the system. $\dot{C}_P$ is the flow rate cost corresponding to the product exergy for the same system. The average cost of fuel for the system (c_F) represents the average cost of each unit of fuel exergy provided for the system:

$$c_F=\frac{\dot{C}_F}{\dot{E}x_F} \quad (8.8)$$

There is a similar case for c_P.

$$c_P=\frac{\dot{C}_P}{\dot{E}x_P} \quad (8.9)$$

Using the definition of fuel and product, the cost equation of a system can be written as follows:

$$\dot{C}_F+\dot{Z}=\dot{C}_P+\dot{C}_L \quad (8.10)$$

8.1.2.2 Auxiliary Equations

In applying cost balance equations to system components, there is usually more than one inlet and outlet flow for each component. Therefore, the number of unknown costs is more than the existing equations. In other words, for n outlet flow from each component, $n - 1$ auxiliary equations are needed. To solve this problem, the p and f rules of SPECO are used.

The F rule states that the total cost associated with the removal of exergy from an exergy stream in a component must be equal to the cost at which the removed exergy was supplied to the same stream in upstream components. The exergy difference of this stream between inlet and outlet is considered in the definition of fuel for the component. The P rule states that each exergy unit is supplied to any stream associated with the product of a component at the same average cost c_P. This cost is calculated from the cost balance and the F rule (Lazzaretto & Tsatsaronis, 1999).

8.1.2.3 Cost of Exergy Waste and Destruction

Calculating the cost associated with exergy destruction stands out as a crucial facet within exergy costing. Specifically, determining the exergy destruction cost within individual components of a system is paramount. This encompasses the exergy destruction cost rate (referred to as "c_{des}") and the quantity of exergy destroyed (termed "ex_{des}"). These costs, though concealed, become apparent exclusively through exergy-economic analysis. The extent of irreversibility or thermodynamic inefficiency inherent in each system component is quantified by expressing it as exergy destruction cost. To compute the cost of exergy destruction, the following equation is employed:

$$\dot{C}_{des} = c_F \dot{E}x_{des} \tag{8.11}$$

The cost of exergy loss can be calculated from the following equation:

$$\dot{C}_L = c_F \dot{E}x_L \tag{8.12}$$

8.1.2.4 Total Cost Rate of Operation

The total cost rate of operation for a system is defined as the sum of equipment cost, exergy destruction cost, and exergy waste cost:

$$\dot{C}_{total} = \dot{Z} + \dot{C}_{des} + \dot{C}_L \tag{8.13}$$

8.2 HEAT EXCHANGERS

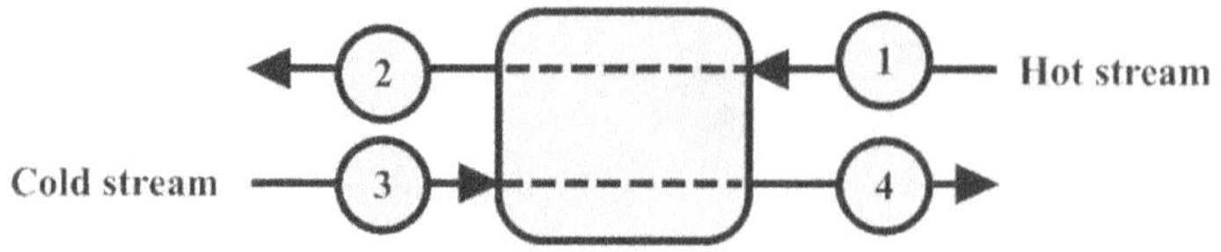

FIGURE 8.1 Schematic of a heat exchanger.

Cost of mass flows:

$$\dot{C}_1 = c_1\dot{E}x_1 \quad (8.14)$$

$$\dot{C}_2 = c_2\dot{E}x_2$$

$$\dot{C}_3 = c_3\dot{E}x_3$$

$$\dot{C}_4 = c_4\dot{E}x_4$$

Fuel exergy:

$$\dot{C}_F = c_1\dot{E}x_1 - c_2\dot{E}x_2 \quad (8.15)$$

Auxiliary equation (F rule):

$$c_1 = c_2 \quad (8.16)$$

Fuel exergy unit cost:

$$c_F = \frac{c_1\dot{E}x_1 - c_2\dot{E}x_2}{\dot{E}x_1 - \dot{E}x_2} \quad (8.17)$$

Product exergy:

$$\dot{C}_P = c_4\dot{E}x_4 - c_3\dot{E}x_3 \quad (8.18)$$

✐Example 8.1: In Example 6.1, if the cost of the inlet juice to the heat exchanger is 1 USD/s and the cost of the inlet condensate is 0.1 USD/s, calculate the exergy destruction cost and the total cost rate of operation. Consider the equipment cost of 2500 USD, maintenance cost of 5% of the equipment cost ($\phi = 1.05$), factory working period of 6 months, discount rate of 20% (0.2), and number of operation years of the equipment of 20 years.

✍Solution

The equipment cost rate is determined using Equation 6.8. For this purpose, the rate of return and the number of annual operating hours are determined:

$$\begin{cases} n = 20 \\ i = 0.2 \end{cases} \rightarrow CRF = \frac{i(1+i)^n}{(1+i)^n - 1} = \frac{0.2(1+0.2)^{20}}{(1+0.2)^{20} - 1} = 0.20535653$$

$$N = 6\,\text{month} \times \frac{30\,\text{day}}{1\,\text{month}} \times \frac{24\,\text{hour}}{1\,\text{day}} = 4320\,\text{hour}$$

$$\begin{cases} Z = 2500 \\ \varphi = 1.05 \end{cases} \rightarrow \dot{Z} = \frac{Z \times CRF \times \varphi}{N \times 3600} = \frac{2500 \times 0.20535653 \times 1.05}{4320 \times 3600} = 0.00003466 \text{ USD}/s$$

Cost of flows:

$$\dot{C}_1 = c_1\dot{E}x_1 \rightarrow c_1 = \frac{\dot{C}_1}{\dot{E}x_1} = \frac{\dot{C}_1}{\dot{m}_1 ex_1} = \frac{1}{15.12} = 0.06611872 \text{ USD/kJ}$$

$$\dot{C}_2 = c_2\dot{E}x_2 = c_2\dot{m}_2 ex_2 = 43.55c_2$$

$$\dot{C}_3 = c_3\dot{E}x_3 \rightarrow c_3 = \frac{\dot{C}_3}{\dot{E}x_3} = \frac{\dot{C}_1}{\dot{m}_3 ex_3} = \frac{0.1}{126.17} = 0.00079260 \text{ USD/kJ}$$

$$\dot{C}_4 = c_4\dot{E}x_4 = c_4\dot{m}_4 ex_4 = 73.55c_4$$

Auxiliary equation:

$$c_4 = c_3 = 0.00079260 \frac{\text{USD}}{\text{kJ}}$$

Cost of fuel and product:

$$\dot{C}_F = \dot{C}_3 - \dot{C}_4 = 0.1 - 73.55(0.00079260) = 0.04170353 \text{ USD}/s$$

$$\dot{C}_P = \dot{C}_2 - \dot{C}_1 = 43.55c_2 - 1$$

Cost balance:

$$\dot{C}_F + \dot{Z} = \dot{C}_P$$

$$\rightarrow 0.04170353 + 0.00003466 = 43.55c_2 - 1$$

$$\rightarrow c_2 = 0.02391853 \text{ USD/kJ}$$

Fuel exergy unit cost:

$$c_F = \frac{\dot{C}_F}{\dot{E}x_F} = \frac{0.0417}{52.62} = 0.00079260 \text{ USD/kJ}$$

Exergy destruction cost:

$$\dot{C}_{des} = c_F\dot{E}x_{des} = 0.00079260 \times 24.19 = 0.01917051 \text{USD/s}$$

Total cost rate of operation:

$$\dot{C}_{des} = c_F\dot{E}x_{des} = 0.00079260 \times 24.19 = 0.01917051 \text{USD/s}$$

✐Example 8.2: In Example 6.3, if the cost of the inlet syrup to the heat exchanger is 1 USD/s and the cost of the inlet steam is 4 USD/s, calculate the exergy destruction cost and the total cost rate of operation. Consider the equipment cost of 8570 USD, maintenance cost of 5% of the equipment cost ($\phi = 1.05$), factory working period of 3 months, discount rate of 20% (0.2), and number of operation years of the equipment of 20 years.

✍Solution

$$\begin{cases} n = 20 \\ i = 0.2 \end{cases} \rightarrow CRF = \frac{i(1+i)^n}{(1+i)^n - 1} = \frac{0.2(1+0.2)^{20}}{(1+0.2)^{20} - 1} = 0.20535653$$

$$N = 3\,\text{month} \times \frac{30\,\text{day}}{1\,\text{month}} \times \frac{24\,\text{hour}}{1\,\text{day}} = 2160\,\text{hour}$$

$$\begin{cases} Z = 8570 \\ \varphi = 1.05 \end{cases} \rightarrow \dot{Z} = \frac{Z \times CRF \times \varphi}{N \times 3600} = \frac{8570 \times 0.20535653 \times 1.05}{2160 \times 3600} = 0.00023764\,\text{USD/s}$$

Cost of flows:

$$\dot{C}_1 = c_1 \dot{E}x_1 \rightarrow c_1 = \frac{\dot{C}_1}{\dot{E}x_1} = \frac{\dot{C}_1}{\dot{m}_1 ex_1} = \frac{1}{2269.27} = 0.00044067\,\text{USD/kJ}$$

$$\dot{C}_2 = c_2 \dot{E}x_2 = c_2 \dot{m}_2 ex_2 = 2542.25 c_2$$

$$\dot{C}_3 = c_3 \dot{E}x_3 \rightarrow c_3 = \frac{\dot{C}_3}{\dot{E}x_3} = \frac{\dot{C}_1}{\dot{m}_3 ex_3} = \frac{(4/3600)}{360.19} = 0.00000308\,\text{USD/kJ}$$

$$\dot{C}_4 = c_4 \dot{E}x_4 = c_4 \dot{m}_4 ex_4 = 40.61 c_4$$

Auxiliary equation (F rule):

$$c_4 = c_3 = 0.00000308 \frac{\text{USD}}{\text{kJ}}$$

Cost of fuel and product:

$$\dot{C}_F = \dot{C}_3 - \dot{C}_4 = (4/3600) - 40.61(0.00000308) = 0.00098584\,\text{USD}/s$$

$$\dot{C}_P = \dot{C}_2 - \dot{C}_1 = 2542.25 c_2 - 1$$

Fuel exergy unit cost:

$$c_F = \frac{\dot{C}_F}{\dot{E}x_F} = \frac{0.00098584}{319.58} = 0.00000308 \frac{\text{USD}}{\text{kJ}}$$

Cost of exergy waste:

$$\dot{C}_L = c_F \dot{E}x_L = 0.00000308 \times 0.38 = 0.00000116\ \text{USD}/s$$

Cost of balance:

$$\dot{C}_F + \dot{Z} = \dot{C}_P + \dot{C}_L$$
$$\rightarrow 0.00098584 + 0.00023764 = 2542.25c_2 - 1 + 0.00000116$$
$$\rightarrow c_2 = 0.00039383\ \text{USD}/\text{kJ}$$

Exergy destruction cost:

$$\dot{C}_{des} = c_F \dot{E}x_{des} = 0.00000308 \times 46.22 = 0.00014257\ \text{USD}/\text{s}$$

Total cost rate of operation:

$$\dot{C}_{total} = \dot{C}_{des} + \dot{Z} + \dot{C}_L = 0.00014257 + 0.00023764 + 0.00000116 = 0.00038138\ \text{USD}/\text{s}$$

8.3 PUMPS

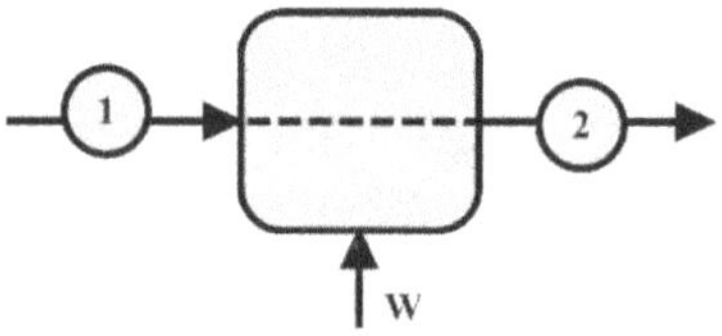

FIGURE 8.2 Schematic of a pump.

The cost of mass flows:

$$\dot{C}_1 = c_1 \dot{E}x_1 \quad (8.19)$$
$$\dot{C}_2 = c_2 \dot{E}x_2$$

Fuel exergy:

$$\dot{C}_F = c_W \dot{W} \quad (8.20)$$

Fuel exergy unit cost:

$$c_F = c_W \quad (8.21)$$

Product exergy:

$$\dot{C}_P = c_2 \dot{E}x_2 - c_1 \dot{E}x_1 \quad (8.22)$$

✐**Example 8.3: In Example 6.5, if the cost of the inlet syrup to the pump is 1 USD/s and the cost of the unit power is 1.22 USD/s, calculate the exergy destruction cost and the total cost rate of operation cost. Consider the equipment cost of 1430 USD, maintenance cost of 5% of the equipment cost ($\phi = 1.05$), factory working period of 3 months, discount rate of 20% (0.2), and number of operation years of the equipment of 20 years.**

✍**Solution**

The equipment cost rate is determined using Equation 6.8. For this purpose, the rate of return and the number of annual operating hours are determined:

$$\begin{cases} n = 20 \\ i = 0.2 \end{cases} \rightarrow CRF = \frac{i(1+i)^n}{(1+i)^n - 1} = \frac{0.2(1+0.2)^{20}}{(1+0.2)^{20} - 1} = 0.20535653$$

$$N = 3\text{month} \times \frac{30\,\text{day}}{1\,\text{month}} \times \frac{24\,\text{hour}}{1\,\text{day}} = 2160\text{ hour}$$

$$\begin{cases} Z = 1430 \\ \varphi = 1.05 \end{cases} \rightarrow \dot{Z} = \frac{Z \times CRF \times \varphi}{N \times 3600} = \frac{1430 \times 0.20535653 \times 1.05}{2160 \times 3600} = 0.00003965\text{ USD}/s$$

Cost of flows:

$$\dot{C}_1 = c_1 \dot{E}x_1 \rightarrow c_1 = \frac{\dot{C}_1}{\dot{E}x_1} = \frac{\dot{C}_1}{\dot{m}_1 ex_1} = \frac{1}{1154.32} = 0.00086631\,\text{USD}/\text{kJ}$$

$$\dot{C}_2 = c_2 \dot{E}x_2 = c_2 \dot{m}_2 ex_2 = 1179.05 c_2$$

Cost of fuel and product:

$$\dot{C}_F = c_W \dot{W} = 0.00000122(28.51) = 0.00003478\text{ USD}/s$$

$$\dot{C}_P = \dot{C}_2 - \dot{C}_1 = 1179.05 c_2 - 1$$

Cost balance:

$$\dot{C}_F + \dot{Z} = \dot{C}_P$$

$$\rightarrow 0.00003478 + 0.00003466 = 43.55 c_2 - 1$$

$$\rightarrow c_2 = 0.02391853\text{ USD}/\text{kJ}$$

Fuel exergy unit cost:

$$c_F = \frac{\dot{C}_F}{\dot{E}x_F} = \frac{0.0417}{52.62} = 0.00079260\text{ USD}/\text{kJ}$$

Exergy destruction cost:

$$\dot{C}_{des} = c_F \dot{E}x_{des} = 0.00079260 \times 24.19 = 0.01917051 \text{ USD}/s$$

Total cost rate of operation:

$$\dot{C}_{total} = \dot{C}_{des} + \dot{Z} = 0.1917051 + 0.00003466 = 0.01920517 \text{ USD}/\text{s}$$

8.4 MIXING

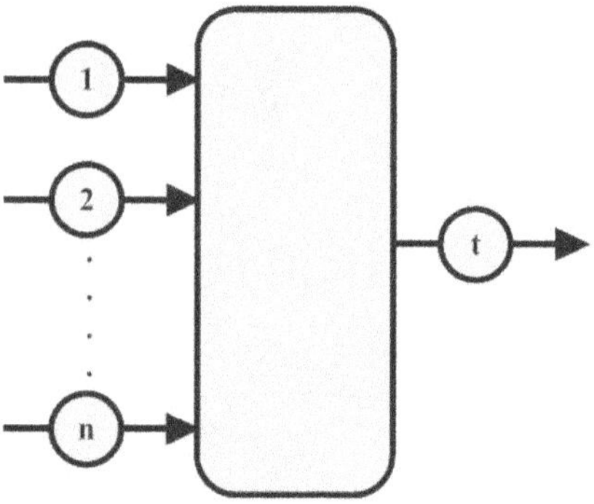

FIGURE 8.3 Schematic of a mixing chamber.

The cost of mass flows:

$$\begin{aligned} \dot{C}_1 &= c_1 \dot{E}x_1 \\ \dot{C}_2 &= c_2 \dot{E}x_2 \\ \dot{C}_n &= c_n \dot{E}x_n \end{aligned} \tag{8.23}$$

Fuel exergy:

$$\dot{C}_F = \dot{C}_1 + \dot{C}_2 + \ldots + \dot{C}_n \tag{8.24}$$

Fuel exergy unit cost:

$$c_F = \frac{c_1 \dot{E}x_1 + c_2 \dot{E}x_2 \ldots + c_n \dot{E}x_n}{\dot{E}x_1 + \dot{E}x_2 + \ldots + \dot{E}x_n} \tag{8.25}$$

Product exergy:

$$\dot{C}_P = c_t \dot{E}x_t \tag{8.26}$$

✐**Example 8.4: In Example 6.6, if the cost of the inlet steam to the barometric condenser is 3 USD/s, and the cost of the cooling water is 0.25 USD/s. Calculate the exergy destruction cost and the total cost rate of operation. Consider the equipment cost of 89,000 USD, maintenance cost of 5% of the equipment cost ($\phi = 1.05$), factory working period of 6 months, interest rate of 20% (0.2), and number of operation years of the equipment of 20 years.**

✍**Solution**

The equipment cost rate is determined using Equation 6.8. For this purpose, the rate of return and the number of annual operating hours are determined:

$$\begin{cases} n = 20 \\ i = 0.2 \end{cases} \rightarrow CRF = \frac{i(1+i)^n}{(1+i)^n - 1} = \frac{0.2(1+0.2)^{20}}{(1+0.2)^{20} - 1} = 0.20535653$$

$$N = 3\,\text{month} \times \frac{30\,\text{day}}{1\,\text{month}} \times \frac{24\,\text{hour}}{1\,\text{day}} = 2160\,\text{hour}$$

$$\begin{cases} Z = 8900 \\ \varphi = 1.05 \end{cases} \rightarrow \dot{Z} = \frac{Z \times CRF \times \varphi}{N \times 3600} = \frac{8900 \times 0.20535653 \times 1.05}{2160 \times 3600} = 0.00012340\,\text{USD}/s$$

Cost of flows:

$$\dot{C}_1 = c_1\dot{E}x_1 \rightarrow c_1 = \frac{\dot{C}_1}{\dot{E}x_1} = \frac{\dot{C}_1}{\dot{m}_1 ex_1} = \frac{3}{373.66} = 0.00802878\,\text{USD}/\text{kJ}$$

$$\dot{C}_2 = c_2\dot{E}x_2 \rightarrow c_2 = \frac{\dot{C}_2}{\dot{E}x_2} = \frac{\dot{C}_2}{\dot{m}_2 ex_2} = \frac{0.25}{16.50} = 0.01515049\,\text{USD}/\text{kJ}$$

$$\dot{C}_3 = c_3\dot{E}x_3 = c_3\dot{m}_3 ex_3 = 52.15c_3$$

Cost of fuel and product:

$$\dot{C}_F = \dot{C}_1 + \dot{C}_3 = 3.25\,\text{USD}/s$$

$$\dot{C}_P = \dot{C}_3 = 52.15c_2$$

Cost balance:

$$\dot{C}_F + \dot{Z} = \dot{C}_P$$

$$\rightarrow 3.25 + 0.00012340 = 52.15c_3$$

$$\rightarrow c_2 = 0.06232608\,\text{USD}/\text{kJ}$$

Fuel exergy unit cost:

$$c_F = \frac{\dot{C}_F}{\dot{E}x_F} = \frac{3.25}{390.16} = 0.00832998\ \text{USD/kJ}$$

Exergy destruction cost:

$$\dot{C}_{des} = c_F \dot{E}x_{des} = 0.00832998 \times 338.01 = 2.81561585\ \text{USD/s}$$

Total cost rate of operation:

$$\dot{C}_{total} = \dot{C}_{des} + \dot{Z} = 2.81561585 + 0.00012340 = 2.81573924\ \text{USD/s}$$

8.5 EVAPORATORS

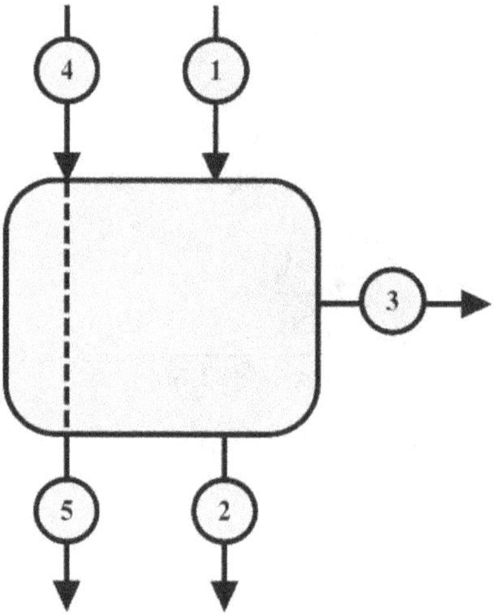

FIGURE 8.4 Schematic of an evaporator.

The cost of mass flows:

$$\begin{aligned} \dot{C}_1 &= c_1 \dot{E}x_1 \\ \dot{C}_2 &= c_2 \dot{E}x_2 \\ \dot{C}_3 &= c_3 \dot{E}x_3 \\ \dot{C}_4 &= c_4 \dot{E}x_4 \\ \dot{C}_5 &= c_5 \dot{E}x_5 \end{aligned} \tag{8.27}$$

Fuel exergy:

$$\dot{C}_F = c_1\dot{E}x_1 + \left(c_4\dot{E}x_4 - c_5\dot{E}x_5\right) \tag{8.28}$$

Auxiliary equation (F rule):

$$c_4 = c_5 \tag{8.29}$$

Fuel exergy unit cost:

$$c_F = \frac{c_1\dot{E}x_1 + \left(c_4\dot{E}x_4 - c_5\dot{E}x_5\right)}{\dot{E}x_1 + \left(\dot{E}x_4 - \dot{E}x_5\right)} \tag{8.30}$$

Product exergy:

$$\dot{C}_P = c_2\dot{E}x_2 + c_3\dot{E}x_3 \tag{8.31}$$

Auxiliary equation (P rule):

$$c_2 = c_3 \tag{8.32}$$

✐**Example 8.5: In Example 6.10, if the cost of the inlet syrup to the evaporator is 0.85 USD/s and the cost of the inlet steam is 0.03 USD/s, calculate the exergy destruction cost and the total cost rate of operation. Consider the equipment cost of 86,000 USD, maintenance cost of 5% of the equipment cost (ϕ = 1.05), factory working period of 6 months, discount rate of 20% (0.2), and number of operation years of the equipment of 20 years.**

✍**Solution**

The equipment cost rate is determined using Equation 6.8. For this purpose, the rate of return and the number of annual operating hours are determined:

$$\begin{cases} n = 20 \\ i = 0.2 \end{cases} \rightarrow CRF = \frac{i(1+i)^n}{(1+i)^n - 1} = \frac{0.2(1+0.2)^{20}}{(1+0.2)^{20} - 1} = 0.20535653$$

$$N = 6\ \text{month} \times \frac{30\ \text{day}}{1\ \text{month}} \times \frac{24\ \text{hour}}{1\ \text{day}} = 4320\ \text{hour}$$

$$\begin{cases} Z = 86000 \\ \varphi = 1.05 \end{cases} \rightarrow \dot{Z} = \frac{Z \times CRF \times \varphi}{N \times 3600} = \frac{86000 \times 0.20535653 \times 1.05}{4320 \times 3600} = 0.00012340\frac{\text{USD}}{\text{s}}$$

Cost of flows:

$$\dot{C}_1 = c_1\dot{E}x_1 \rightarrow c_1 = \frac{\dot{C}_1}{\dot{E}x_1} = \frac{\dot{C}_1}{\dot{m}_1 ex_1} = \frac{0.85}{1916.10} = 0.00044361\frac{\text{USD}}{\text{kJ}}$$

$$\dot{C}_2 = c_2\dot{E}x_2 = c_2\dot{m}_2 ex_2 = 1232.55c_2$$

$$\dot{C}_3 = c_3\dot{E}x_3 = c_3\dot{m}_3 ex_3 = 7897.81c_3$$

$$\dot{C}_4 = c_4\dot{E}x_4 \rightarrow c_4 = \frac{\dot{C}_4}{\dot{E}x_4} = \frac{\dot{C}_4}{\dot{m}_4 ex_4} = \frac{0.03}{8982.47} = 0.00000334\frac{\text{USD}}{\text{kJ}}$$

$$\dot{C}_5 = c_5\dot{E}x_5 = c_5\dot{m}_5 ex_5 = 1012.71c_5$$

Fuel and product cost:

$$\dot{C}_F = \dot{C}_1 + \left(\dot{C}_4 - \dot{C}_5\right)$$

$$\dot{C}_P = \dot{C}_2 + \dot{C}_3$$

Auxiliary equations:

$$c_5 = c_4 = 0.00000334\frac{\text{USD}}{\text{kJ}}\ \{\text{F rule}\}$$

$$c_2 = c_3\ \{\text{P rule}\}$$

Fuel exergy unit cost:

$$\begin{aligned} c_F &= \frac{\dot{C}_F}{\dot{E}x_F} = \frac{\dot{C}_1 + \left(\dot{C}_4 - \dot{C}_5\right)}{\dot{E}x_F} = \frac{\dot{C}_1 + \left(\dot{C}_4 - c_5\dot{E}x_5\right)}{\dot{E}x_F} \\ &= \frac{0.85 + 0.03 - (0.00000334 \times 1012.71)}{9885.86} \\ &= 0.00008867\frac{\text{USD}}{\text{kJ}} \end{aligned}$$

Cost of exergy loss:

$$\dot{C}_L = c_F\dot{E}x_L = 0.00008867 \times 3.15 = 0.00027948\frac{\text{USD}}{\text{s}}$$

Cost balance:

$$\dot{C}_F + \dot{Z} = \dot{C}_P + \dot{C}_L$$

$$\begin{aligned} &\rightarrow \left[0.85 + 0.03 - (0.00000334 \times 1012.71)\right] + 0.00012340 \\ &= 1232.55c_2 + 7897.81c_3 = c_2(1232.55 + 7897.81) \end{aligned}$$

$$\rightarrow c_2 = c_3 = 0.00009624\frac{\text{USD}}{\text{kJ}}$$

Exergy destruction cost:

$$\dot{C}_{des} = c_F \dot{E}x_{des} = 0.00008867 \times 752.35 = 0.06671363 \frac{\text{USD}}{\text{s}}$$

Total cost rate of operation:

$$\dot{C}_{total} = \dot{C}_{des} + \dot{Z} + \dot{C}_L = 0.06671363 + 0.00012340 + 0.00027948$$
$$= 0.06937784 \frac{\text{USD}}{\text{s}}$$

8.6 SEPARATORS

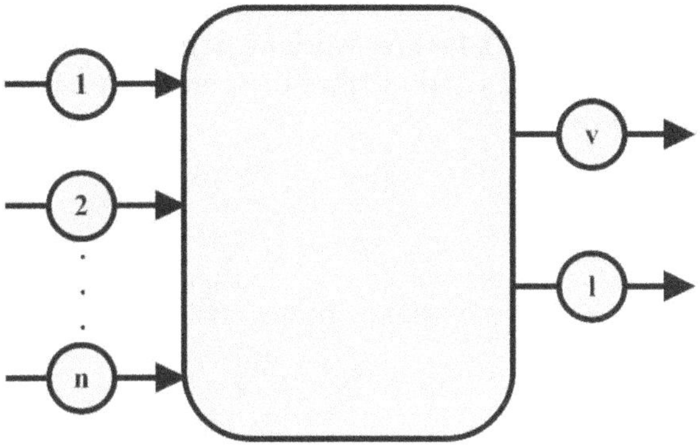

FIGURE 8.5 Schematic of a separator.

The cost of mass flows:

$$\begin{aligned} \dot{C}_1 &= c_1 \dot{E}x_1 \\ \dot{C}_2 &= c_2 \dot{E}x_2 \\ \dot{C}_n &= c_n \dot{E}x_n \\ \dot{C}_v &= c_v \dot{E}x_v \\ \dot{C}_l &= c_l \dot{E}x_l \end{aligned} \quad (8.33)$$

Fuel exergy:

$$\dot{C}_F = \dot{C}_1 + \dot{C}_2 + \ldots + \dot{C}_n \quad (8.34)$$

Fuel exergy unit cost:

$$c_F = \frac{c_1 \dot{E}x_1 + c_2 \dot{E}x_2 \ldots + c_n \dot{E}x_n}{\dot{E}x_1 + \dot{E}x_2 + \ldots + \dot{E}x_n} \quad (8.35)$$

Product exergy:

$$\dot{C}_P = c_v \dot{E}x_v + c_l \dot{E}x_l \tag{8.36}$$

Auxiliary equation (P rule):

$$c_v = c_l \tag{8.37}$$

Example 8.6: In Example 6.12, if the cost of condensates 1, 2, and 3 is 0.3, 0.001, and 0.0001, respectively, calculate the exergy destruction cost and the total cost rate of operation. Consider the equipment cost of 3500 USD, maintenance cost of 5% of the equipment cost (ϕ = 1.05), factory working period of 3 months, interest rate of 20% (0.2), and number of operation years of the equipment of 20 years.

Solution

The equipment cost rate is determined using Equation 6.8. For this purpose, the rate of return and the number of annual operating hours are determined:

$$\begin{cases} n = 20 \\ i = 0.2 \end{cases} \rightarrow CRF = \frac{i(1+i)^n}{(1+i)^n - 1} = \frac{0.2(1+0.2)^{20}}{(1+0.2)^{20} - 1} = 0.20535653$$

$$N = 3\,\text{month} \times \frac{30\,\text{day}}{1\,\text{month}} \times \frac{24\,\text{hour}}{1\,\text{day}} = 2160\,\text{hour}$$

$$\begin{cases} Z = 3500 \\ \varphi = 1.05 \end{cases} \rightarrow \dot{Z} = \frac{Z \times CRF \times \varphi}{N \times 3600} = \frac{3500 \times 0.20535653 \times 1.05}{2160 \times 3600} = 0.00009705 \frac{\text{USD}}{\text{s}}$$

Cost of flows:

$$\dot{C}_1 = c_1 \dot{E}x_1 \rightarrow c_1 = \frac{\dot{C}_1}{\dot{E}x_1} = \frac{\dot{C}_1}{\dot{m}_1 ex_1} = \frac{0.003}{1020.63} = 0.00000294 \frac{\text{USD}}{\text{kJ}}$$

$$\dot{C}_2 = c_2 \dot{E}x_2 \rightarrow c_2 = \frac{\dot{C}_2}{\dot{E}x_2} = \frac{\dot{C}_2}{\dot{m}_2 ex_2} = \frac{0.001}{334.58} = 0.00000299 \frac{\text{USD}}{\text{kJ}}$$

$$\dot{C}_3 = c_3 \dot{E}x_3 \rightarrow c_3 = \frac{\dot{C}_3}{\dot{E}x_3} = \frac{\dot{C}_3}{\dot{m}_3 ex_3} = \frac{0.0001}{39.90} = 0.00000251 \frac{\text{USD}}{\text{kJ}}$$

$$\dot{C}_4 = c_4 \dot{E}x_4 = c_4 \dot{m}_4 ex_4 = 1001.16 c_4$$

$$\dot{C}_5 = c_5 \dot{E}x_5 = c_5 \dot{m}_5 ex_5 = 370.37 c_5$$

Fuel and product cost:

$$\dot{C}_F = \dot{C}_1 + \dot{C}_2 + \dot{C}_3$$
$$\dot{C}_P = \dot{C}_4 + \dot{C}_5$$

Auxiliary equations:

$$c_4 = c_5 \ \{\text{P rule}\}$$

Cost balance:

$$\dot{C}_F + \dot{Z} = \dot{C}_P + \dot{C}_L$$
$$\rightarrow \left[0.85 + 0.03 - (0.00000334 \times 1012.71)\right] + 0.00012340$$
$$= 1232.55c_2 + 7897.81c_3 = c_2(1232.55 + 7897.81)$$
$$\rightarrow c_2 = c_3 = 0.00009624 \frac{\text{USD}}{\text{kJ}}$$

Fuel exergy unit cost:

$$c_F = \frac{\dot{C}_F}{\dot{E}x_F} = \frac{\dot{C}_1 + \dot{C}_2 + \dot{C}_3}{\dot{E}x_F} = \frac{0.003 + 0.001 + 0.0001}{1395.12} = 0.00000294 \frac{\text{USD}}{\text{kJ}}$$

Exergy destruction cost:

$$\dot{C}_{des} = c_F \dot{E}x_{des} = 0.00000294 \times 23.59 = 0.00006933 \frac{\text{USD}}{s}$$

Total cost rate of operation:

$$\dot{C}_{total} = \dot{C}_{des} + \dot{Z} = 0.00006933 + 0.00009705 = 0.00016638 \frac{\text{USD}}{s}$$

8.7 THROTTLING VALVES

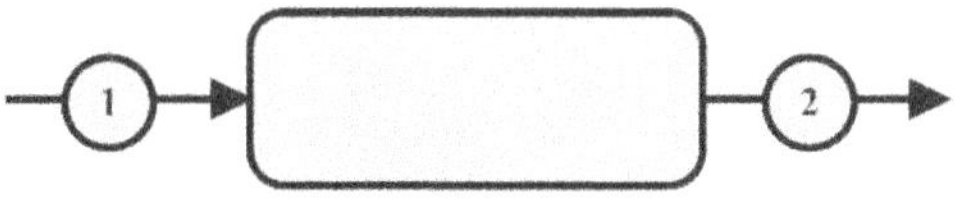

FIGURE 8.6 Schematic of a throttling valve.

The cost of mass flows

$$\dot{C}_1 = c_1 \dot{E}x_1$$

$$\dot{C}_2 = c_2\dot{E}x_2 \tag{8.38}$$

Fuel exergy:

$$\dot{C}_F = c_1\dot{E}x_1 \tag{8.39}$$

Fuel exergy unit cost:

$$c_F = c_1 \tag{8.40}$$

Product exergy:

$$\dot{C}_P = c_2\dot{E}x_2 \tag{8.41}$$

Example 8.7: In Example 6.13, if the unit cost of exergy of steam entering the pressure relief valve is 0.00005 USD/kJ and the steam mass flow rate is 1.42 kg/s, calculate the cost of exergy destruction and the total cost rate of operation. Consider the equipment cost of 180 USD, maintenance cost of 5% of the equipment cost (ϕ = 1.05), factory working period of 6 months, discount rate of 20% (0.2), and number of operation years of the equipment of 20 years.

Solution

$$\begin{cases} n = 20 \\ i = 0.2 \end{cases} \rightarrow CRF = \frac{i(1+i)^n}{(1+i)^n - 1} = \frac{0.2(1+0.2)^{20}}{(1+0.2)^{20} - 1} = 0.20535653$$

$$N = 6\text{ month} \times \frac{30\text{ day}}{1\text{ month}} \times \frac{24\text{ hour}}{1\text{ day}} = 4320\text{ hour}$$

$$\begin{cases} Z = 1800 \\ \varphi = 1.05 \end{cases} \rightarrow \dot{Z} = \frac{Z \times CRF \times \varphi}{N \times 3600} = \frac{180 \times 0.20535653 \times 1.05}{4320 \times 3600} = 0.00000250\text{ USD/s}$$

Cost of flows:

$$\dot{C}_1 = c_1\dot{E}x_1 = c_1\dot{m}_1 ex_1 = 0.00005(1.42)(794.79) = 0.05642987\text{ USD/s}$$

$$\dot{C}_2 = c_2\dot{E}x_2 = c_2\dot{m}_2 ex_2 = 664.70c_2$$

Fuel and product cost:

$$\dot{C}_F = \dot{C}_1 = 0.05642987\text{USD/s}$$

$$\dot{C}_P = \dot{C}_2 = 664.70c_2$$

Cost balance:

$$\dot{C}_F + \dot{Z} = \dot{C}_P$$
$$\rightarrow 0.05642987 + 0.00000250 = 664.70c_2$$
$$\rightarrow c_2 = 0.00005979\ \text{USD/kJ}$$

Fuel exergy unit cost:

$$c_F = c_1 = 0.00005\ \text{USD/kJ}$$

Exergy destruction cost:

$$\dot{C}_{des} = c_F \dot{E}x_{des} = 0.00005 \times 184.72 = 0.00923587\ \text{USD/s}$$

Total cost rate of operation:

$$\dot{C}_{total} = \dot{C}_{des} + \dot{Z} = 0.00923587 + 0.00000250 = 0.00923837\ \text{USD/s}$$

References

Abott, M., & Van Ness, H. C. (1989). *Schaum's outline of thermodynamics with chemical applications*. McGraw Hill Professional, p. 362.

Aboud, S. A., Altemimi, A. B., RS Al-Hilphy, A., Yi-Chen, L., & Cacciola, F. (2019). A comprehensive review on infrared heating applications in food processing. *Molecules, 24*(22), 4125.

Aghbashlo, M., Tabatabaei, M., & Karimi, K. (2016). Exergy-based sustainability assessment of ethanol production via Mucor indicus from fructose, glucose, sucrose, and molasses. *Energy, 98*, 240–252.

Balmer, R. T. (2011). *Modern engineering thermodynamics-textbook with tables booklet*. Academic Press.

Borgnakke, C., & Sonntag, R. E. (2022). *Fundamentals of thermodynamics*. John Wiley & Sons.

Bühler, F., Nguyen, T. V., Jensen, J. K., Holm, F. M., & Elmegaard, B. (2018). Energy, exergy and advanced exergy analysis of a milk processing factory. *Energy, 162*, 576–592.

Cengel, Y. A., & Boles, M. A. (2011). *Thermodynamics: An engineering approach*. McGraw-Hill.

Cimbala, J. M., & Cengel, Y. A. (2006). *Fluid mechanics: Fundamentals and applications*. McGraw-Hill Higher Education.

Dinçer, I. (Ed.). (2018). *Comprehensive energy systems*. Elsevier.

Dinçer, I., & Rosen, M. A. (2015). *Exergy analysis of heating, refrigerating and air conditioning: Methods and applications*. Academic Press.

Dinçer, İ., & Zamfirescu, C. (2016). *Drying phenomena: Theory and applications*. John Wiley & Sons.

Hazervazifeh, A., Nikbakht, A. M., Moghaddam, P. A., & Sharifian, F. (2018). Energy economy and kinetic investigation of sugar cube dehydration using microwave supplemented with thermal imaging. *Journal of Food Processing and Preservation, 42*(2), e13504.

Herfeh, N. S., Mobli, H., Nikbakht, A. M., Keyhani, A., & Piri, A. (2022). Integration of thermal performance of sour cherry concentration plant with 3E procedures: Energy, exergy and exergoeconomy. *Journal of Thermal Analysis and Calorimetry, 147*(19), 10419–10437.

Jacobs, P. (2013). *Thermodynamics*. World Scientific Publishing Company.

Kanoğlu, M., Çengel, Y. A., & Dinçer, İ. (2012). *Efficiency evaluation of energy systems*. Springer Science & Business Media.

Klepzig, L. S., Juckers, A., Knerr, P., Harms, F., & Strube, J. (2020). Digital twin for lyophilization by process modeling in manufacturing of biologics. *Processes, 8*(10), 1325.

Lazzaretto, A., & Tsatsaronis, G. (1999, November). On the calculation of efficiencies and costs in thermal systems. In *ASME international mechanical engineering congress and exposition* (Vol. 16509, pp. 421–430). American Society of Mechanical Engineers.

Lazzaretto, A., & Tsatsaronis, G. (2006). SPECO: A systematic and general methodology for calculating efficiencies and costs in thermal systems. *Energy, 31*(8–9), 1257–1289.

Moran, M. J., Shapiro, H. N., Boettner, D. D., & Bailey, M. B. (2010). *Fundamentals of engineering thermodynamics*. John Wiley & Sons.

Mujumdar, A. S. (2006). Principles, classification, and selection of dryers. In A. S. Mujumdar (Ed.), *Handbook of industrial drying* (3rd ed.). CRC Press.

Piri, A., Hazervazifeh, A., & Nikbakht, A. M. (2022). A comprehensive analysis of uncertainty in thermal utilities: 3E case study of a sugar plant. *Sustainable Energy Technologies and Assessments, 53*, 102498.

Piri, A., Nikbakht, A. M., Hazervazifeh, A., Sarnavi, H. J., & Fanaei, A. R. (2021). Role of 3E analysis in detection of thermodynamic losses in the evaporation line and steam and power production unit in the sugar processing plant. *Biomass Conversion and Biorefinery*, 1–21.

Potter, M. C., & Somerton, C. W. (2014). *Schaum's outline of thermodynamics for engineers.* McGraw-Hill Education.

Qiu, J. (2019). *Mild conductive drying of foods* [Doctoral dissertation, Wageningen University and Research].

Shavit, A., & Gutfinger, C. (2008). *Thermodynamics: From concepts to applications.* CRC Press.

Shukla, A., & Kumar, D. S. Y. (2017). A review on exergy, life cycle and thermo economic analysis of sugar industry. *International Journal of Mechanical Engineering and Technology, 8*(10).

Singh, G., Singh, P. J., Tyagi, V. V., Barnwal, P., & Pandey, A. K. (2019). Exergy and thermoeconomic analysis of cream pasteurisation plant. *Journal of Thermal Analysis and Calorimetry, 137,* 1381–1400.

Singh, R. P., & Heldman, D. R. (2001). *Introduction to food engineering.* Gulf Professional Publishing.

Varzakas, T., & Tzia, C. (Eds.). (2014). *Food engineering handbook: Food engineering fundamentals.* CRC Press.

Yildirim, N., & Genc, S. (2017). Energy and exergy analysis of a milk powder production system. *Energy Conversion and Management, 149*, 698–705.

Appendix 1: Thermodynamic Properties Used in the Examples

TABLE A.1

Saturated Water Characteristics from Temperature

		Specific Volume, m^3/kg		Internal Energy, kJ/kg			Enthalpy, kJ/kg			Entropy, kJ/kg.K		
Temp.,	**Sat Press.,**	**Liquid,**	**Steam,**	**Sat Liquid,**	**Evap.,**	**Sat Steam,**	**Sat Liquid,**	**Evap.,**	**Sat Steam,**	**Sat Liquid,**	**Evap.,**	**Sat Steam,**
T °C	*Psat kPa*	*vf*	*vg*	*uf*	*ufg*	*ug*	*hf*	*hfg*	*hg*	*sf*	*sfg*	*sg*
0.01	0.6117	0.001	206	0	2374.9	2374.9	0.001	2500.9	2500.9	0	9.1556	9.1556
5	0.8725	0.001	147.03	21.019	2360.8	2381.8	21.02	2489.1	2510.1	0.0763	8.9487	9.0249
10	1.2281	0.001	106.32	42.02	2346.6	2388.7	42.022	2477.2	2519.2	0.1511	8.7488	8.8999
15	1.7057	0.001001	77.885	62.98	2332.5	2395.5	62.982	2465.4	2528.3	0.2245	8.5559	8.7803
20	2.3392	0.001002	57.762	83.913	2318.4	2402.3	83.915	2453.5	2537.4	0.2965	8.3696	8.6661
25	3.1698	0.001003	43.34	104.83	2304.3	2409.1	104.83	2441.7	2546.5	0.3672	8.1895	8.5567
30	4.2469	0.001004	32.879	125.73	2290.2	2415.9	125.74	2429.8	2555.6	0.4368	8.0152	8.452
35	5.6291	0.001006	25.205	146.63	2276	2422.7	146.64	2417.9	2564.6	0.5051	7.8466	8.3517
40	7.3851	0.001008	19.515	167.53	2261.9	2429.4	167.53	2406	2573.5	0.5724	7.6832	8.2556
45	9.5953	0.00101	15.251	188.43	2247.7	2436.1	188.44	2394	2582.4	0.6386	7.5247	8.1633
50	12.352	0.001012	12.026	209.33	2233.4	2442.7	209.34	2382	2591.3	0.7038	7.371	8.0748

(Continued)

TABLE A.1 (*Continued*)

Saturated Water Characteristics from Temperature

Temp.,	Sat Press.,	Specific Volume, m³/kg		Internal Energy, kJ/kg			Enthalpy, kJ/kg			Entropy, kJ/kg.K		
		Liquid,	Steam,	Sat Liquid,	Evap.,	Sat Steam,	Sat Liquid,	Evap.,	Sat Steam,	Sat Liquid,	Evap.,	Sat Steam,
T °C	*Psat kPa*	*vf*	*vg*	*uf*	*ufg*	*ug*	*hf*	*hfg*	*hg*	*sf*	*sfg*	*sg*
55	15.763	0.001015	9.5639	230.24	2219.1	2449.3	230.26	2369.8	2600.1	0.768	7.2218	7.9898
60	19.947	0.001017	7.667	251.16	2204.7	2455.9	251.18	2357.7	2608.8	0.8313	7.0769	7.9082
65	25.043	0.00102	6.1935	272.09	2190.3	2462.4	272.12	2345.4	2617.5	0.8937	6.936	7.8296
70	31.202	0.001023	5.0396	293.04	2175.8	2468.9	293.07	2333	2626.1	0.9551	6.7989	7.754
75	38.597	0.001026	4.1291	313.99	2161.3	2475.3	314.03	2320.6	2634.6	1.0158	6.6655	7.6812
80	47.416	0.001029	3.4053	334.97	2146.6	2481.6	335.02	2308	2643	1.0756	6.5355	7.6111
85	57.868	0.001032	2.8261	355.96	2131.9	2487.8	356.02	2295.3	2651.4	1.1346	6.4089	7.5435
90	70.183	0.001036	2.3593	376.97	2117	2494	377.04	2282.5	2659.6	1.1929	6.2853	7.4782
95	84.609	0.00104	1.9808	398	2102	2500.1	398.09	2269.6	2667.6	1.2504	6.1647	7.4151
100	101.42	0.001043	1.672	419.06	2087	2506	419.17	2256.4	2675.6	1.3072	6.047	7.3542
105	120.9	0.001047	1.4186	440.15	2071.8	2511.9	440.28	2243.1	2683.4	1.3634	5.9319	7.2952
110	143.38	0.001052	1.2094	461.27	2056.4	2517.7	461.42	2229.7	2691.1	1.4188	5.8193	7.2382
115	169.18	0.001056	1.036	482.42	2040.9	2523.3	482.59	2216	2698.6	1.4737	5.7092	7.1829
120	198.67	0.00106	0.89133	503.6	2025.3	2528.9	503.81	2202.1	2706	1.5279	5.6013	7.1292
125	232.23	0.001065	0.77012	524.83	2009.5	2534.3	525.07	2188.1	2713.1	1.5816	5.4956	7.0771
130	270.28	0.00107	0.66808	546.1	1993.4	2539.5	546.38	2173.7	2720.1	1.6346	5.3919	7.0265
135	313.22	0.001075	0.58179	567.41	1977.3	2544.7	567.75	2159.1	2726.9	1.6872	5.2901	6.9773
140	361.53	0.00108	0.5085	588.77	1960.9	2549.6	589.16	2144.3	2733.5	1.7392	5.1901	6.9294
145	415.68	0.001085	0.446	610.19	1944.2	2554.4	610.64	2129.2	2739.8	1.7908	5.0919	6.8827
150	476.16	0.001091	0.39248	631.66	1927.4	2559.1	632.18	2113.8	2745.9	1.8418	4.9953	6.8371
155	543.49	0.001096	0.34648	653.19	1910.3	2563.5	653.79	2098	2751.8	1.8924	4.9002	6.7927
160	618.23	0.001102	0.3068	674.79	1893	2567.8	675.47	2082	2757.5	1.9426	4.8066	6.7492

165	700.93	0.001108	0.27244	696.46	1875.4	2571.9	697.24	2065.6	2762.8	1.9923	4.7143	6.7067
170	792.18	0.001114	0.2426	718.2	1857.5	2575.7	719.08	2048.8	2767.9	2.0417	4.6233	6.665
175	892.6	0.001121	0.21659	740.02	1839.4	2579.4	741.02	2031.7	2772.7	2.0906	4.5335	6.6242
180	1002.8	0.001127	0.19384	761.92	1820.9	2582.8	763.05	2014.2	2777.2	2.1392	4.4448	6.5841
185	1123.5	0.001134	0.1739	783.91	1802.1	2586	785.19	1996.2	2781.4	2.1875	4.3572	6.5447
190	1255.2	0.001141	0.15636	806	1783	2589	807.43	1977.9	2785.3	2.2355	4.2705	6.5059
195	1398.8	0.001149	0.14089	828.18	1763.6	2591.7	829.78	1959	2788.8	2.2831	4.1847	6.4678
200	1554.9	0.001157	0.12721	850.46	1743.7	2594.2	852.26	1939.8	2792	2.3305	4.0997	6.4302
205	1724.3	0.001164	0.11508	872.86	1723.5	2596.4	874.87	1920	2794.8	2.3776	4.0154	6.393
210	1907.7	0.001173	0.10429	895.38	1702.9	2598.3	897.61	1899.7	2797.3	2.4245	3.9318	6.3563
215	2105.9	0.001181	0.09468	918.02	1681.9	2599.9	920.5	1878.8	2799.3	2.4712	3.8489	6.32
220	2319.6	0.00119	0.086094	940.79	1660.5	2601.3	943.55	1857.4	2801	2.5176	3.7664	6.284
225	2549.7	0.001199	0.078405	963.7	1638.6	2602.3	966.76	1835.4	2802.2	2.5639	3.6844	6.2483
230	2797.1	0.001209	0.071505	986.76	1616.1	2602.9	990.14	1812.8	2802.9	2.61	3.6028	6.2128
235	3062.6	0.001219	0.0653	1010	1593.2	2603.2	1013.7	1789.5	2803.2	2.656	3.5216	6.1775
240	3347	0.001229	0.059707	1033.4	1569.8	2603.1	1037.5	1765.5	2803	2.7018	3.4405	6.1424
245	3651.2	0.00124	0.054656	1056.9	1545.7	2602.7	1061.5	1740.8	2802.2	2.7476	3.3596	6.1072
250	3976.2	0.001252	0.050085	1080.7	1521.1	2601.8	1085.7	1715.3	2801	2.7933	3.2788	6.0721
255	4322.9	0.001263	0.045941	1104.7	1495.8	2600.5	1110.1	1689	2799.1	2.839	3.1979	6.0369
260	4692.3	0.001276	0.042175	1128.8	1469.9	2598.7	1134.8	1661.8	2796.6	2.8847	3.1169	6.0017
265	5085.3	0.001289	0.038748	1153.3	1443.2	2596.5	1159.8	1633.7	2793.5	2.9304	3.0358	5.9662
270	5503	0.001303	0.035622	1177.9	1415.7	2593.7	1185.1	1604.6	2789.7	2.9762	2.9542	5.9305
275	5946.4	0.001317	0.032767	1202.9	1387.4	2590.3	1210.7	1574.5	2785.2	3.0221	2.8723	5.8944
280	6416.6	0.001333	0.030153	1228.2	1358.2	2586.4	1236.7	1543.2	2779.9	3.0681	2.7898	5.8579
285	6914.6	0.001349	0.027756	1253.7	1328.1	2581.8	1263.1	1510.7	2773.7	3.1144	2.7066	5.821
290	7441.8	0.001366	0.025554	1279.7	1296.9	2576.5	1289.8	1476.9	2766.7	3.1608	2.6225	5.7834
295	7999	0.001384	0.023528	1306	1264.5	2570.5	1317.1	1441.6	2758.7	3.2076	2.5374	5.745
300	8587.9	0.001404	0.021659	1332.7	1230.9	2563.6	1344.8	1404.8	2749.6	3.2548	2.4511	5.7059

(Continued)

TABLE A.1 (*Continued*)

Saturated Water Characteristics from Temperature

		Specific Volume, m^3/kg		Internal Energy, kJ/kg			Enthalpy, kJ/kg			Entropy, kJ/kg.K		
Temp.,	Sat Press.,	Liquid,	Steam,	Sat Liquid,	Evap.,	Sat Steam,	Sat Liquid,	Evap.,	Sat Steam,	Sat Liquid,	Evap.,	Sat Steam,
T °C	*Psat kPa*	*vf*	*vg*	*uf*	*ufg*	*ug*	*hf*	*hfg*	*hg*	*sf*	*sfg*	*sg*
305	9209.4	0.001425	0.019932	1360	1195.9	2555.8	1373.1	1366.3	2739.4	3.3024	2.3633	5.6657
310	9865	0.001447	0.018333	1387.7	1159.3	2547.1	1402	1325.9	2727.9	3.3506	2.2737	5.6243
315	10.556	0.001472	0.016849	1416.1	1121.1	2537.2	1431.6	1283.4	2715	3.3994	2.1821	5.5816
320	11.284	0.001499	0.01547	1445.1	1080.9	2526	1462	1238.5	2700.6	3.4491	2.0881	5.5372
325	12.051	0.001528	0.014183	1475	1038.5	2513.4	1493.4	1191	2684.3	3.4998	1.9911	5.4908
330	12.858	0.00156	0.012979	1505.7	993.5	2499.2	1525.8	1140.3	2666	3.5516	1.8906	5.4422
335	13.707	0.001597	0.011848	1537.5	945.5	2483	1559.4	1086	2645.4	3.605	1.7857	5.3907
340	14.601	0.001638	0.010783	1570.7	893.8	2464.5	1594.6	1027.4	2622	3.6602	1.6756	5.3358
345	15.541	0.001685	0.009772	1605.5	837.7	2443.2	1631.7	963.4	2595.1	3.7179	1.5585	5.2765
350	16.529	0.001741	0.008806	1642.4	775.9	2418.3	1671.2	892.7	2563.9	3.7788	1.4326	5.2114
355	17.570	0.001808	0.007872	1682.2	706.4	2388.6	1714	812.9	2526.9	3.8442	1.2942	5.1384
360	18.666	0.001895	0.00695	1726.2	625.7	2351.9	1761.5	720.1	2481.6	3.9165	1.1373	5.0537
365	19.822	0.002015	0.006009	1777.2	526.4	2303.6	1817.2	605.5	2422.7	4.0004	0.9489	4.9493
370	21.044	0.002217	0.004953	1844.5	385.6	2230.1	1891.2	443.1	2334.3	4.1119	0.689	4.8009
373.95	22.064	0.003106	0.003106	2015.7	0	2015.7	0	2084.3	2084.3	4.407	0	4.407

TABLE A.2

Saturated Water Characteristics from Pressure

		Specific Volume, m³/kg		Internal Energy, kJ/kg			Enthalpy, kJ/kg			Entropy, kJ/kg.K		
Press.,	**Temp.,**	**Liquid,**	**Steam,**	**Sat Liquid,**	**Evap.,**	**Sat Steam,**	**Sat Liquid,**	**Evap.,**	**Sat Steam,**	**Sat Liquid,**	**Evap.,**	**Sat Steam,**
P kPa	***Tsat °C***	***vf***	***vg***	***uf***	***ufg***	***ug***	***hf***	***hfg***	***hg***	***sf***	***sfg***	***sg***
1	6.97	0.001	129.19	29.302	2355.2	2384.5	29.303	2484.4	2513.7	0.1059	8.869	8.9749
1.5	13.02	0.001001	87.964	54.686	2338.1	2392.8	54.688	2470.1	2524.7	0.1956	8.6314	8.827
2	17.5	0.001001	66.99	73.431	2325.5	2398.9	73.433	2459.5	2532.9	0.2606	8.4621	8.7227
2.5	21.08	0.001002	54.242	88.422	2315.4	2403.8	88.424	2451	2539.4	0.3118	8.3302	8.6421
3	24.08	0.001003	45.654	100.98	2306.9	2407.9	100.98	2443.9	2544.8	0.3543	8.2222	8.5765
4	28.96	0.001004	34.791	121.39	2293.1	2414.5	121.39	2432.3	2553.7	0.4224	8.051	8.4734
5	32.87	0.001005	28.185	137.75	2282.1	2419.8	137.75	2423	2560.7	0.4762	7.9176	8.3938
7.5	40.29	0.001008	19.233	168.74	2261.1	2429.8	168.75	2405.3	2574	0.5763	7.6738	8.2501
10	45.81	0.00101	14.67	191.79	2245.4	2437.2	191.81	2392.1	2583.9	0.6492	7.4996	8.1488
15	53.97	0.001014	10.02	225.93	2222.1	2448	225.94	2372.3	2598.3	0.7549	7.2522	8.0071
20	60.06	0.001017	7.6481	251.4	2204.6	2456	251.42	2357.5	2608.9	0.832	7.0752	7.9073
25	64.96	0.00102	6.2034	271.93	2190.4	2462.4	271.96	2345.5	2617.5	0.8932	6.937	7.8302
30	69.09	0.001022	5.2287	289.24	2178.5	2467.7	289.27	2335.3	2624.6	0.9441	6.8234	7.7675
40	75.86	0.001026	3.9933	317.58	2158.8	2476.3	317.62	2318.4	2636.1	1.0261	6.643	7.6691
50	81.32	0.00103	3.2403	340.49	2142.7	2483.2	340.54	2304.7	2645.2	1.0912	6.5019	7.5931
75	91.76	0.001037	2.2172	384.36	2111.8	2496.1	384.44	2278	2662.4	1.2132	6.2426	7.4558
100	99.61	0.001043	1.6941	417.4	2088.2	2505.6	417.51	2257.5	2675	1.3028	6.0562	7.3589
101.325	99.97	0.001043	1.6734	418.95	2087	2506	419.06	2256.5	2675.6	1.3069	6.0476	7.3545
125	105.97	0.001048	1.375	444.23	2068.8	2513	444.36	2240.6	2684.9	1.3741	5.91	7.2841
150	111.35	0.001053	1.1594	466.97	2052.3	2519.2	467.13	2226	2693.1	1.4337	5.7894	7.2231
175	116.04	0.001057	1.0037	486.82	2037.7	2524.5	487.01	2213.1	2700.2	1.485	5.6865	7.1716

(Continued)

TABLE A.2 *(Continued)*

Saturated Water Characteristics from Pressure

		Specific Volume, m^3/kg		Internal Energy, kJ/kg			Enthalpy, kJ/kg			Entropy, kJ/kg.K		
Press.,	Temp.,	Liquid,	Steam,	Sat Liquid,	Evap.,	Sat Steam,	Sat Liquid,	Evap.,	Sat Steam,	Sat Liquid,	Evap.,	Sat Steam,
P kPa	*Tsat °C*	*vf*	*vg*	*uf*	*ufg*	*ug*	*hf*	*hfg*	*hg*	*sf*	*sfg*	*sg*
200	120.21	0.001061	0.88578	504.5	2024.6	2529.1	504.71	2201.6	2706.3	1.5302	5.5968	7.127
225	123.97	0.001064	0.79329	520.47	2012.7	2533.2	520.71	2191	2711.7	1.5706	5.5171	7.0877
250	127.41	0.001067	0.71873	535.08	2001.8	2536.8	535.35	2181.2	2716.5	1.6072	5.4453	7.0525
275	130.58	0.00107	0.65732	548.57	1991.6	2540.1	548.86	2172	2720.9	1.6408	5.38	7.0207
300	133.52	0.001073	0.60582	561.11	1982.1	2543.2	561.43	2163.5	2724.9	1.6717	5.32	6.9917
325	136.27	0.001076	0.56199	572.84	1973.1	2545.9	573.19	2155.4	2728.6	1.7005	5.2645	6.965
350	138.86	0.001079	0.52422	583.89	1964.6	2548.5	584.26	2147.7	2732	1.7274	5.2128	6.9402
375	141.3	0.001081	0.49133	594.32	1956.6	2550.9	594.73	2140.4	2735.1	1.7526	5.1645	6.9171
400	143.61	0.001084	0.46242	604.22	1948.9	2553.1	604.66	2133.4	2738.1	1.7765	5.1191	6.8955
450	147.9	0.001088	0.41392	622.65	1934.5	2557.1	623.14	2120.3	2743.4	1.8205	5.0356	6.8561
500	151.83	0.001093	0.37483	639.54	1921.2	2560.7	640.09	2108	2748.1	1.8604	4.9603	6.8207
550	155.46	0.001097	0.34261	655.16	1908.8	2563.9	655.77	2096.6	2752.4	1.897	4.8916	6.7886
600	158.83	0.001101	0.3156	669.72	1897.1	2566.8	670.38	2085.8	2756.2	1.9308	4.8285	6.7593
650	161.98	0.001104	0.2926	683.37	1886.1	2569.4	684.08	2075.5	2759.6	1.9623	4.7699	6.7322
700	164.95	0.001108	0.27278	696.23	1875.6	2571.8	697	2065.8	2762.8	1.9918	4.7153	6.7071
750	167.75	0.001111	0.25552	708.40	1865.6	2574.0	709.24	2056.4	2765.7	2.0195	4.6642	6.6837
800	170.41	0.001115	0.24035	719.97	1856.1	2576	720.87	2047.5	2768.3	2.0457	4.616	6.6616
850	172.94	0.001118	0.2269	731	1846.9	2577.9	731.95	2038.8	2770.8	2.0705	4.5705	6.6409
900	175.35	0.001121	0.21489	741.55	1838.1	2579.6	742.56	2030.5	2773	2.0941	4.5273	6.6213
950	177.66	0.001124	0.20411	751.67	1829.6	2581.3	752.74	2022.4	2775.2	2.1166	4.4862	6.6027
1000	179.88	0.001127	0.19436	761.39	1821.4	2582.8	762.51	2014.6	2777.1	2.1381	4.447	6.585
1100	184.06	0.001133	0.17745	779.78	1805.7	2585.5	781.03	1999.6	2780.7	2.1785	4.3735	6.552
1200	187.96	0.001138	0.16326	796.96	1790.9	2587.8	798.33	1985.4	2783.8	2.2159	4.3058	6.5217
1300	191.6	0.001144	0.15119	813.1	1776.8	2589.9	814.59	1971.9	2786.5	2.2508	4.2428	6.4936

1400	195.04	0.001149	0.14078	828.35	1763.4	2591.8	829.96	1958.9	2788.9	2.2835	4.184	6.4675
1500	198.29	0.001154	0.13171	842.82	1750.6	2593.4	844.55	1946.4	2791	2.3143	4.1287	6.443
1750	205.72	0.001166	0.11344	876.12	1720.6	2596.7	878.16	1917.1	2795.2	2.3844	4.0033	6.3877
2000	212.38	0.001177	0.099587	906.12	1693	2599.1	908.47	1889.8	2798.3	2.4467	3.8923	6.339
2250	218.41	0.001187	0.088717	933.54	1667.3	2600.9	936.21	1864.3	2800.5	2.5029	3.7926	6.2954
2500	223.95	0.001197	0.079952	958.87	1643.2	2602.1	961.87	1840.1	2801.9	2.5542	3.7016	6.2558
3000	233.85	0.001217	0.066667	1004.6	1598.5	2603.2	1008.3	1794.9	2803.2	2.6454	3.5402	6.1856
3500	242.56	0.001235	0.057061	1045.4	1557.6	2603	1049.7	1753	2802.7	2.7253	3.3991	6.1244
4000	250.35	0.001252	0.049779	1082.4	1519.3	2601.7	1087.4	1713.5	2800.8	2.7966	3.2731	6.0696
5000	263.94	0.001286	0.039448	1148.1	1448.9	2597	1154.5	1639.7	2794.2	2.9207	3.053	5.9737
6000	275.59	0.001319	0.032449	1205.8	1384.1	2589.9	1213.8	1570.9	2784.6	3.0275	2.8627	5.8902
7000	285.83	0.001352	0.027378	1258	1323	2581	1267.5	1505.2	2772.6	3.122	2.6927	5.8148
8000	295.01	0.001384	0.023525	1306	1264.5	2570.5	1317.1	1441.6	2758.7	3.2077	2.5373	5.745
9000	303.35	0.001418	0.020489	1350.9	1207.6	2558.5	1363.7	1379.3	2742.9	3.2866	2.3925	5.6791
10.000	311	0.001452	0.018028	1393.3	1151.8	2545.2	1407.8	1317.6	2725.5	3.3603	2.2556	5.6159
11.000	318.08	0.001488	0.015988	1433.9	1096.6	2530.4	1450.2	1256.1	2706.3	3.4299	2.1245	5.5544
12.000	324.68	0.001526	0.014264	1473	1041.3	2514.3	1491.3	1194.1	2685.4	3.4964	1.9975	5.4939
13.000	330.85	0.001566	0.012781	1511	985.5	2496.6	1531.4	1131.3	2662.7	3.5606	1.873	5.4336
14.000	336.67	0.00161	0.011487	1548.4	928.7	2477.1	1571	1067	2637.9	3.6232	1.7497	5.3728
15.000	342.16	0.001657	0.010341	1585.5	870.3	2455.7	1610.3	1000.5	2610.8	3.6848	1.6261	5.3108
16.000	347.36	0.00171	0.009312	1622.6	809.4	2432	1649.9	931.1	2581	3.7461	1.5005	5.2466
17.000	352.29	0.00177	0.008374	1660.2	745.1	2405.4	1690.3	857.4	2547.7	3.8082	1.3709	5.1791
18.000	356.99	0.00184	0.007504	1699.1	675.9	2375	1732.2	777.8	2510	3.872	1.2343	5.1064
19.000	361.47	0.001926	0.006677	1740.3	598.9	2339.2	1776.8	689.2	2466	3.9396	1.086	5.0256
20.000	365.75	0.002038	0.005862	1785.8	509	2294.8	1826.6	585.5	2412.1	4.0146	0.9164	4.931
21.000	369.83	0.002207	0.004994	1841.6	391.9	2233.5	1888	450.4	2338.4	4.1071	0.7005	4.8076
22.000	373.71	0.002703	0.003644	1951.7	140.8	2092.4	2011.1	161.5	2172.6	4.2942	0.2496	4.5439
22.064	373.95	0.003106	0.003106	2015.7	0	2015.7	2084.3	0	2084.3	4.407	0	4.407

TABLE A.3
Superheated Water

T	v	u	h	s	v	U	h	s	v	u	h	s
°C	m³/kg	kJ/kg	kJ/kg	kJ/kg K	m³/kg	kJ/kg	kJ/kg	kJ/kg K	m3/kg	kJ/kg	kJ/kg	kJ/kg K
		P = 0.01 MPa (45.81°C)				P = 0.05 MPa (81.32°C)				P = 0.10 MPa (99.61°C)		
Sat.[†]	14.67	2437.2	2583.9	8.1488	3.2403	2483.2	2645.2	7.5931	1.6941	2505.6	2675	7.3589
50	14.867	2443.3	2592	8.1741								
100	17.196	2515.5	2687.5	8.4489	3.4187	2511.5	2682.4	7.6953	1.6959	2506.2	2675.8	7.3611
150	19.513	2587.9	2783	8.6893	3.8897	2585.7	2780.2	7.9413	1.9367	2582.9	2776.6	7.6148
200	21.826	2661.4	2879.6	8.9049	4.3562	2660	2877.8	8.1592	2.1724	2658.2	2875.5	7.8356
250	24.136	2736.1	2977.5	9.1015	4.8206	2735.1	2976.2	8.3568	2.4062	2733.9	2974.5	8.0346
300	26.446	2812.3	3076.7	9.2827	5.2841	2811.6	3075.8	8.5387	2.6389	2810.7	3074.5	8.2172
400	31.063	2969.3	3280	9.6094	6.2094	2968.9	3279.3	8.8659	3.1027	2968.3	3278.6	8.5452
500	35.68	3132.9	3489.7	9.8998	7.1338	3132.6	3489.3	9.1566	3.5655	3132.2	3488.7	8.8362
600	40.296	3303.3	3706.3	10.1631	8.0577	3303.1	3706	9.4201	4.0279	3302.8	3705.6	9.0999
700	44.911	3480.8	3929.9	10.4056	8.9813	3480.6	3929.7	9.6626	4.49	3480.4	3929.4	9.3424
800	49.527	3665.4	4160.6	10.6312	9.9047	3665.2	4160.4	9.8883	4.9519	3665	4160.2	9.5682
900	54.143	3856.9	4398.3	10.8429	10.828	3856.8	4398.2	10.1	5.4137	3856.7	4398	9.78
1000	58.758	4055.3	4642.8	11.0429	11.7513	4055.2	4642.7	10.3	5.8755	4055	4642.6	9.98
1100	63.373	4260	4893.8	11.2326	12.6745	4259.9	4893.7	10.4897	6.3372	4259.8	4893.6	10.1698
1200	67.989	4470.9	5150.8	11.4132	13.5977	4470.8	5150.7	10.6704	6.7988	4470.7	5150.6	10.3504
1300	72.604	4687.4	5413.4	11.5857	14.5209	4687.3	5413.3	10.8429	7.2605	4687.2	5413.3	10.5229
		P 0.20 MPa (120.21°C)				P 0.30 MPa (133.52°C)				P 0.40 MPa (143.61°C)		
Sat.	0.88578	2529.1	2706.3	7.127	0.60582	2543.2	2724.9	6.9917	0.46242	2553.1	2738.1	6.8955
150	0.95986	2577.1	2769.1	7.281	0.63402	2571	2761.2	7.0792	0.47088	2564.4	2752.8	6.9306
200	1.08049	2654.6	2870.7	7.5081	0.71643	2651	2865.9	7.3132	0.53434	2647.2	2860.9	7.1723
250	1.1989	2731.4	2971.2	7.71	0.79645	2728.9	2967.9	7.518	0.5952	2726.4	2964.5	7.3804

300	1.31623	2808.8	3072.1	7.8941	0.87535	2807	3069.6	7.7037	0.65489	2805.1	3067.1	7.5677
400	1.54934	2967.2	3277	8.2236	1.03155	2966	3275.5	8.0347	0.77265	2964.9	3273.9	7.9003
500	1.78142	3131.4	3487.7	8.5153	1.18672	3130.6	3486.6	8.3271	0.88936	3129.8	3485.5	8.1933
600	2.01302	3302.2	3704.8	8.7793	1.34139	3301.6	3704	8.5915	1.00558	3301	3703.3	8.458
700	2.24434	3479.9	3928.8	9.0221	1.4958	3479.5	3928.2	8.8345	1.12152	3479	3927.6	8.7012
800	2.4755	3664.7	4159.8	9.2479	1.65004	3664.3	4159.3	9.0605	1.2373	3663.9	4158.9	8.9274
900	2.70656	3856.3	4397.7	9.4598	1.80417	3856	4397.3	9.2725	1.35298	3855.7	4396.9	9.1394
1000	2.93755	4054.8	4642.3	9.6599	1.95824	4054.5	4642	9.4726	1.46859	4054.3	4641.7	9.3396
1100	3.16848	4259.6	4893.3	9.8497	2.11226	4259.4	4893.1	9.6624	1.58414	4259.2	4892.9	9.5295
1200	3.39938	4470.5	5150.4	10.0304	2.26624	4470.3	5150.2	9.8431	1.69966	4470.2	5150	9.7102
1300	3.63026	4687.1	5413.1	10.2029	2.42019	4686.9	5413	10.0157	1.81516	4686.7	5412.8	9.8828
		P 0.50 MPa (151.83°C)				*P* 0.60 MPa (158.83°C)				*P* 0.80 MPa (170.41°C)		
Sat.	0.37483	2560.7	2748.1	6.8207	0.3156	2566.8	2756.2	6.7593	0.24035	2576	2768.3	6.6616
200	0.42503	2643.3	2855.8	7.061	0.35212	2639.4	2850.6	6.9683	0.26088	2631.1	2839.8	6.8177
250	0.47443	2723.8	2961	7.2725	0.3939	2721.2	2957.6	7.1833	0.29321	2715.9	2950.4	7.0402
300	0.52261	2803.3	3064.6	7.4614	0.43442	2801.4	3062	7.374	0.32416	2797.5	3056.9	7.2345
350	0.57015	2883	3168.1	7.6346	0.47428	2881.6	3166.1	7.5481	0.35442	2878.6	3162.2	7.4107
400	0.61731	2963.7	3272.4	7.7956	0.51374	2962.5	3270.8	7.7097	0.38429	2960.2	3267.7	7.5735
500	0.71095	3129	3484.5	8.0893	0.592	3128.2	3483.4	8.0041	0.44332	3126.6	3481.3	7.8692
600	0.80409	3300.4	3702.5	8.3544	0.66976	3299.8	3701.7	8.2695	0.50186	3298.7	3700.1	8.1354
700	0.89696	3478.6	3927	8.5978	0.74725	3478.1	3926.4	8.5132	0.56011	3477.2	3925.3	8.3794
800	0.98966	3663.6	4158.4	8.824	0.82457	3663.2	4157.9	8.7395	0.6182	3662.5	4157	8.6061
900	1.08227	3855.4	4396.6	9.0362	0.90179	3855.1	4396.2	8.9518	0.67619	3854.5	4395.5	8.8185
1000	1.1748	4054	4641.4	9.2364	0.97893	4053.8	4641.1	9.1521	0.73411	4053.3	4640.5	9.0189
1100	1.26728	4259	4892.6	9.4263	1.05603	4258.8	4892.4	9.342	0.79197	4258.3	4891.9	9.209
1200	1.35972	4470	5149.8	9.6071	1.13309	4469.8	5149.6	9.5229	0.8498	4469.4	5149.3	9.3898
1300	1.45214	4686.6	5412.6	9.7797	1.21012	4686.4	5412.5	9.6955	0.90761	4686.1	5412.2	9.5625

(Continued)

TABLE A.3 (*Continued*)

Superheated water

T	*v*	*u*	*h*	*s*	*v*	*u*	*h*	*s*	*v*	*u*	*h*	*s*
°C	m³/kg	kJ/kg	kJ/kg	kJ/kg.K	m³/kg	kJ/kg	kJ/kg	kJ/kg.K	m3/kg	kJ/kg	kJ/kg	kJ/kg.K
		P 1.00 MPa (179.88C)				P 1.20 MPa (187.96C)				P 1.40 MPa (195.04C)		
Sat.	0.19437	2582.8	2777.1	6.585	0.16326	2587.8	2783.8	6.5217	0.14078	2591.8	2788.9	6.4675
200	0.20602	2622.3	2828.3	6.6956	0.16934	2612.9	2816.1	6.5909	0.14303	2602.7	2803	6.4975
250	0.23275	2710.4	2943.1	6.9265	0.19241	2704.7	2935.6	6.8313	0.16356	2698.9	2927.9	6.7488
300	0.25799	2793.7	3051.6	7.1246	0.21386	2789.7	3046.3	7.0335	0.18233	2785.7	3040.9	6.9553
350	0.2825	2875.7	3158.2	7.3029	0.23455	2872.7	3154.2	7.2139	0.20029	2869.7	3150.1	7.1379
400	0.30661	2957.9	3264.5	7.467	0.25482	2955.5	3261.3	7.3793	0.21782	2953.1	3258.1	7.3046
500	0.35411	3125	3479.1	7.7642	0.29464	3123.4	3477	7.6779	0.25216	3121.8	3474.8	7.6047
600	0.40111	3297.5	3698.6	8.0311	0.33395	3296.3	3697	7.9456	0.28597	3295.1	3695.5	7.873
700	0.44783	3476.3	3924.1	8.2755	0.37297	3475.3	3922.9	8.1904	0.31951	3474.4	3921.7	8.1183
800	0.49438	3661.7	4156.1	8.5024	0.41184	3661	4155.2	8.4176	0.35288	3660.3	4154.3	8.3458
900	0.54083	3853.9	4394.8	8.715	0.45059	3853.3	4394	8.6303	0.38614	3852.7	4393.3	8.5587
1000	0.58721	4052.7	4640	8.9155	0.48928	4052.2	4639.4	8.831	0.41933	4051.7	4638.8	8.7595
1100	0.63354	4257.9	4891.4	9.1057	0.52792	4257.5	4891	9.0212	0.45247	4257	4890.5	8.9497
1200	0.67983	4469	5148.9	9.2866	0.56652	4468.7	5148.5	9.2022	0.48558	4468.3	5148.1	9.1308
1300	0.7261	4685.8	5411.9	9.4593	0.60509	4685.5	5411.6	9.375	0.51866	4685.1	5411.3	9.3036
		P 1.60 MPa (201.37C)				P 1.80 MPa (207.11C)				P 2.00 MPa (212.38C)		
Sat.	0.12374	2594.8	2792.8	6.42	0.11037	2597.3	2795.9	6.3775	0.09959	2599.1	2798.3	6.339
225	0.13293	2645.1	2857.8	6.5537	0.11678	2637	2847.2	6.4825	0.10381	2628.5	2836.1	6.416
250	0.1419	2692.9	2919.9	6.6753	0.12502	2686.7	2911.7	6.6088	0.1115	2680.3	2903.3	6.5475
300	0.15866	2781.6	3035.4	6.8864	0.14025	2777.4	3029.9	6.8246	0.12551	2773.2	3024.2	6.7684
350	0.17459	2866.6	3146	7.0713	0.1546	2863.6	3141.9	7.012	0.1386	2860.5	3137.7	6.9583
400	0.19007	2950.8	3254.9	7.2394	0.16849	2948.3	3251.6	7.1814	0.15122	2945.9	3248.4	7.1292

500	0.22029	3120.1	3472.6	7.541	0.19551	3118.5	3470.4	7.4845	0.17568	3116.9	3468.3	7.4337
600	0.24999	3293.9	3693.9	7.8101	0.222	3292.7	3692.3	7.7543	0.19962	3291.5	3690.7	7.7043
700	0.27941	3473.5	3920.5	8.0558	0.24822	3472.6	3919.4	8.0005	0.22326	3471.7	3918.2	7.9509
800	0.30865	3659.5	4153.4	8.2834	0.27426	3658.8	4152.4	8.2284	0.24674	3658	4151.5	8.1791
900	0.3378	3852.1	4392.6	8.4965	0.3002	3851.5	4391.9	8.4417	0.27012	3850.9	4391.1	8.3925
1000	0.36687	4051.2	4638.2	8.6974	0.32606	4050.7	4637.6	8.6427	0.29342	4050.2	4637.1	8.5936
1100	0.39589	4256.6	4890	8.8878	0.35188	4256.2	4889.6	8.8331	0.31667	4255.7	4889.1	8.7842
1200	0.42488	4467.9	5147.7	9.0689	0.37766	4467.6	5147.3	9.0143	0.33989	4467.2	5147	8.9654
1300	0.45383	4684.8	5410.9	9.2418	0.40341	4684.5	5410.6	9.1872	0.36308	4684.2	5410.3	9.1384
		P 2.50 MPa (223.95C)				P 3.00 MPa (233.85C)				P 3.50 MPa (242.56C)		
Sat.	0.07995	2602.1	2801.9	6.2558	0.06667	2603.2	2803.2	6.1856	0.05706	2603	2802.7	6.1244
225	0.08026	2604.8	2805.5	6.2629								
250	0.08705	2663.3	2880.9	6.4107	0.07063	2644.7	2856.5	6.2893	0.05876	2624	2829.7	6.1764
300	0.09894	2762.2	3009.6	6.6459	0.08118	2750.8	2994.3	6.5412	0.06845	2738.8	2978.4	6.4484
350	0.10979	2852.5	3127	6.8424	0.09056	2844.4	3116.1	6.745	0.0768	2836	3104.9	6.6601
400	0.12012	2939.8	3240.1	7.017	0.09938	2933.6	3231.7	6.9235	0.08456	2927.2	3223.2	6.8428
450	0.13015	3026.2	3351.6	7.1768	0.10789	3021.2	3344.9	7.0856	0.09198	3016.1	3338.1	7.0074
500	0.13999	3112.8	3462.8	7.3254	0.1162	3108.6	3457.2	7.2359	0.09919	3104.5	3451.7	7.1593
600	0.15931	3288.5	3686.8	7.5979	0.13245	3285.5	3682.8	7.5103	0.11325	3282.5	3678.9	7.4357
700	0.17835	3469.3	3915.2	7.8455	0.14841	3467	3912.2	7.759	0.12702	3464.7	3909.3	7.6855
800	0.19722	3656.2	4149.2	8.0744	0.1642	3654.3	4146.9	7.9885	0.14061	3652.5	4144.6	7.9156
900	0.21597	3849.4	4389.3	8.2882	0.17988	3847.9	4387.5	8.2028	0.1541	3846.4	4385.7	8.1304
1000	0.23466	4049	4635.6	8.4897	0.19549	4047.7	4634.2	8.4045	0.16751	4046.4	4632.7	8.3324
1100	0.2533	4254.7	4887.9	8.6804	0.21105	4253.6	4886.7	8.5955	0.18087	4252.5	4885.6	8.5236
1200	0.2719	4466.3	5146	8.8618	0.22658	4465.3	5145.1	8.7771	0.1942	4464.4	5144.1	8.7053
1300	0.29048	4683.4	5409.5	9.0349	0.24207	4682.6	5408.8	8.9502	0.2075	4681.8	5408	8.8786

(Continued)

TABLE A.3 (*Continued*)

Superheated water

T	*v*	*u*	*h*	*s*	*v*	*u*	*h*	*s*	*v*	*u*	*h*	*s*
°C	m³/kg	kJ/kg	kJ/kg	kJ/kg.K	m³/kg	kJ/kg	kJ/kg	kJ/kg.K	m3/kg	kJ/kg	kJ/kg	kJ/kg.K
		P 4.0 MPa (250.35C)				P 4.5 MPa (257.44C)				P 5.0 MPa (263.94C)		
Sat.	0.04978	2601.7	2800.8	6.0696	0.04406	2599.7	2798	6.0198	0.03945	2597	2794.2	5.9737
275	0.05461	2668.9	2887.3	6.2312	0.04733	2651.4	2864.4	6.1429	0.04144	2632.3	2839.5	6.0571
300	0.05887	2726.2	2961.7	6.3639	0.05138	2713	2944.2	6.2854	0.04535	2699	2925.7	6.2111
350	0.06647	2827.4	3093.3	6.5843	0.05842	2818.6	3081.5	6.5153	0.05197	2809.5	3069.3	6.4516
400	0.07343	2920.8	3214.5	6.7714	0.06477	2914.2	3205.7	6.7071	0.05784	2907.5	3196.7	6.6483
450	0.08004	3011	3331.2	6.9386	0.07076	3005.8	3324.2	6.877	0.06332	3000.6	3317.2	6.821
500	0.08644	3100.3	3446	7.0922	0.07652	3096	3440.4	7.0323	0.06858	3091.8	3434.7	6.9781
600	0.09886	3279.4	3674.9	7.3706	0.08766	3276.4	3670.9	7.3127	0.0787	3273.3	3666.9	7.2605
700	0.11098	3462.4	3906.3	7.6214	0.0985	3460	3903.3	7.5647	0.08852	3457.7	3900.3	7.5136
800	0.12292	3650.6	4142.3	7.8523	0.10916	3648.8	4140	7.7962	0.09816	3646.9	4137.7	7.7458
900	0.13476	3844.8	4383.9	8.0675	0.11972	3843.3	4382.1	8.0118	0.10769	3841.8	4380.2	7.9619
1000	0.14653	4045.1	4631.2	8.2698	0.1302	4043.9	4629.8	8.2144	0.11715	4042.6	4628.3	8.1648
1100	0.15824	4251.4	4884.4	8.4612	0.14064	4250.4	4883.2	8.406	0.12655	4249.3	4882.1	8.3566
1200	0.16992	4463.5	5143.2	8.643	0.15103	4462.6	5142.2	8.588	0.13592	4461.6	5141.3	8.5388
1300	0.18157	4680.9	5407.2	8.8164	0.1614	4680.1	5406.5	8.7616	0.14527	4679.3	5405.7	8.7124
		P 6.0 MPa (275.59C)				P 7.0 MPa (285.83C)				P 8.0 MPa (295.01C)		
Sat.	0.03245	2589.9	2784.6	5.8902	0.027378	2581	2772.6	5.8148	0.023525	2570.5	2758.7	5.745
300	0.03619	2668.4	2885.6	6.0703	0.029492	2633.5	2839.9	5.9337	0.024279	2592.3	2786.5	5.7937
350	0.04225	2790.4	3043.9	6.3357	0.035262	2770.1	3016.9	6.2305	0.029975	2748.3	2988.1	6.1321
400	0.04742	2893.7	3178.3	6.5432	0.039958	2879.5	3159.2	6.4502	0.034344	2864.6	3139.4	6.3658
450	0.05217	2989.9	3302.9	6.7219	0.044187	2979	3288.3	6.6353	0.038194	2967.8	3273.3	6.5579
500	0.05667	3083.1	3423.1	6.8826	0.048157	3074.3	3411.4	6.8	0.041767	3065.4	3399.5	6.7266

550	0.06102	3175.2	3541.3	7.0308	0.051966	3167.9	3531.6	6.9507	0.045172	3160.5	3521.8	6.88
600	0.06527	3267.2	3658.8	7.1693	0.055665	3261	3650.6	7.091	0.048463	3254.7	3642.4	7.0221
700	0.07355	3453	3894.3	7.4247	0.06285	3448.3	3888.3	7.3487	0.054829	3443.6	3882.2	7.2822
800	0.08165	3643.2	4133.1	7.6582	0.069856	3639.5	4128.5	7.5836	0.061011	3635.7	4123.8	7.5185
900	0.08964	3838.8	4376.6	7.8751	0.07675	3835.7	4373	7.8014	0.067082	3832.7	4369.3	7.7372
1000	0.09756	4040.1	4625.4	8.0786	0.083571	4037.5	4622.5	8.0055	0.073079	4035	4619.6	7.9419
1100	0.10543	4247.1	4879.7	8.2709	0.090341	4245	4877.4	8.1982	0.079025	4242.8	4875	8.135
1200	0.11326	4459.8	5139.4	8.4534	0.097075	4457.9	5137.4	8.381	0.084934	4456.1	5135.5	8.3181
1300	0.12107	4677.7	5404.1	8.6273	0.103781	4676.1	5402.6	8.5551	0.090817	4674.5	5401	8.4925
		P 9.0 MPa (303.35C)				P 10.0 MPa (311.00C)				P 12.5 MPa (327.81C)		
Sat.	0.020489	2558.5	2742.9	5.6791	0.018028	2545.2	2725.5	5.6159	0.013496	2505.6	2674.3	5.4638
325	0.023284	2647.6	2857.1	5.8738	0.019877	2611.6	2810.3	5.7596				
350	0.025816	2725	2957.3	6.038	0.02244	2699.6	2924	5.946	0.016138	2624.9	2826.6	5.713
400	0.02996	2849.2	3118.8	6.2876	0.026436	2833.1	3097.5	6.2141	0.02003	2789.6	3040	6.0433
450	0.033524	2956.3	3258	6.4872	0.029782	2944.5	3242.4	6.4219	0.023019	2913.7	3201.5	6.2749
500	0.036793	3056.3	3387.4	6.6603	0.032811	3047	3375.1	6.5995	0.02563	3023.2	3343.6	6.4651
550	0.039885	3153	3512	6.8164	0.035655	3145.4	3502	6.7585	0.028033	3126.1	3476.5	6.6317
600	0.042861	3248.4	3634.1	6.9605	0.038378	3242	3625.8	6.9045	0.030306	3225.8	3604.6	6.7828
650	0.045755	3343.4	3755.2	7.0954	0.041018	3338	3748.1	7.0408	0.032491	3324.1	3730.2	6.9227
700	0.048589	3438.8	3876.1	7.2229	0.043597	3434	3870	7.1693	0.034612	3422	3854.6	7.054
800	0.054132	3632	4119.2	7.4606	0.048629	3628.2	4114.5	7.4085	0.038724	3618.8	4102.8	7.2967
900	0.059562	3829.6	4365.7	7.6802	0.053547	3826.5	4362	7.629	0.04272	3818.9	4352.9	7.5195
1000	0.064919	4032.4	4616.7	7.8855	0.058391	4029.9	4613.8	7.8349	0.046641	4023.5	4606.5	7.7269
1100	0.070224	4240.7	4872.7	8.0791	0.063183	4238.5	4870.3	8.0289	0.05051	4233.1	4864.5	7.922
1200	0.075492	4454.2	5133.6	8.2625	0.067938	4452.4	5131.7	8.2126	0.054342	4447.7	5127	8.1065
1300	0.080733	4672.9	5399.5	8.4371	0.072667	4671.3	5398	8.3874	0.058147	4667.3	5394.1	8.2819

(Continued)

TABLE A.3 (*Continued*)

Superheated water

T °C	*v* m³/kg	*u* kJ/kg	*h* kJ/kg	*s* kJ/kg.K	*vu* m³/kg kJ/kg	*h* kJ/kg	*s* kJ/kg.K	*v* m3/kg	*u* kJ/kg	*h* kJ/kg	*s* kJ/kg.K	
		P 15.0 MPa (342.16C)				P 17.5 MPa (354.67C)				P 20.0 MPa (365.75C)		
Sat.	0.010341	2455.7	2610.8	5.3108	0.007932	2390.7	2529.5	5.1435	0.005862	2294.8	2412.1	4.931
350	0.011481	2520.9	2693.1	5.4438								
400	0.015671	2740.6	2975.7	5.8819	0.012463	2684.3	2902.4	5.7211	0.00995	2617.9	2816.9	5.5526
450	0.018477	2880.8	3157.9	6.1434	0.015204	2845.4	3111.4	6.0212	0.012721	2807.3	3061.7	5.9043
500	0.020828	2998.4	3310.8	6.348	0.017385	2972.4	3276.7	6.2424	0.014793	2945.3	3241.2	6.1446
550	0.022945	3106.2	3450.4	6.523	0.019305	3085.8	3423.6	6.4266	0.016571	3064.7	3396.2	6.339
600	0.024921	3209.3	3583.1	6.6796	0.021073	3192.5	3561.3	6.589	0.018185	3175.3	3539	6.5075
650	0.026804	3310.1	3712.1	6.8233	0.022742	3295.8	3693.8	6.7366	0.019695	3281.4	3675.3	6.6593
700	0.028621	3409.8	3839.1	6.9573	0.024342	3397.5	3823.5	6.8735	0.021134	3385.1	3807.8	6.7991
800	0.032121	3609.3	4091.1	7.2037	0.027405	3599.7	4079.3	7.1237	0.02387	3590.1	4067.5	7.0531
900	0.035503	3811.2	4343.7	7.4288	0.030348	3803.5	4334.6	7.3511	0.026484	3795.7	4325.4	7.2829
1000	0.038808	4017.1	4599.2	7.6378	0.033215	4010.7	4592	7.5616	0.02902	4004.3	4584.7	7.495
1100	0.042062	4227.7	4858.6	7.8339	0.036029	4222.3	4852.8	7.7588	0.031504	4216.9	4847	7.6933
1200	0.045279	4443.1	5122.3	8.0192	0.038806	4438.5	5117.6	7.9449	0.033952	4433.8	5112.9	7.8802
1300	0.048469	4663.3	5390.3	8.1952	0.041556	4659.2	5386.5	8.1215	0.036371	4655.2	5382.7	8.0574
		P 25.0 MPa				P 30.0 MPa				P 35.0 MPa		
375	0.001978	1799.9	1849.4	4.0345	0.001792	1738.1	1791.9	3.9313	0.001701	1702.8	1762.4	3.8724
400	0.006005	2428.5	2578.7	5.14	0.002798	2068.9	2152.8	4.4758	0.002105	1914.9	1988.6	4.2144
425	0.007886	2607.8	2805	5.4708	0.005299	2452.9	2611.8	5.1473	0.003434	2253.3	2373.5	4.7751
450	0.009176	2721.2	2950.6	5.6759	0.006737	2618.9	2821	5.4422	0.004957	2497.5	2671	5.1946
500	0.011143	2887.3	3165.9	5.9643	0.008691	2824	3084.8	5.7956	0.006933	2755.3	2997.9	5.6331

550	0.012736	3020.8	3339.2	6.1816	0.010175	2974.5	3279.7	6.0403	0.008348	2925.8	3218	5.9093
600	0.01414	3140	3493.5	6.3637	0.011445	3103.4	3446.8	6.2373	0.009523	3065.6	3399	6.1229
650	0.01543	3251.9	3637.7	6.5243	0.01259	3221.7	3599.4	6.4074	0.010565	3190.9	3560.7	6.303
700	0.016643	3359.9	3776	6.6702	0.013654	3334.3	3743.9	6.5599	0.011523	3308.3	3711.6	6.4623
800	0.018922	3570.7	4043.8	6.9322	0.015628	3551.2	4020	6.8301	0.013278	3531.6	3996.3	6.7409
900	0.021075	3780.2	4307.1	7.1668	0.017473	3764.6	4288.8	7.0695	0.014904	3749	4270.6	6.9853
1000	0.02315	3991.5	4570.2	7.3821	0.01924	3978.6	4555.8	7.288	0.01645	3965.8	4541.5	7.2069
1100	0.025172	4206.1	4835.4	7.5825	0.020954	4195.2	4823.9	7.4906	0.017942	4184.4	4812.4	7.4118
1200	0.027157	4424.6	5103.5	7.771	0.02263	4415.3	5094.2	7.6807	0.019398	4406.1	5085	7.6034
1300	0.029115	4647.2	5375.1	7.9494	0.024279	4639.2	5367.6	7.8602	0.020827	4631.2	5360.2	7.7841
		P 40.0 MPa				P 50.0 MPa				P 60.0 MPa		
375	0.001641	1677	1742.6	3.829	0.00156	1638.6	1716.6	3.7642	0.001503	1609.7	1699.9	3.7149
400	0.001911	1855	1931.4	4.1145	0.001731	1787.8	1874.4	4.0029	0.001633	1745.2	1843.2	3.9317
425	0.002538	2097.5	2199	4.5044	0.002009	1960.3	2060.7	4.2746	0.001816	1892.9	2001.8	4.163
450	0.003692	2364.2	2511.8	4.9449	0.002487	2160.3	2284.7	4.5896	0.002086	2055.1	2180.2	4.414
500	0.005623	2681.6	2906.5	5.4744	0.00389	2528.1	2722.6	5.1762	0.002952	2393.2	2570.3	4.9356
550	0.006985	2875.1	3154.4	5.7857	0.005118	2769.5	3025.4	5.5563	0.003955	2664.6	2901.9	5.3517
600	0.008089	3026.8	3350.4	6.017	0.006108	2947.1	3252.6	5.8245	0.004833	2866.8	3156.8	5.6527
650	0.009053	3159.5	3521.6	6.2078	0.006957	3095.6	3443.5	6.0373	0.005591	3031.3	3366.8	5.8867
700	0.00993	3282	3679.2	6.374	0.007717	3228.7	3614.6	6.2179	0.006265	3175.4	3551.3	6.0814
800	0.011521	3511.8	3972.6	6.6613	0.009073	3472.2	3925.8	6.5225	0.007456	3432.6	3880	6.4033
900	0.01298	3733.3	4252.5	6.9107	0.010296	3702	4216.8	6.7819	0.008519	3670.9	4182.1	6.6725
1000	0.01436	3952.9	4527.3	7.1355	0.011441	3927.4	4499.4	7.0131	0.009504	3902	4472.2	6.9099
1100	0.015686	4173.7	4801.1	7.3425	0.012534	4152.2	4778.9	7.2244	0.010439	4130.9	4757.3	7.1255
1200	0.016976	4396.9	5075.9	7.5357	0.01359	4378.6	5058.1	7.4207	0.011339	4360.5	5040.8	7.3248
1300	0.018239	4623.3	5352.8	7.7175	0.01462	4607.5	5338.5	7.6048	0.012213	4591.8	5324.5	7.5111

TABLE A.4
Compressed Water

T	*v*	*u*	*h*	*s*	*v*	*u*	*h*	*s*	*v*	*u*	*h*	*s*
°C	m³/kg	kJ/kg	kJ/kg	kJ/kg.K	m³/kg	kJ/kg	kJ/kg	kJ/kg.K	m³/kg	kJ/kg	kJ/kg	kJ/kg.K
		P 5 MPa (263.94C)				*P* 10 MPa (311.00C)				*P* 15 MPa (342.16C)		
Sat.	0.0012862	1148.1	1154.5	2.9207	0.0014522	1393.3	1407.9	3.3603	0.0016572	1585.5	1610.3	3.6848
0	0.0009977	0.04	5.03	0.0001	0.0009952	0.12	10.07	0.0003	0.0009928	0.18	15.07	0.0004
20	0.0009996	83.61	88.61	0.2954	0.0009973	83.31	93.28	0.2943	0.0009951	83.01	97.93	0.2932
40	0.0010057	166.92	171.95	0.5705	0.0010035	166.33	176.37	0.5685	0.0010013	165.75	180.77	0.5666
60	0.0010149	250.29	255.36	0.8287	0.0010127	249.43	259.55	0.826	0.0010105	248.58	263.74	0.8234
80	0.0010267	333.82	338.96	1.0723	0.0010244	332.69	342.94	1.0691	0.0010221	331.59	346.92	1.0659
100	0.001041	417.65	422.85	1.3034	0.0010385	416.23	426.62	1.2996	0.0010361	414.85	430.39	1.2958
120	0.0010576	501.91	507.19	1.5236	0.0010549	500.18	510.73	1.5191	0.0010522	498.5	514.28	1.5148
140	0.0010769	586.8	592.18	1.7344	0.0010738	584.72	595.45	1.7293	0.0010708	582.69	598.75	1.7243
160	0.0010988	672.55	678.04	1.9374	0.0010954	670.06	681.01	1.9316	0.001092	667.63	684.01	1.9259
180	0.001124	759.47	765.09	2.1338	0.00112	756.48	767.68	2.1271	0.001116	753.58	770.32	2.1206
200	0.0011531	847.92	853.68	2.3251	0.0011482	844.32	855.8	2.3174	0.0011435	840.84	858	2.31
220	0.0011868	938.39	944.32	2.5127	0.0011809	934.01	945.82	2.5037	0.0011752	929.81	947.43	2.4951
240	0.0012268	1031.6	1037.7	2.6983	0.0012192	1026.2	1038.3	2.6876	0.0012121	1021	1039.2	2.6774
260	0.0012755	1128.5	1134.9	2.8841	0.0012653	1121.6	1134.3	2.871	0.001256	1115.1	1134	2.8586
280					0.0013226	1221.8	1235	3.0565	0.0013096	1213.4	1233	3.041
300					0.001398	1329.4	1343.3	3.2488	0.0013783	1317.6	1338.3	3.2279
320									0.0014733	1431.9	1454	3.4263
340									0.0016311	1567.9	1592.4	3.6555

	P 20 MPa (365.75C)				P 30 MPa				P 50 MPa			
Sat.	0.0020378	1785.8	1826.6	4.0146								
0	0.0009904	0.23	20.03	0.0005	0.0009857	0.29	29.86	0.0003	0.0009767	0.29	49.13	−0.001
20	0.0009929	82.71	102.57	0.2921	0.0009886	82.11	111.77	0.2897	0.0009805	80.93	129.95	0.2845
40	0.0009992	165.17	185.16	0.5646	0.0009951	164.05	193.9	0.5607	0.0009872	161.9	211.25	0.5528
60	0.0010084	247.75	267.92	0.8208	0.0010042	246.14	276.26	0.8156	0.0009962	243.08	292.88	0.8055
80	0.0010199	330.5	350.9	1.0627	0.0010155	328.4	358.86	1.0564	0.0010072	324.42	374.78	1.0442
100	0.0010337	413.5	434.17	1.292	0.001029	410.87	441.74	1.2847	0.0010201	405.94	456.94	1.2705
120	0.0010496	496.85	517.84	1.5105	0.0010445	493.66	525	1.502	0.0010349	487.69	539.43	1.4859
140	0.0010679	580.71	602.07	1.7194	0.0010623	576.9	608.76	1.7098	0.0010517	569.77	622.36	1.6916
160	0.0010886	665.28	687.05	1.9203	0.0010823	660.74	693.21	1.9094	0.0010704	652.33	705.85	1.8889
180	0.0011122	750.78	773.02	2.1143	0.0011049	745.4	778.55	2.102	0.0010914	735.49	790.06	2.079
200	0.001139	837.49	860.27	2.3027	0.0011304	831.11	865.02	2.2888	0.0011149	819.45	875.19	2.2628
220	0.0011697	925.77	949.16	2.4867	0.0011595	918.15	952.93	2.4707	0.0011412	904.39	961.45	2.4414
240	0.0012053	1016.1	1040.2	2.6676	0.0011927	1006.9	1042.7	2.6491	0.0011708	990.55	1049.1	2.6156
260	0.0012472	1109	1134	2.8469	0.0012314	1097.8	1134.7	2.825	0.0012044	1078.2	1138.4	2.7864
280	0.0012978	1205.6	1231.5	3.0265	0.001277	1191.5	1229.8	3.0001	0.001243	1167.7	1229.9	2.9547
300	0.0013611	1307.2	1334.4	3.2091	0.0013322	1288.9	1328.9	3.1761	0.0012879	1259.6	1324	3.1218
320	0.001445	1416.6	1445.5	3.3996	0.0014014	1391.7	1433.7	3.3558	0.0013409	1354.3	1421.4	3.2888
340	0.0015693	1540.2	1571.6	3.6086	0.0014932	1502.4	1547.1	3.5438	0.0014049	1452.9	1523.1	3.4575
360	0.0018248	1703.6	1740.1	3.8787	0.0016276	1626.8	1675.6	3.7499	0.0014848	1556.5	1630.7	3.6301
380					0.0018729	1782	1838.2	4.0026	0.0015884	1667.1	1746.5	3.8102

Appendix 2: EES Codes of Sample Examples

2.1 Example 6.4

```
{0}
P[0]=101.325
T[0]=25
T_K[0]=T[0]+273.15
h[0]=enthalpy(Water,T=T[0],P=P[0])
s[0]=entropy(Water,T=T[0],P=P[0])

T_ref=0
C_P_Protein_T_ref=2.0082
C_P_Fat_T_ref=1.9842
C_P_Carbohydrate_T_ref=1.5488
C_P_Ash_T_ref=1.0926
C_P_Water_T_ref=4.1762

C_P_Protein_T_0=2.0082+1.2089*0.001*T[0]-1.3129*0.000001*T[0]^2
C_P_Fat_T_0=1.9842+1.4733*0.001*T[0]-4.8008*0.000001*T[0]^2
C_P_Carbohydrate_T_0=1.5488+1.9625*0.001*T[0]-5.9399*0.000001*T[0]^2
C_P_Ash_T_0=1.0926+1.8896*0.001*T[0]-3.6817*0.000001*T[0]^2
C_P_Water_T_0=4.1762-9.0864*0.00001*T[0]+5.4731*0.000001*T[0]^2

X_Ash=0.0048
X_Fat=0.4
X_Carbohydrate=0.0285
X_Protein=0.0202
X_Water=0.5465

{1}
P[1]=300
T[1]=54
T_K[1]=T[1]+273.15
C_P_Protein[1]=2.0082+1.2089*0.001*T[1]-1.3129*0.000001*T[1]^2
C_P_Fat[1]=1.9842+1.4733*0.001*T[1]-4.8008*0.000001*T[1]^2
C_P_Carbohydrate[1]=1.5488+1.9625*0.001*T[1]-5.9399*0.000001*T[1]^2
C_P_Ash[1]=1.0926+1.8896*0.001*T[1]-3.6817*0.000001*T[1]^2
C_P_Water[1]=4.1762-9.0864*0.00001*T[1]+5.4731*0.000001*T[1]^2

C_P[1]=X_Protein*C_P_Protein[1]+X_Ash*C_P_Ash[1]+X_Fat*C_P_
Fat[1]+X_Carbohydrate*C_P_Carbohydrate[1]+X_Water*C_P_Water[1]
C_P_0[1]=X_Protein*C_P_Protein_T_0+X_Ash*C_P_Ash_T_0+X_Fat*C_P_
Fat_T_0+X_Carbohydrate*C_P_Carbohydrate_T_0+X_Water*C_P_Water_T_0
C_P_ref[1]=X_Protein*C_P_Protein_T_ref+X_Ash*C_P_Ash_T_ref+X_
Fat*C_P_Fat_T_ref+X_Carbohydrate*C_P_Carbohydrate_T_ref+X_
Water*C_P_Water_T_ref
C_P_ave_en[1]=(C_P[1]+C_P_ref[1])/2
C_P_ave_ex[1]=(C_P[1]+C_P_0[1])/2
```

```
RHO_Protein[1]=1.3299*1000-0.5184*T[1]
RHO_Fat[1]=9.2559*100-0.41757*T[1]
RHO_Carbohydrate[1]=1.5991*1000-0.31046*T[1]
RHO_Ash[1]=2.4238*1000-0.28063*T[1]
RHO_Water[1]=9.9718*100+3.1439*0.001*T[1]-3.7574*0.001*T[1]^2
v[1]=X_Protein/RHO_Protein[1]+X_Ash/RHO_Ash[1]+X_Fat/RHO_
Fat[1]+X_Carbohydrate/RHO_Carbohydrate[1]+X_Water/RHO_Water[1]

h[1]=C_P_ave_en[1]*T[1]+v[1]*(P[1]-P[0])
ex[1]=C_P_ave_ex[1]*(T[1]-T[0]-T_K[0]*ln(T_K[1]/T_K[0]))+v[1]
*(P[1]-P[0])

{2}
P[2]=235
T[2]=78.12
T_K[2]=T[2]+273.15

C_P_Protein[2]=2.0082+1.2089*0.001*T[2]-1.3129*0.000001*T[2]^2
C_P_Fat[2]=1.9842+1.4733*0.001*T[2]-4.8008*0.000001*T[2]^2
C_P_Carbohydrate[2]=1.5488+1.9625*0.001*T[2]-5.9399*0.000001*T[2]^2
C_P_Ash[2]=1.0926+1.8896*0.001*T[2]-3.6817*0.000001*T[2]^2
C_P_Water[2]=4.1762-9.0864*0.00001*T[2]+5.4731*0.000001*T[2]^2

C_P[2]=X_Protein*C_P_Protein[2]+X_Ash*C_P_Ash[2]+X_Fat*C_P_
Fat[2]+X_Carbohydrate*C_P_Carbohydrate[2]+X_Water*C_P_Water[2]
C_P_0[2]=X_Protein*C_P_Protein_T_0+X_Ash*C_P_Ash_T_0+X_Fat*C_P_
Fat_T_0+X_Carbohydrate*C_P_Carbohydrate_T_0+X_Water*C_P_Water_T_0
C_P_ref[2]=X_Protein*C_P_Protein_T_ref+X_Ash*C_P_Ash_T_ref+X_
Fat*C_P_Fat_T_ref+X_Carbohydrate*C_P_Carbohydrate_T_ref+X_
Water*C_P_Water_T_ref
C_P_ave_en[2]=(C_P[2]+C_P_ref[2])/2
C_P_ave_ex[2]=(C_P[2]+C_P_0[2])/2

RHO_Protein[2]=1.3299*1000-0.5184*T[2]
RHO_Fat[2]=9.2559*100-0.41757*T[2]
RHO_Carbohydrate[2]=1.5991*1000-0.31046*T[2]
RHO_Ash[2]=2.4238*1000-0.28063*T[2]
RHO_Water[2]=9.9718*100+3.1439*0.001*T[2]-3.7574*0.001*T[2]^2
v[2]=X_Protein/RHO_Protein[2]+X_Ash/RHO_Ash[2]+X_Fat/RHO_
Fat[2]+X_Carbohydrate/RHO_Carbohydrate[2]+X_Water/RHO_Water[2]

h[2]=C_P_ave_en[2]*T[2]+v[2]*(P[2]-P[0])
ex[2]=C_P_ave_ex[2]*(T[2]-T[0]-T_K[0]*ln(T_K[2]/T_K[0]))+v[2]
*(P[2]-P[0])

{3}
P[3]=450
T[3]=78.12
T_K[3]=T[3]+273.15

C_P_Protein[3]=2.0082+1.2089*0.001*T[3]-1.3129*0.000001*T[3]^2
C_P_Fat[3]=1.9842+1.4733*0.001*T[3]-4.8008*0.000001*T[3]^2
C_P_Carbohydrate[3]=1.5488+1.9625*0.00
1*T[3]-5.9399*0.000001*T[3]^2
```

```
C_P_Ash[3]=1.0926+1.8896*0.001*T[3]-3.6817*0.000001*T[3]^2
C_P_Water[3]=4.1762-9.0864*0.00001*T[3]+5.4731*0.000001*T[3]^2

C_P[3]=X_Protein*C_P_Protein[3]+X_Ash*C_P_Ash[3]+X_Fat*C_P_
Fat[3]+X_Carbohydrate*C_P_Carbohydrate[3]+X_Water*C_P_Water[3]
C_P_0[3]=X_Protein*C_P_Protein_T_0+X_Ash*C_P_Ash_T_0+X_Fat*C_P_
Fat_T_0+X_Carbohydrate*C_P_Carbohydrate_T_0+X_Water*C_P_Water_T_0
C_P_ref[3]=X_Protein*C_P_Protein_T_ref+X_Ash*C_P_Ash_T_ref+X_
Fat*C_P_Fat_T_ref+X_Carbohydrate*C_P_Carbohydrate_T_ref+X_
Water*C_P_Water_T_ref
C_P_ave_en[3]=(C_P[3]+C_P_ref[3])/2
C_P_ave_ex[3]=(C_P[3]+C_P_0[3])/2

RHO_Protein[3]=1.3299*1000-0.5184*T[3]
RHO_Fat[3]=9.2559*100-0.41757*T[3]
RHO_Carbohydrate[3]=1.5991*1000-0.31046*T[3]
RHO_Ash[3]=2.4238*1000-0.28063*T[3]
RHO_Water[3]=9.9718*100+3.1439*0.001*T[3]-3.7574*0.001*T[3]^2
v[3]=X_Protein/RHO_Protein[3]+X_Ash/RHO_Ash[3]+X_Fat/RHO_
Fat[3]+X_Carbohydrate/RHO_Carbohydrate[3]+X_Water/RHO_Water[3]

h[3]=C_P_ave_en[3]*T[3]+v[3]*(P[3]-P[0])
ex[3]=C_P_ave_ex[3]*(T[3]-T[0]-T_K[0]*ln(T_K[3]/T_K[0]))+v[3]
*(P[3]-P[0])

{4}
P[4]=425
T[4]=90
T_K[4]=T[4]+273.15

C_P_Protein[4]=2.0082+1.2089*0.001*T[4]-1.3129*0.000001*T[4]^2
C_P_Fat[4]=1.9842+1.4733*0.001*T[4]-4.8008*0.000001*T[4]^2
C_P_Carbohydrate[4]=1.5488+1.9625*0.001*T[4]-5.9399*0.000001*T[4]^2
C_P_Ash[4]=1.0926+1.8896*0.001*T[4]-3.6817*0.000001*T[4]^2
C_P_Water[4]=4.1762-9.0864*0.00001*T[4]+5.4731*0.000001*T[4]^2

C_P[4]=X_Protein*C_P_Protein[4]+X_Ash*C_P_Ash[4]+X_Fat*C_P_
Fat[4]+X_Carbohydrate*C_P_Carbohydrate[4]+X_Water*C_P_Water[4]
C_P_0[4]=X_Protein*C_P_Protein_T_0+X_Ash*C_P_Ash_T_0+X_Fat*C_P_
Fat_T_0+X_Carbohydrate*C_P_Carbohydrate_T_0+X_Water*C_P_Water_T_0
C_P_ref[4]=X_Protein*C_P_Protein_T_ref+X_Ash*C_P_Ash_T_ref+X_
Fat*C_P_Fat_T_ref+X_Carbohydrate*C_P_Carbohydrate_T_ref+X_
Water*C_P_Water_T_ref
C_P_ave_en[4]=(C_P[4]+C_P_ref[4])/2
C_P_ave_ex[4]=(C_P[4]+C_P_0[4])/2

RHO_Protein[4]=1.3299*1000-0.5184*T[4]
RHO_Fat[4]=9.2559*100-0.41757*T[4]
RHO_Carbohydrate[4]=1.5991*1000-0.31046*T[4]
RHO_Ash[4]=2.4238*1000-0.28063*T[4]
RHO_Water[4]=9.9718*100+3.1439*0.001*T[4]-3.7574*0.001*T[4]^2
v[4]=X_Protein/RHO_Protein[4]+X_Ash/RHO_Ash[4]+X_Fat/RHO_
Fat[4]+X_Carbohydrate/RHO_Carbohydrate[4]+X_Water/RHO_Water[4]
```

```
h[4]=C_P_ave_en[4]*T[4]+v[4]*(P[4]-P[0])
ex[4]=C_P_ave_ex[4]*(T[4]-T[0]-T_K[0]*ln(T_K[4]/T_K[0]))+v[4]
*(P[4]-P[0])

{5}
P[5]=400
T[5]=62
T_K[5]=T[5]+273.15

C_P_Protein[5]=2.0082+1.2089*0.001*T[5]-1.3129*0.000001*T[5]^2
C_P_Fat[5]=1.9842+1.4733*0.001*T[5]-4.8008*0.000001*T[5]^2
C_P_Carbohydrate[5]=1.5488+1.9625*0.001*T[5]-5.9399*0.000001*T[5]^2
C_P_Ash[5]=1.0926+1.8896*0.001*T[5]-3.6817*0.000001*T[5]^2
C_P_Water[5]=4.1762-9.0864*0.00001*T[5]+5.4731*0.000001*T[5]^2

C_P[5]=X_Protein*C_P_Protein[5]+X_Ash*C_P_Ash[5]+X_Fat*C_P_
Fat[5]+X_Carbohydrate*C_P_Carbohydrate[5]+X_Water*C_P_Water[5]
C_P_0[5]=X_Protein*C_P_Protein_T_0+X_Ash*C_P_Ash_T_0+X_Fat*C_P_
Fat_T_0+X_Carbohydrate*C_P_Carbohydrate_T_0+X_Water*C_P_Water_T_0
C_P_ref[5]=X_Protein*C_P_Protein_T_ref+X_Ash*C_P_Ash_T_ref+X_
Fat*C_P_Fat_T_ref+X_Carbohydrate*C_P_Carbohydrate_T_ref+X_
Water*C_P_Water_T_ref
C_P_ave_en[5]=(C_P[5]+C_P_ref[5])/2
C_P_ave_ex[5]=(C_P[5]+C_P_0[5])/2

RHO_Protein[5]=1.3299*1000-0.5184*T[5]
RHO_Fat[5]=9.2559*100-0.41757*T[5]
RHO_Carbohydrate[5]=1.5991*1000-0.31046*T[5]
RHO_Ash[5]=2.4238*1000-0.28063*T[5]
RHO_Water[5]=9.9718*100+3.1439*0.001*T[5]-3.7574*0.001*T[5]^2
v[5]=X_Protein/RHO_Protein[5]+X_Ash/RHO_Ash[5]+X_Fat/RHO_
Fat[5]+X_Carbohydrate/RHO_Carbohydrate[5]+X_Water/RHO_Water[5]

h[5]=C_P_ave_en[5]*T[5]+v[5]*(P[5]-P[0])
ex[5]=C_P_ave_ex[5]*(T[5]-T[0]-T_K[0]*ln(T_K[5]/
T_K[0]))+v[5]*(P[5]-P[0])

{6}
P[6]=350
T[6]=35.5
T_K[6]=T[6]+273.15

C_P_Protein[6]=2.0082+1.2089*0.001*T[6]-1.3129*0.000001*T[6]^2
C_P_Fat[6]=1.9842+1.4733*0.001*T[6]-4.8008*0.000001*T[6]^2
C_P_Carbohydrate[6]=1.5488+1.9625*0.001*T[6]-5.9399*0.000001*T[6]^2
C_P_Ash[6]=1.0926+1.8896*0.001*T[6]-3.6817*0.000001*T[6]^2
C_P_Water[6]=4.1762-9.0864*0.00001*T[6]+5.4731*0.000001*T[6]^2

C_P[6]=X_Protein*C_P_Protein[6]+X_Ash*C_P_Ash[6]+X_Fat*C_P_
Fat[6]+X_Carbohydrate*C_P_Carbohydrate[6]+X_Water*C_P_Water[6]
C_P_0[6]=X_Protein*C_P_Protein_T_0+X_Ash*C_P_Ash_T_0+X_Fat*C_P_
Fat_T_0+X_Carbohydrate*C_P_Carbohydrate_T_0+X_Water*C_P_Water_T_0
C_P_ref[6]=X_Protein*C_P_Protein_T_ref+X_Ash*C_P_Ash_T_ref+X_
Fat*C_P_Fat_T_ref+X_Carbohydrate*C_P_Carbohydrate_T_ref+X_
Water*C_P_Water_T_ref
C_P_ave_en[6]=(C_P[6]+C_P_ref[6])/2
C_P_ave_ex[6]=(C_P[6]+C_P_0[6])/2
```

```
RHO_Protein[6]=1.3299*1000-0.5184*T[6]
RHO_Fat[6]=9.2559*100-0.41757*T[6]
RHO_Carbohydrate[6]=1.5991*1000-0.31046*T[6]
RHO_Ash[6]=2.4238*1000-0.28063*T[6]
RHO_Water[6]=9.9718*100+3.1439*0.001*T[6]-3.7574*0.001*T[6]^2
v[6]=X_Protein/RHO_Protein[6]+X_Ash/RHO_Ash[6]+X_Fat/RHO_
Fat[6]+X_Carbohydrate/RHO_Carbohydrate[6]+X_Water/RHO_Water[6]

h[6]=C_P_ave_en[6]*T[6]+v[6]*(P[6]-P[0])
ex[6]=C_P_ave_ex[6]*(T[6]-T[0]-T_K[0]*ln(T_K[6]/T_K[0]))+v[6]
*(P[6]-P[0])

{7}
P[7]=325
T[7]=7
T_K[7]=T[7]+273.15

C_P_Protein[7]=2.0082+1.2089*0.001*T[7]-1.3129*0.000001*T[7]^2
C_P_Fat[7]=1.9842+1.4733*0.001*T[7]-4.8008*0.000001*T[7]^2
C_P_Carbohydrate[7]=1.5488+1.9625*0.001*T[7]-5.9399*0.000001*T[7]^2
C_P_Ash[7]=1.0926+1.8896*0.001*T[7]-3.6817*0.000001*T[7]^2
C_P_Water[7]=4.1762-9.0864*0.00001*T[7]+5.4731*0.000001*T[7]^2
C_P[7]=X_Protein*C_P_Protein[7]+X_Ash*C_P_Ash[7]+X_Fat*C_P_
Fat[7]+X_Carbohydrate*C_P_Carbohydrate[7]+X_Water*C_P_Water[7]
C_P_0[7]=X_Protein*C_P_Protein_T_0+X_Ash*C_P_Ash_T_0+X_Fat*C_P_
Fat_T_0+X_Carbohydrate*C_P_Carbohydrate_T_0+X_Water*C_P_Water_T_0
C_P_ref[7]=X_Protein*C_P_Protein_T_ref+X_Ash*C_P_Ash_T_ref+X_
Fat*C_P_Fat_T_ref+X_Carbohydrate*C_P_Carbohydrate_T_ref+X_
Water*C_P_Water_T_ref
C_P_ave_en[7]=(C_P[7]+C_P_ref[7])/2
C_P_ave_ex[7]=(C_P[7]+C_P_0[7])/2

RHO_Protein[7]=1.3299*1000-0.5184*T[7]
RHO_Fat[7]=9.2559*100-0.41757*T[7]
RHO_Carbohydrate[7]=1.5991*1000-0.31046*T[7]
RHO_Ash[7]=2.4238*1000-0.28063*T[7]
RHO_Water[7]=9.9718*100+3.1439*0.001*T[7]-3.7574*0.001*T[7]^2
v[7]=X_Protein/RHO_Protein[7]+X_Ash/RHO_Ash[7]+X_Fat/RHO_
Fat[7]+X_Carbohydrate/RHO_Carbohydrate[7]+X_Water/RHO_Water[7]

h[7]=C_P_ave_en[7]*T[7]+v[7]*(P[7]-P[0])
ex[7]=C_P_ave_ex[7]*(T[7]-T[0]-T_K[0]*ln(T_K[7]/T_K[0]))+v[7]
*(P[7]-P[0])

{8}
P[8]=150
T[8]=99
T_K[8]=T[8]+273.15
h[8]=enthalpy(Water,T=T[8],P=P[8])
s[8]=entropy(Water,T=T[8],P=P[8])
ex[8]=h[8]-h[0]-T_K[0]*(s[8]-s[0])

{9}
P[9]=125
T[9]=56.12
T_K[9]=T[9]+273.15
h[9]=enthalpy(Water,T=T[9],P=P[9])
```

```
s[9]=entropy(Water,T=T[9],P=P[9])
ex[9]=h[9]-h[0]-T_K[0]*(s[9]-s[0])

{10}
P[10]=250
T[10]=1
T_K[10]=T[10]+273.15
h[10]=enthalpy(Water,T=T[10],P=P[10])
s[10]=entropy(Water,T=T[10],P=P[10])
ex[10]=h[10]-h[0]-T_K[0]*(s[10]-s[0])

{11}
P[11]=200
T[11]=8.5
T_K[11]=T[11]+273.15
h[11]=enthalpy(Water,T=T[11],P=P[11])
s[11]=entropy(Water,T=T[11],P=P[11])
ex[11]=h[11]-h[0]-T_K[0]*(s[11]-s[0])

{12}
P[12]=300
T[12]=27
T_K[12]=T[12]+273.15
h[12]=enthalpy(Water,T=T[12],P=P[12])
s[12]=entropy(Water,T=T[12],P=P[12])
ex[12]=h[12]-h[0]-T_K[0]*(s[12]-s[0])

{13}
P[13]=250
T[13]=32
T_K[13]=T[13]+273.15
h[13]=enthalpy(Water,T=T[13],P=P[13])
s[13]=entropy(Water,T=T[13],P=P[13])
ex[13]=h[13]-h[0]-T_K[0]*(s[13]-s[0])

{mass balance}
m_dot[1]=0.67
m_dot[2]=m_dot[1]
m_dot[3]=m_dot[2]
m_dot[4]=m_dot[3]
m_dot[5]=m_dot[4]
m_dot[6]=m_dot[5]
m_dot[7]=m_dot[6]
m_dot[8]=0.2
m_dot[9]=m_dot[8]
m_dot[10]=2.02
m_dot[11]=m_dot[10]
m_dot[12]=2.02
m_dot[13]=m_dot[12]

{energy balance}
m_dot[1]*h[1]+m_dot[3]*h[3]+m_dot[8]*h[8]+m_dot[10]*h[10]+m_
dot[12]*h[12]=m_dot[2]*h[2]+m_dot[7]*h[7]+m_dot[9]*h[9]+m_
dot[13]*h[13]+m_dot[11]*h[11]+Q_dot_L

{exergy balance}
Ex_dot[1]=m_dot[1]*ex[1]
```

```
Ex_dot[2]=m_dot[2]*ex[2]
Ex_dot[3]=m_dot[3]*ex[3]
Ex_dot[4]=m_dot[4]*ex[4]
Ex_dot[5]=m_dot[5]*ex[5]
Ex_dot[6]=m_dot[6]*ex[6]
Ex_dot[7]=m_dot[7]*ex[7]
Ex_dot[8]=m_dot[8]*ex[8]
Ex_dot[9]=m_dot[9]*ex[9]
Ex_dot[10]=m_dot[10]*ex[10]
Ex_dot[11]=m_dot[11]*ex[11]
Ex_dot[12]=m_dot[12]*ex[12]
Ex_dot[13]=m_dot[13]*ex[13]

T_b=40+273.15
Ex_dot_L=Q_dot_L*(1-((T_K[0])/(T_b)))
Ex_dot_F=(Ex_dot[8]-Ex_dot[9])+(Ex_dot[4]-Ex_dot[5])+(Ex_
dot[5]-Ex_dot[6])+(Ex_dot[6]-(-Ex_dot[7]))
Ex_dot_P=(Ex_dot[2]-Ex_dot[1])+(Ex_dot[4]-Ex_dot[3])+((-Ex_
dot[11])-(-Ex_dot[10]))+(Ex_dot[13]-Ex_dot[12])
Ex_dot_des=Ex_dot_F-Ex_dot_P-Ex_dot_L
```

2.2 Example 6.21

```
{0}
P[0]=101.325
T[0]=25
T_K[0]=T[0]+273.15
h[0]=enthalpy(Water,T=T[0],P=P[0])
s[0]=entropy(Water,T=T[0],P=P[0])

T_ref=0
C_P_Protein_T_ref=2.0082
C_P_Fat_T_ref=1.9842
C_P_Carbohydrate_T_ref=1.5488
C_P_Ash_T_ref=1.0926
C_P_Water_T_ref=4.1762

C_P_Protein_T_0=2.0082+1.2089*0.001*T[0]-1.3129*0.000001*T[0]^2
C_P_Fat_T_0=1.9842+1.4733*0.001*T[0]-4.8008*0.000001*T[0]^2
C_P_Carbohydrate_T_0=1.5488+1.9625*0.001*T[0]-5.9399*0.000001
*T[0]^2
C_P_Ash_T_0=1.0926+1.8896*0.001*T[0]-3.6817*0.000001*T[0]^2
C_P_Water_T_0=4.1762-9.0864*0.00001*T[0]+5.4731*0.000001*T[0]^2

{1}
P[1]=360
T[1]=87
T_K[1]=T[1]+273.15
TS[1]=0.1051
X_Ash[1]=0.00320
X_Fat[1]=0.00008
X_Carbohydrate[1]=0.09892
X_Protein[1]=0.00290
X_Water[1]=0.89490
```

```
C_P_Protein[1]=2.0082+1.2089*0.001*T[1]-1.3129*0.000001*T[1]^2
C_P_Fat[1]=1.9842+1.4733*0.001*T[1]-4.8008*0.000001*T[1]^2
C_P_Carbohydrate[1]=1.5488+1.9625*0.001*T[1]-5.9399*0.000001*T[1]^2
C_P_Ash[1]=1.0926+1.8896*0.001*T[1]-3.6817*0.000001*T[1]^2
C_P_Water[1]=4.1762-9.0864*0.00001*T[1]+5.4731*0.000001*T[1]^2

C_P[1]=X_Protein[1]*C_P_Protein[1]+X_Ash[1]*C_P_Ash[1]+X_
Fat[1]*C_P_Fat[1]+X_Carbohydrate[1]*C_P_Carbohydrate[1]+X_
Water[1]*C_P_Water[1]
C_P_0[1]=X_Protein[1]*C_P_Protein_T_0+X_Ash[1]*C_P_Ash_T_0+X_
Fat[1]*C_P_Fat_T_0+X_Carbohydrate[1]*C_P_Carbohydrate_T_0+X_
Water[1]*C_P_Water_T_0
C_P_ref[1]=X_Protein[1]*C_P_Protein_T_ref+X_Ash[1]*C_P_Ash_T_
ref+X_Fat[1]*C_P_Fat_T_ref+X_Carbohydrate[1]*C_P_Carbohydrate_T_
ref+X_Water[1]*C_P_Water_T_ref
C_P_ave_en[1]=(C_P[1]+C_P_ref[1])/2
C_P_ave_ex[1]=(C_P[1]+C_P_0[1])/2

RHO_Protein[1]=1.3299*1000-0.5184*T[1]
RHO_Fat[1]=9.2559*100-0.41757*T[1]
RHO_Carbohydrate[1]=1.5991*1000-0.31046*T[1]
RHO_Ash[1]=2.4238*1000-0.28063*T[1]
RHO_Water[1]=9.9718*100+3.1439*0.001*T[1]-3.7574*0.001*T[1]^2
v[1]=X_Protein[1]/RHO_Protein[1]+X_Ash[1]/RHO_Ash[1]+X_Fat[1]/
RHO_Fat[1]+X_Carbohydrate[1]/RHO_Carbohydrate[1]+X_Water[1]/
RHO_Water[1]

h[1]=C_P_ave_en[1]*T[1]+v[1]*(P[1]-P[0])
ex[1]=C_P_ave_ex[1]*(T[1]-T[0]-T_K[0]*ln(T_K[1]/
T_K[0]))+v[1]*(P[1]-P[0])

{2}
P[2]=170
T[2]=115.4
T_K[2]=T[2]+273.15
TS[2]=0.1404
X_Water[2]=0.85958
X_Protein[2]=0.00388
X_Fat[2]=0.00011
X_Carbohydrate[2]=0.13215
X_Ash[2]=0.00428

C_P_Protein[2]=2.0082+1.2089*0.001*T[2]-1.3129*0.000001*T[2]^2
C_P_Fat[2]=1.9842+1.4733*0.001*T[2]-4.8008*0.000001*T[2]^2
C_P_Carbohydrate[2]=1.5488+1.9625*0.001*T[2]-5.9399*0.000001*T[2]^2
C_P_Ash[2]=1.0926+1.8896*0.001*T[2]-3.6817*0.000001*T[2]^2
C_P_Water[2]=4.1762-9.0864*0.00001*T[2]+5.4731*0.000001*T[2]^2

C_P[2]=X_Protein[2]*C_P_Protein[2]+X_Ash[2]*C_P_Ash[2]+X_
Fat[2]*C_P_Fat[2]+X_Carbohydrate[2]*C_P_Carbohydrate[2]+X_
Water[2]*C_P_Water[2]
C_P_0[2]=X_Protein[2]*C_P_Protein_T_0+X_Ash[2]*C_P_Ash_T_0+X_
Fat[2]*C_P_Fat_T_0+X_Carbohydrate[2]*C_P_Carbohydrate_T_0+X_
Water[2]*C_P_Water_T_0
C_P_ref[2]=X_Protein[2]*C_P_Protein_T_ref+X_Ash[2]*C_P_Ash_T_
ref+X_Fat[2]*C_P_Fat_T_ref+X_Carbohydrate[2]*C_P_Carbohydrate_T_
ref+X_Water[2]*C_P_Water_T_ref
```

```
C_P_ave_en[2]=(C_P[2]+C_P_ref[2])/2
C_P_ave_ex[2]=(C_P[2]+C_P_0[2])/2

RHO_Protein[2]=1.3299*1000-0.5184*T[2]
RHO_Fat[2]=9.2559*100-0.41757*T[2]
RHO_Carbohydrate[2]=1.5991*1000-0.31046*T[2]
RHO_Ash[2]=2.4238*1000-0.28063*T[2]
RHO_Water[2]=9.9718*100+3.1439*0.001*T[2]-3.7574*0.001*T[2]^2
v[2]=X_Protein[2]/RHO_Protein[2]+X_Ash[2]/RHO_Ash[2]+X_Fat[2]/
RHO_Fat[2]+X_Carbohydrate[2]/RHO_Carbohydrate[2]+X_Water[2]/
RHO_Water[2]
h[2]=C_P_ave_en[2]*T[2]+v[2]*(P[2]-P[0])
ex[2]=C_P_ave_ex[2]*(T[2]-T[0]-T_K[0]*ln(T_K[2]/T_K[0]))+v[2]
*(P[2]-P[0])

{3}
P[3]=170
x[3]=1
h[3]=enthalpy(Water,x=x[3],P=P[3])
s[3]=entropy(Water,x=x[3],P=P[3])
ex[3]=h[3]-h[0]-T_K[0]*(s[3]-s[0])

{4}
P[4]=240
x[4]=1
h[4]=enthalpy(Water,x=x[4],P=P[4])
s[4]=entropy(Water,x=x[4],P=P[4])
ex[4]=h[4]-h[0]-T_K[0]*(s[4]-s[0])

{5}
P[5]=240
x[5]=0
h[5]=enthalpy(Water,x=x[5],P=P[5])
s[5]=entropy(Water,x=x[5],P=P[5])
ex[5]=h[5]-h[0]-T_K[0]*(s[5]-s[0])

{6}
P[6]=400
T[6]=74
T_K[6]=T[6]+273.15
TS[6]=0.1287
X_Ash[6]=0.00433
X_Fat[6]=0.00011
X_Carbohydrate[6]=0.12029
X_Protein[6]=0.00393
X_Water[6]=0.87134

C_P_Protein[6]=2.0082+1.2089*0.001*T[6]-1.3129*0.000001*T[6]^2
C_P_Fat[6]=1.9842+1.4733*0.001*T[6]-4.8008*0.000001*T[6]^2
C_P_Carbohydrate[6]=1.5488+1.9625*0.001*T[6]-5.9399*0.000001*T[6]^2
C_P_Ash[6]=1.0926+1.8896*0.001*T[6]-3.6817*0.000001*T[6]^2
C_P_Water[6]=4.1762-9.0864*0.00001*T[6]+5.4731*0.000001*T[6]^2

C_P[6]=X_Protein[6]*C_P_Protein[6]+X_Ash[6]*C_P_Ash[6]+X_
Fat[6]*C_P_Fat[6]+X_Carbohydrate[6]*C_P_Carbohydrate[6]+X_
Water[6]*C_P_Water[6]
```

```
C_P_0[6]=X_Protein[6]*C_P_Protein_T_0+X_Ash[6]*C_P_Ash_T_0+X_
Fat[6]*C_P_Fat_T_0+X_Carbohydrate[6]*C_P_Carbohydrate_T_0+X_
Water[6]*C_P_Water_T_0
C_P_ref[6]=X_Protein[6]*C_P_Protein_T_ref+X_Ash[6]*C_P_Ash_T_
ref+X_Fat[6]*C_P_Fat_T_ref+X_Carbohydrate[6]*C_P_Carbohydrate_T_
ref+X_Water[6]*C_P_Water_T_ref
C_P_ave_en[6]=(C_P[6]+C_P_ref[6])/2
C_P_ave_ex[6]=(C_P[6]+C_P_0[6])/2

RHO_Protein[6]=1.3299*1000-0.5184*T[6]
RHO_Fat[6]=9.2559*100-0.41757*T[6]
RHO_Carbohydrate[6]=1.5991*1000-0.31046*T[6]
RHO_Ash[6]=2.4238*1000-0.28063*T[6]
RHO_Water[6]=9.9718*100+3.1439*0.001*T[6]-3.7574*0.001*T[6]^2
v[6]=X_Protein[6]/RHO_Protein[6]+X_Ash[6]/RHO_Ash[6]+X_Fat[6]/
RHO_Fat[6]+X_Carbohydrate[6]/RHO_Carbohydrate[6]+X_Water[6]/
RHO_Water[6]

h[6]=C_P_ave_en[6]*T[6]+v[6]*(P[6]-P[0])
ex[6]=C_P_ave_ex[6]*(T[6]-T[0]-T_K[0]*ln(T_K[6]/T_K[0]))+v[6]
*(P[6]-P[0])

{7}
P[7]=350
T[7]=79
T_K[7]=T[7]+273.15
TS[7]=TS[6]
X_Ash[7]=X_Ash[6]
X_Fat[7]=X_Fat[6]
X_Carbohydrate[7]=X_Carbohydrate[6]
X_Protein[7]=X_Protein[6]
X_Water[7]=X_Water[6]

C_P_Protein[7]=2.0082+1.2089*0.001*T[7]-1.3129*0.000001*T[7]^2
C_P_Fat[7]=1.9842+1.4733*0.001*T[7]-4.8008*0.000001*T[7]^2
C_P_Carbohydrate[7]=1.5488+1.9625*0.001*T[7]-5.9399*0.000001*T[7]^2
C_P_Ash[7]=1.0926+1.8896*0.001*T[7]-3.6817*0.000001*T[7]^2
C_P_Water[7]=4.1762-9.0864*0.00001*T[7]+5.4731*0.000001*T[7]^2

C_P[7]=X_Protein[7]*C_P_Protein[7]+X_Ash[7]*C_P_Ash[7]+X_
Fat[7]*C_P_Fat[7]+X_Carbohydrate[7]*C_P_Carbohydrate[7]+X_
Water[7]*C_P_Water[7]
C_P_0[7]=X_Protein[7]*C_P_Protein_T_0+X_Ash[7]*C_P_Ash_T_0+X_
Fat[7]*C_P_Fat_T_0+X_Carbohydrate[7]*C_P_Carbohydrate_T_0+X_
Water[7]*C_P_Water_T_0
C_P_ref[7]=X_Protein[7]*C_P_Protein_T_ref+X_Ash[7]*C_P_Ash_T_
ref+X_Fat[7]*C_P_Fat_T_ref+X_Carbohydrate[7]*C_P_Carbohydrate_T_
ref+X_Water[7]*C_P_Water_T_ref
C_P_ave_en[7]=(C_P[7]+C_P_ref[7])/2
C_P_ave_ex[7]=(C_P[7]+C_P_0[7])/2

RHO_Protein[7]=1.3299*1000-0.5184*T[7]
RHO_Fat[7]=9.2559*100-0.41757*T[7]
RHO_Carbohydrate[7]=1.5991*1000-0.31046*T[7]
RHO_Ash[7]=2.4238*1000-0.28063*T[7]
RHO_Water[7]=9.9718*100+3.1439*0.001*T[7]-3.7574*0.001*T[7]^2
```

```
v[7]=X_Protein[7]/RHO_Protein[7]+X_Ash[7]/RHO_Ash[7]+X_Fat[7]/
RHO_Fat[7]+X_Carbohydrate[7]/RHO_Carbohydrate[7]+X_Water[7]/
RHO_Water[7]

h[7]=C_P_ave_en[7]*T[7]+v[7]*(P[7]-P[0])
ex[7]=C_P_ave_ex[7]*(T[7]-T[0]-T_K[0]*ln(T_K[7]/T_K[0]))+v[7]
*(P[7]-P[0])

{8}
P[8]=600
T[8]=62
T_K[8]=T[8]+273.15
TS[8]=TS[1]
X_Ash[8]=X_Ash[1]
X_Fat[8]=X_Fat[1]
X_Carbohydrate[8]=X_Carbohydrate[1]
X_Protein[8]=X_Protein[1]
X_Water[8]=X_Water[1]

C_P_Protein[8]=2.0082+1.2089*0.001*T[8]-1.3129*0.000001*T[8]^2
C_P_Fat[8]=1.9842+1.4733*0.001*T[8]-4.8008*0.000001*T[8]^2
C_P_Carbohydrate[8]=1.5488+1.9625*0.001*T[8]-5.9399*0.000001*T[8]^2
C_P_Ash[8]=1.0926+1.8896*0.001*T[8]-3.6817*0.000001*T[8]^2
C_P_Water[8]=4.1762-9.0864*0.00001*T[8]+5.4731*0.000001*T[8]^2

C_P[8]=X_Protein[8]*C_P_Protein[8]+X_Ash[8]*C_P_Ash[8]+X_
Fat[8]*C_P_Fat[8]+X_Carbohydrate[8]*C_P_Carbohydrate[8]+X_
Water[8]*C_P_Water[8]
C_P_0[8]=X_Protein[8]*C_P_Protein_T_0+X_Ash[8]*C_P_Ash_T_0+X_
Fat[8]*C_P_Fat_T_0+X_Carbohydrate[8]*C_P_Carbohydrate_T_0+X_
Water[8]*C_P_Water_T_0
C_P_ref[8]=X_Protein[8]*C_P_Protein_T_ref+X_Ash[8]*C_P_Ash_T_
ref+X_Fat[8]*C_P_Fat_T_ref+X_Carbohydrate[8]*C_P_Carbohydrate_T_
ref+X_Water[8]*C_P_Water_T_ref
C_P_ave_en[8]=(C_P[8]+C_P_ref[8])/2
C_P_ave_ex[8]=(C_P[8]+C_P_0[8])/2

RHO_Protein[8]=1.3299*1000-0.5184*T[8]
RHO_Fat[8]=9.2559*100-0.41757*T[8]
RHO_Carbohydrate[8]=1.5991*1000-0.31046*T[8]
RHO_Ash[8]=2.4238*1000-0.28063*T[8]
RHO_Water[8]=9.9718*100+3.1439*0.001*T[8]-3.7574*0.001*T[8]^2
v[8]=X_Protein[8]/RHO_Protein[8]+X_Ash[8]/RHO_Ash[8]+X_Fat[8]/
RHO_Fat[8]+X_Carbohydrate[8]/RHO_Carbohydrate[8]+X_Water[8]/
RHO_Water[8]

h[8]=C_P_ave_en[8]*T[8]+v[8]*(P[8]-P[0])
ex[8]=C_P_ave_ex[8]*(T[8]-T[0]-T_K[0]*ln(T_K[8]/T_K[0]))+v[8]
*(P[8]-P[0])

{9}
P[9]=550
T[9]=67
T_K[9]=T[9]+273.15
TS[9]=TS[1]
X_Ash[9]=X_Ash[1]
```

```
X_Fat[9]=X_Fat[1]
X_Carbohydrate[9]=X_Carbohydrate[1]
X_Protein[9]=X_Protein[1]
X_Water[9]=X_Water[1]

C_P_Protein[9]=2.0082+1.2089*0.001*T[9]-1.3129*0.000001*T[9]^2
C_P_Fat[9]=1.9842+1.4733*0.001*T[9]-4.8008*0.000001*T[9]^2
C_P_Carbohydrate[9]=1.5488+1.9625*0.001*T[9]-5.9399*0.000001*T[9]^2
C_P_Ash[9]=1.0926+1.8896*0.001*T[9]-3.6817*0.000001*T[9]^2
C_P_Water[9]=4.1762-9.0864*0.00001*T[9]+5.4731*0.000001*T[9]^2

C_P[9]=X_Protein[9]*C_P_Protein[9]+X_Ash[9]*C_P_Ash[9]+X_
Fat[9]*C_P_Fat[9]+X_Carbohydrate[9]*C_P_Carbohydrate[9]+X_
Water[9]*C_P_Water[9]
C_P_0[9]=X_Protein[9]*C_P_Protein_T_0+X_Ash[9]*C_P_Ash_T_0+X_
Fat[9]*C_P_Fat_T_0+X_Carbohydrate[9]*C_P_Carbohydrate_T_0+X_
Water[9]*C_P_Water_T_0
C_P_ref[9]=X_Protein[9]*C_P_Protein_T_ref+X_Ash[9]*C_P_Ash_T_
ref+X_Fat[9]*C_P_Fat_T_ref+X_Carbohydrate[9]*C_P_Carbohydrate_T_
ref+X_Water[9]*C_P_Water_T_ref
C_P_ave_en[9]=(C_P[9]+C_P_ref[9])/2
C_P_ave_ex[9]=(C_P[9]+C_P_0[9])/2

RHO_Protein[9]=1.3299*1000-0.5184*T[9]
RHO_Fat[9]=9.2559*100-0.41757*T[9]
RHO_Carbohydrate[9]=1.5991*1000-0.31046*T[9]
RHO_Ash[9]=2.4238*1000-0.28063*T[9]
RHO_Water[9]=9.9718*100+3.1439*0.001*T[9]-3.7574*0.001*T[9]^2
v[9]=X_Protein[9]/RHO_Protein[9]+X_Ash[9]/RHO_Ash[9]+X_Fat[9]/
RHO_Fat[9]+X_Carbohydrate[9]/RHO_Carbohydrate[9]+X_Water[9]/
RHO_Water[9]

h[9]=C_P_ave_en[9]*T[9]+v[9]*(P[9]-P[0])
ex[9]=C_P_ave_ex[9]*(T[9]-T[0]-T_K[0]*ln(T_K[9]/T_K[0]))+v[9]
*(P[9]-P[0])

{mass balance}
m_dot[1]=3.5
m_dot[6]=3.5
m_dot[8]=3.5
m_dot[4]=m_dot[5]
m_dot[6]=m_dot[7]
m_dot[8]=m_dot[9]
m_dot[1]=m_dot[2]+m_dot[3]
TS[1]*m_dot[1]=TS[2]*m_dot[2]

{energy balance}
m_dot[1]*h[1]+m_dot[4]*h[4]+m_dot[6]*h[6]+m_dot[8]*h[8]+W_dot=m_
dot[2]*h[2]+m_dot[3]*h[3]+m_dot[5]*h[5]+m_dot[7]*h[7]+m_
dot[9]*h[9]
W_dot=1
{exergy balance}
Ex_dot[1]=m_dot[1]*ex[1]
Ex_dot[2]=m_dot[2]*ex[2]
Ex_dot[3]=m_dot[3]*ex[3]
Ex_dot[4]=m_dot[4]*ex[4]
Ex_dot[5]=m_dot[5]*ex[5]
```

```
Ex_dot[6]=m_dot[6]*ex[6]
Ex_dot[7]=m_dot[7]*ex[7]
Ex_dot[8]=m_dot[8]*ex[8]
Ex_dot[9]=m_dot[9]*ex[9]

Ex_dot_F=Ex_dot[1]+(Ex_dot[4]-Ex_dot[5])+W_dot
Ex_dot_P=Ex_dot[2]+Ex_dot[3]+(Ex_dot[7]-Ex_dot[6])+(Ex_dot[9]-Ex_
dot[8])
Ex_dot_L=0
Ex_dot_des=Ex_dot_F-Ex_dot_P-Ex_dot_L
ETA_ex=(Ex_dot_P*100)/Ex_dot_F
```

2.3 Example 6.22

```
{0}
P[0]=101.325
T[0]=25
T_K[0]=T[0]+273.15
h[0]=enthalpy(Water,T=T[0],P=P[0])
s[0]=entropy(Water,T=T[0],P=P[0])

T_ref=0
C_P_Protein_T_ref=2.0082
C_P_Fat_T_ref=1.9842
C_P_Carbohydrate_T_ref=1.5488
C_P_Ash_T_ref=1.0926
C_P_Water_T_ref=4.1762

C_P_Protein_T_0=2.0082+1.2089*0.001*T[0]-1.3129*0.000001*T[0]^2
C_P_Fat_T_0=1.9842+1.4733*0.001*T[0]-4.8008*0.000001*T[0]^2
C_P_Carbohydrate_T_0=1.5488+1.9625*0.001*T[0]-5.9399*0.000001*T[0]^2
C_P_Ash_T_0=1.0926+1.8896*0.001*T[0]-3.6817*0.000001*T[0]^2
C_P_Water_T_0=4.1762-9.0864*0.00001*T[0]+5.4731*0.000001*T[0]^2

X_Ash=0
X_Fat=0
X_Carbohydrate=0.142
X_Protein=0
X_Water=0.858

{1}
P[1]=551.325
T[1]=90.10
T_K[1]=T[1]+273.15

C_P_Protein[1]=2.0082+1.2089*0.001*T[1]-1.3129*0.000001*T[1]^2
C_P_Fat[1]=1.9842+1.4733*0.001*T[1]-4.8008*0.000001*T[1]^2
C_P_Carbohydrate[1]=1.5488+1.9625*0.001*T[1]-5.9399*0.000001*T[1]^2
C_P_Ash[1]=1.0926+1.8896*0.001*T[1]-3.6817*0.000001*T[1]^2
C_P_Water[1]=4.1762-9.0864*0.00001*T[1]+5.4731*0.000001*T[1]^2

C_P[1]=X_Protein*C_P_Protein[1]+X_Ash*C_P_Ash[1]+X_Fat*C_P_
Fat[1]+X_Carbohydrate*C_P_Carbohydrate[1]+X_Water*C_P_Water[1]
C_P_0[1]=X_Protein*C_P_Protein_T_0+X_Ash*C_P_Ash_T_0+X_Fat*C_P_
Fat_T_0+X_Carbohydrate*C_P_Carbohydrate_T_0+X_Water*C_P_Water_T_0
```

```
C_P_ref[1]=X_Protein*C_P_Protein_T_ref+X_Ash*C_P_Ash_T_ref+X_
Fat*C_P_Fat_T_ref+X_Carbohydrate*C_P_Carbohydrate_T_ref+X_
Water*C_P_Water_T_ref
C_P_ave_en[1]=(C_P[1]+C_P_ref[1])/2
C_P_ave_ex[1]=(C_P[1]+C_P_0[1])/2

RHO_Protein[1]=1.3299*1000-0.5184*T[1]
RHO_Fat[1]=9.2559*100-0.41757*T[1]
RHO_Carbohydrate[1]=1.5991*1000-0.31046*T[1]
RHO_Ash[1]=2.4238*1000-0.28063*T[1]
RHO_Water[1]=9.9718*100+3.1439*0.001*T[1]-3.7574*0.001*T[1]^2
v[1]=X_Protein/RHO_Protein[1]+X_Ash/RHO_Ash[1]+X_Fat/RHO_
Fat[1]+X_Carbohydrate/RHO_Carbohydrate[1]+X_Water/RHO_Water[1]

h[1]=C_P_ave_en[1]*T[1]+v[1]*(P[1]-P[0])
ex[1]=C_P_ave_ex[1]*(T[1]-T[0]-T_K[0]*ln(T_K[1]/T_K[0]))+v[1]
*(P[1]-P[0])

{2}
P[2]=491.325
T[2]=108
T_K[2]=T[2]+273.15

C_P_Protein[2]=2.0082+1.2089*0.001*T[2]-1.3129*0.000001*T[2]^2
C_P_Fat[2]=1.9842+1.4733*0.001*T[2]-4.8008*0.000001*T[2]^2
C_P_Carbohydrate[2]=1.5488+1.9625*0.001*T[2]-5.9399*0.000001*T[2]^2
C_P_Ash[2]=1.0926+1.8896*0.001*T[2]-3.6817*0.000001*T[2]^2
C_P_Water[2]=4.1762-9.0864*0.00001*T[2]+5.4731*0.000001*T[2]^2

C_P[2]=X_Protein*C_P_Protein[2]+X_Ash*C_P_Ash[2]+X_Fat*C_P_
Fat[2]+X_Carbohydrate*C_P_Carbohydrate[2]+X_Water*C_P_Water[2]
C_P_0[2]=X_Protein*C_P_Protein_T_0+X_Ash*C_P_Ash_T_0+X_Fat*C_P_
Fat_T_0+X_Carbohydrate*C_P_Carbohydrate_T_0+X_Water*C_P_Water_T_0
C_P_ref[2]=X_Protein*C_P_Protein_T_ref+X_Ash*C_P_Ash_T_ref+X_
Fat*C_P_Fat_T_ref+X_Carbohydrate*C_P_Carbohydrate_T_ref+X_
Water*C_P_Water_T_ref
C_P_ave_en[2]=(C_P[2]+C_P_ref[2])/2
C_P_ave_ex[2]=(C_P[2]+C_P_0[2])/2

RHO_Protein[2]=1.3299*1000-0.5184*T[2]
RHO_Fat[2]=9.2559*100-0.41757*T[2]
RHO_Carbohydrate[2]=1.5991*1000-0.31046*T[2]
RHO_Ash[2]=2.4238*1000-0.28063*T[2]
RHO_Water[2]=9.9718*100+3.1439*0.001*T[2]-3.7574*0.001*T[2]^2
v[2]=X_Protein/RHO_Protein[2]+X_Ash/RHO_Ash[2]+X_Fat/RHO_
Fat[2]+X_Carbohydrate/RHO_Carbohydrate[2]+X_Water/RHO_Water[2]

h[2]=C_P_ave_en[2]*T[2]+v[2]*(P[2]-P[0])
ex[2]=C_P_ave_ex[2]*(T[2]-T[0]-T_K[0]*ln(T_K[2]/T_K[0]))+v[2]
*(P[2]-P[0])

{3}
P[3]=441.325
T[3]=118
T_K[3]=T[3]+273.15

C_P_Protein[3]=2.0082+1.2089*0.001*T[3]-1.3129*0.000001*T[3]^2
C_P_Fat[3]=1.9842+1.4733*0.001*T[3]-4.8008*0.000001*T[3]^2
```

```
C_P_Carbohydrate[3]=1.5488+1.9625*0.001*T[3]-5.9399*0.000001*T[3]^2
C_P_Ash[3]=1.0926+1.8896*0.001*T[3]-3.6817*0.000001*T[3]^2
C_P_Water[3]=4.1762-9.0864*0.00001*T[3]+5.4731*0.000001*T[3]^2

C_P[3]=X_Protein*C_P_Protein[3]+X_Ash*C_P_Ash[3]+X_Fat*C_P_
Fat[3]+X_Carbohydrate*C_P_Carbohydrate[3]+X_Water*C_P_Water[3]
C_P_0[3]=X_Protein*C_P_Protein_T_0+X_Ash*C_P_Ash_T_0+X_Fat
*C_P_Fat_T_0+X_Carbohydrate*C_P_Carbohydrate_T_0+X_Water
*C_P_Water_T_0
C_P_ref[3]=X_Protein*C_P_Protein_T_ref+X_Ash*C_P_Ash_T_ref+X_
Fat*C_P_Fat_T_ref+X_Carbohydrate*C_P_Carbohydrate_T_ref+X_
Water*C_P_Water_T_ref
C_P_ave_en[3]=(C_P[3]+C_P_ref[3])/2
C_P_ave_ex[3]=(C_P[3]+C_P_0[3])/2

RHO_Protein[3]=1.3299*1000-0.5184*T[3]
RHO_Fat[3]=9.2559*100-0.41757*T[3]
RHO_Carbohydrate[3]=1.5991*1000-0.31046*T[3]
RHO_Ash[3]=2.4238*1000-0.28063*T[3]
RHO_Water[3]=9.9718*100+3.1439*0.001*T[3]-3.7574*0.001*T[3]^2
v[3]=X_Protein/RHO_Protein[3]+X_Ash/RHO_Ash[3]+X_Fat/RHO_
Fat[3]+X_Carbohydrate/RHO_Carbohydrate[3]+X_Water/RHO_Water[3]

h[3]=C_P_ave_en[3]*T[3]+v[3]*(P[3]-P[0])
ex[3]=C_P_ave_ex[3]*(T[3]-T[0]-T_K[0]*ln(T_K[3]/T_K[0]))+v[3]
*(P[3]-P[0])

{4}
P[4]=391.325
T[4]=124
T_K[4]=T[4]+273.15

C_P_Protein[4]=2.0082+1.2089*0.001*T[4]-1.3129*0.000001*T[4]^2
C_P_Fat[4]=1.9842+1.4733*0.001*T[4]-4.8008*0.000001*T[4]^2
C_P_Carbohydrate[4]=1.5488+1.9625*0.001*T[4]-5.9399*0.000001
*T[4]^2
C_P_Ash[4]=1.0926+1.8896*0.001*T[4]-3.6817*0.000001*T[4]^2
C_P_Water[4]=4.1762-9.0864*0.00001*T[4]+5.4731*0.000001*T[4]^2

C_P[4]=X_Protein*C_P_Protein[4]+X_Ash*C_P_Ash[4]+X_Fat*C_P_
Fat[4]+X_Carbohydrate*C_P_Carbohydrate[4]+X_Water*C_P_Water[4]
C_P_0[4]=X_Protein*C_P_Protein_T_0+X_Ash*C_P_Ash_T_0+X_Fat
*C_P_Fat_T_0+X_Carbohydrate*C_P_Carbohydrate_T_0+X_Water
*C_P_Water_T_0
C_P_ref[4]=X_Protein*C_P_Protein_T_ref+X_Ash*C_P_Ash_T_ref+X_
Fat*C_P_Fat_T_ref+X_Carbohydrate*C_P_Carbohydrate_T_ref+X_
Water*C_P_Water_T_ref
C_P_ave_en[4]=(C_P[4]+C_P_ref[4])/2
C_P_ave_ex[4]=(C_P[4]+C_P_0[4])/2

RHO_Protein[4]=1.3299*1000-0.5184*T[4]
RHO_Fat[4]=9.2559*100-0.41757*T[4]
RHO_Carbohydrate[4]=1.5991*1000-0.31046*T[4]
RHO_Ash[4]=2.4238*1000-0.28063*T[4]
RHO_Water[4]=9.9718*100+3.1439*0.001*T[4]-3.7574*0.001*T[4]^2
v[4]=X_Protein/RHO_Protein[4]+X_Ash/RHO_Ash[4]+X_Fat/RHO_
Fat[4]+X_Carbohydrate/RHO_Carbohydrate[4]+X_Water/RHO_Water[4]
```

```
h[4]=C_P_ave_en[4]*T[4]+v[4]*(P[4]-P[0])
ex[4]=C_P_ave_ex[4]*(T[4]-T[0]-T_K[0]*ln(T_K[4]/T_K[0]))+v[4]
*(P[4]-P[0])

{5}
P[5]=171.325
x[5]=1
h[5]=enthalpy(Water,x=x[5],P=P[5])
s[5]=entropy(Water,x=x[5],P=P[5])
ex[5]=h[5]-h[0]-T_K[0]*(s[5]-s[0])

{6}
P[6]=141.325
x[6]=0
h[6]=enthalpy(Water,x=x[6],P=P[6])
s[6]=entropy(Water,x=x[6],P=P[6])
ex[6]=h[6]-h[0]-T_K[0]*(s[6]-s[0])

{7}
P[7]=241.325
x[7]=1
h[7]=enthalpy(Water,x=x[7],P=P[7])
s[7]=entropy(Water,x=x[7],P=P[7])
ex[7]=h[7]-h[0]-T_K[0]*(s[7]-s[0])

{8}
P[8]=201.325
x[8]=0
h[8]=enthalpy(Water,x=x[8],P=P[8])
s[8]=entropy(Water,x=x[8],P=P[8])
ex[8]=h[8]-h[0]-T_K[0]*(s[8]-s[0])

{9}
P[9]=369.45
x[9]=1
h[9]=enthalpy(Water,x=x[9],P=P[9])
s[9]=entropy(Water,x=x[9],P=P[9])
ex[9]=h[9]-h[0]-T_K[0]*(s[9]-s[0])

{10}
P[10]=340.325
x[10]=0
h[10]=enthalpy(Water,x=x[10],P=P[10])
s[10]=entropy(Water,x=x[10],P=P[10])
ex[10]=h[10]-h[0]-T_K[0]*(s[10]-s[0])

{mass and energy balance}
m_dot[1]=48.60
m_dot[2]=m_dot[1]
m_dot[5]=1.54
m_dot[6]=m_dot[5]
m_dot[1]*h[1]+m_dot[5]*h[5]=m_dot[2]*h[2]+m_dot[6]*h[6]+Q_dot_L_1

m_dot[3]=m_dot[2]
m_dot[7]=0.88
m_dot[8]=m_dot[7]
m_dot[2]*h[2]+m_dot[7]*h[7]=m_dot[3]*h[3]+m_dot[8]*h[8]+Q_dot_L_2
```

```
m_dot[4]=m_dot[3]
m_dot[9]=0.55
m_dot[10]=m_dot[9]
m_dot[3]*h[3]+m_dot[9]*h[9]=m_dot[4]*h[4]+m_dot[10]*h[10]+Q_dot_L_3

{exergy balance}
Ex_dot[1]=m_dot[1]*ex[1]
Ex_dot[2]=m_dot[2]*ex[2]
Ex_dot[3]=m_dot[3]*ex[3]
Ex_dot[4]=m_dot[4]*ex[4]
Ex_dot[5]=m_dot[5]*ex[5]
Ex_dot[6]=m_dot[6]*ex[6]
Ex_dot[7]=m_dot[7]*ex[7]
Ex_dot[8]=m_dot[8]*ex[8]
Ex_dot[9]=m_dot[9]*ex[9]
Ex_dot[10]=m_dot[10]*ex[10]

T_b_1=45+273.15
Ex_dot_L_1=Q_dot_L_1*(1-((T_K[0])/(T_b_1)))
Ex_dot_F_1=(Ex_dot[5]-Ex_dot[6])
Ex_dot_P_1=(Ex_dot[2]-Ex_dot[1])
Ex_dot_des_1=Ex_dot_F_1-Ex_dot_P_1-Ex_dot_L_1
ETA_ex_1=(Ex_dot_P_1*100)/Ex_dot_F_1

T_b_2=49+273.15
Ex_dot_L_2=Q_dot_L_2*(1-((T_K[0])/(T_b_2)))
Ex_dot_F_2=(Ex_dot[7]-Ex_dot[8])
Ex_dot_P_2=(Ex_dot[3]-Ex_dot[2])
Ex_dot_des_2=Ex_dot_F_2-Ex_dot_P_2-Ex_dot_L_2
ETA_ex_2=(Ex_dot_P_2*100)/Ex_dot_F_2

T_b_3=56+273.15
Ex_dot_L_3=Q_dot_L_3*(1-((T_K[0])/(T_b_3)))
Ex_dot_F_3=(Ex_dot[9]-Ex_dot[10])
Ex_dot_P_3=(Ex_dot[4]-Ex_dot[3])
Ex_dot_des_3=Ex_dot_F_3-Ex_dot_P_3-Ex_dot_L_3
ETA_ex_3=(Ex_dot_P_3*100)/Ex_dot_F_3
```

2.4 Example 6.23

```
{0}
P[0]=101.325
T[0]=25
T_K[0]=T[0]+273.15
h[0]=enthalpy(Water,T=T[0],P=P[0])
s[0]=entropy(Water,T=T[0],P=P[0])

T_ref=0
C_P_Protein_T_ref=2.0082
C_P_Fat_T_ref=1.9842
C_P_Carbohydrate_T_ref=1.5488
C_P_Ash_T_ref=1.0926
C_P_Water_T_ref=4.1762

C_P_Protein_T_0=2.0082+1.2089*0.001*T[0]-1.3129*0.000001*T[0]^2
C_P_Fat_T_0=1.9842+1.4733*0.001*T[0]-4.8008*0.000001*T[0]^2
```

```
C_P_Carbohydrate_T_0=1.5488+1.9625*0.001*T[0]-5.9399*0.000001*T[0]^2
C_P_Ash_T_0=1.0926+1.8896*0.001*T[0]-3.6817*0.000001*T[0]^2
C_P_Water_T_0=4.1762-9.0864*0.00001*T[0]+5.4731*0.000001*T[0]^2

RHO_Protein_T_0=1.3299*1000
RHO_Fat_T_0=9.2559*100
RHO_Carbohydrate_T_0=1.5991*1000
RHO_Ash_T_0=2.4238*1000
RHO_Water_T_0=9.9718*100

X_Ash=0
X_Fat=0
X_Carbohydrate=0.8
X_Protein=0
X_Water=0.2

v[0]=X_Protein/RHO_Protein_T_0+X_Ash/RHO_Ash_T_0+X_Fat/RHO_
Fat_T_0+X_Carbohydrate/RHO_Carbohydrate_T_0+X_Water/RHO_Water_T_0

{1}
P[1]=101.325
T[1]=30
T_K[1]=T[1]+273.15

C_P_Protein[1]=2.0082+1.2089*0.001*T[1]-1.3129*0.000001*T[1]^2
C_P_Fat[1]=1.9842+1.4733*0.001*T[1]-4.8008*0.000001*T[1]^2
C_P_Carbohydrate[1]=1.5488+1.9625*0.001*T[1]-5.9399*0.000001*T[1]^2
C_P_Ash[1]=1.0926+1.8896*0.001*T[1]-3.6817*0.000001*T[1]^2
C_P_Water[1]=4.1762-9.0864*0.00001*T[1]+5.4731*0.000001*T[1]^2

C_P[1]=X_Protein*C_P_Protein[1]+X_Ash*C_P_Ash[1]+X_Fat*C_P_
Fat[1]+X_Carbohydrate*C_P_Carbohydrate[1]+X_Water*C_P_Water[1]
C_P_0[1]=X_Protein*C_P_Protein_T_0+X_Ash*C_P_Ash_T_0+X_Fat*C_P_
Fat_T_0+X_Carbohydrate*C_P_Carbohydrate_T_0+X_Water*C_P_Water_T_0
C_P_ref[1]=X_Protein*C_P_Protein_T_ref+X_Ash*C_P_Ash_T_ref+X_
Fat*C_P_Fat_T_ref+X_Carbohydrate*C_P_Carbohydrate_T_ref+X_
Water*C_P_Water_T_ref
C_P_ave_en[1]=(C_P[1]+C_P_ref[1])/2
C_P_ave_ex[1]=(C_P[1]+C_P_0[1])/2

RHO_Protein[1]=1.3299*1000-0.5184*T[1]
RHO_Fat[1]=9.2559*100-0.41757*T[1]
RHO_Carbohydrate[1]=1.5991*1000-0.31046*T[1]
RHO_Ash[1]=2.4238*1000-0.28063*T[1]
RHO_Water[1]=9.9718*100+3.1439*0.001*T[1]-3.7574*0.001*T[1]^2
v[1]=X_Protein/RHO_Protein[1]+X_Ash/RHO_Ash[1]+X_Fat/RHO_
Fat[1]+X_Carbohydrate/RHO_Carbohydrate[1]+X_Water/RHO_Water[1]

u[1]=C_P_ave_en[1]*T[1]
ex[1]=C_P_ave_ex[1]*(T[1]-T[0]-T_K[0]*ln(T_K[1]/T_K[0]))+P[0]
*(v[1]-v[0])

{2}
P[2]=101.325
T[2]=74
T_K[2]=T[2]+273.15
```

```
C_P_Protein[2]=2.0082+1.2089*0.001*T[2]-1.3129*0.000001*T[2]^2
C_P_Fat[2]=1.9842+1.4733*0.001*T[2]-4.8008*0.000001*T[2]^2
C_P_Carbohydrate[2]=1.5488+1.9625*0.001*T[2]-5.9399*0.000001*T[2]^2
C_P_Ash[2]=1.0926+1.8896*0.001*T[2]-3.6817*0.000001*T[2]^2
C_P_Water[2]=4.1762-9.0864*0.00001*T[2]+5.4731*0.000001*T[2]^2

C_P[2]=X_Protein*C_P_Protein[2]+X_Ash*C_P_Ash[2]+X_Fat*C_P_
Fat[2]+X_Carbohydrate*C_P_Carbohydrate[2]+X_Water*C_P_Water[2]
C_P_0[2]=X_Protein*C_P_Protein_T_0+X_Ash*C_P_Ash_T_0+X_Fat*C_P_
Fat_T_0+X_Carbohydrate*C_P_Carbohydrate_T_0+X_Water*C_P_Water_T_0
C_P_ref[2]=X_Protein*C_P_Protein_T_ref+X_Ash*C_P_Ash_T_ref+X_
Fat*C_P_Fat_T_ref+X_Carbohydrate*C_P_Carbohydrate_T_ref+X_
Water*C_P_Water_T_ref
C_P_ave_en[2]=(C_P[2]+C_P_ref[2])/2
C_P_ave_ex[2]=(C_P[2]+C_P_0[2])/2

RHO_Protein[2]=1.3299*1000-0.5184*T[2]
RHO_Fat[2]=9.2559*100-0.41757*T[2]
RHO_Carbohydrate[2]=1.5991*1000-0.31046*T[2]
RHO_Ash[2]=2.4238*1000-0.28063*T[2]
RHO_Water[2]=9.9718*100+3.1439*0.001*T[2]-3.7574*0.001*T[2]^2
v[2]=X_Protein/RHO_Protein[2]+X_Ash/RHO_Ash[2]+X_Fat/RHO_
Fat[2]+X_Carbohydrate/RHO_Carbohydrate[2]+X_Water/RHO_Water[2]

u[2]=C_P_ave_en[2]*T[2]
ex[2]=C_P_ave_ex[2]*(T[2]-T[0]-T_K[0]*ln(T_K[2]/T_K[0]))+P[0]
*(v[2]-v[0])

{3}
P[3]=151.325
T[3]=147.73
h[3]=enthalpy(Water,T=T[3],P=P[3])
s[3]=entropy(Water,T=T[3],P=P[3])
ex[3]=h[3]-h[0]-T_K[0]*(s[3]-s[0])

{4}
P[4]=121.325
T[4]=84.3
h[4]=enthalpy(Water,T=T[4],P=P[4])
s[4]=entropy(Water,T=T[4],P=P[4])
ex[4]=h[4]-h[0]-T_K[0]*(s[4]-s[0])

{mass and energy balance}
m_dot[3]=0.01
m_dot[4]=m_dot[3]
m_honey=120

(m_dot[3]*(h[3]-h[4])+1)*(15*60)-Q_L=m_honey*(u[2]-u[1])

{exergy balance}
Ex_dot[3]=m_dot[3]*ex[3]
Ex_dot[4]=m_dot[4]*ex[4]

T_b=60+273.15
Ex_L=Q_L*(1-((T_K[0])/(T_b)))
(m_dot[3]*(ex[3]-ex[4])+1)*(15*60)-Ex_L-Ex_des=m_
honey*(ex[2]-ex[1])
```

2.5 Example 8.1

```
{0}
P[0]=101.325
T[0]=25
T_K[0]=T[0]+273.15
h[0]=enthalpy(Water,T=T[0],P=P[0])
s[0]=entropy(Water,T=T[0],P=P[0])

T_ref=0
C_P_Protein_T_ref=2.0082
C_P_Fat_T_ref=1.9842
C_P_Carbohydrate_T_ref=1.5488
C_P_Ash_T_ref=1.0926
C_P_Water_T_ref=4.1762

C_P_Protein_T_0=2.0082+1.2089*0.001*T[0]-1.3129*0.000001*T[0]^2
C_P_Fat_T_0=1.9842+1.4733*0.001*T[0]-4.8008*0.000001*T[0]^2
C_P_Carbohydrate_T_0=1.5488+1.9625*0.001*T[0]-5.9399*0.000001
*T[0]^2
C_P_Ash_T_0=1.0926+1.8896*0.001*T[0]-3.6817*0.000001*T[0]^2
C_P_Water_T_0=4.1762-9.0864*0.00001*T[0]+5.4731*0.000001*T[0]^2

{1}
P[1]=551.325
T[1]=50
T_K[1]=T[1]+273.15
X_Ash[1]=0.0032
X_Fat[1]=0.00008
X_Carbohydrate[1]=0.09892
X_Protein[1]=0.0029
X_Water[1]=0.8949

C_P_Protein[1]=2.0082+1.2089*0.001*T[1]-1.3129*0.000001*T[1]^2
C_P_Fat[1]=1.9842+1.4733*0.001*T[1]-4.8008*0.000001*T[1]^2
C_P_Carbohydrate[1]=1.5488+1.9625*0.001*T[1]-5.9399*0.000001
*T[1]^2
C_P_Ash[1]=1.0926+1.8896*0.001*T[1]-3.6817*0.000001*T[1]^2
C_P_Water[1]=4.1762-9.0864*0.00001*T[1]+5.4731*0.000001*T[1]^2

C_P[1]=X_Protein[1]*C_P_Protein[1]+X_Ash[1]*C_P_Ash[1]+X_
Fat[1]*C_P_Fat[1]+X_Carbohydrate[1]*C_P_Carbohydrate[1]+X_
Water[1]*C_P_Water[1]
C_P_0[1]=X_Protein[1]*C_P_Protein_T_0+X_Ash[1]*C_P_Ash_T_0+X_
Fat[1]*C_P_Fat_T_0+X_Carbohydrate[1]*C_P_Carbohydrate_T_0+X_
Water[1]*C_P_Water_T_0
C_P_ref[1]=X_Protein[1]*C_P_Protein_T_ref+X_Ash[1]*C_P_Ash_T_
ref+X_Fat[1]*C_P_Fat_T_ref+X_Carbohydrate[1]*C_P_Carbohydrate_T_
ref+X_Water[1]*C_P_Water_T_ref
C_P_ave_en[1]=(C_P[1]+C_P_ref[1])/2
C_P_ave_ex[1]=(C_P[1]+C_P_0[1])/2

RHO_Protein[1]=1.3299*1000-0.5184*T[1]
RHO_Fat[1]=9.2559*100-0.41757*T[1]
RHO_Carbohydrate[1]=1.5991*1000-0.31046*T[1]
RHO_Ash[1]=2.4238*1000-0.28063*T[1]
RHO_Water[1]=9.9718*100+3.1439*0.001*T[1]-3.7574*0.001*T[1]^2
```

```
v[1]=X_Protein[1]/RHO_Protein[1]+X_Ash[1]/RHO_Ash[1]+X_Fat[1]/
RHO_Fat[1]+X_Carbohydrate[1]/RHO_Carbohydrate[1]+X_Water[1]/
RHO_Water[1]

h[1]=C_P_ave_en[1]*T[1]+v[1]*(P[1]-P[0])
ex[1]=C_P_ave_ex[1]*(T[1]-T[0]-T_K[0]*ln(T_K[1]/T_K[0]))+v[1]
*(P[1]-P[0])

{2}
P[2]=451.325
T[2]=70
T_K[2]=T[2]+273.15
X_Ash[2]=0.0032
X_Fat[2]=0.00008
X_Carbohydrate[2]=0.09892
X_Protein[2]=0.0029
X_Water[2]=0.8949

C_P_Protein[2]=2.0082+1.2089*0.001*T[2]-1.3129*0.000001*T[2]^2
C_P_Fat[2]=1.9842+1.4733*0.001*T[2]-4.8008*0.000001*T[2]^2
C_P_Carbohydrate[2]=1.5488+1.9625*0.001*T[2]-5.9399*0.000001
*T[2]^2
C_P_Ash[2]=1.0926+1.8896*0.001*T[2]-3.6817*0.000001*T[2]^2
C_P_Water[2]=4.1762-9.0864*0.00001*T[2]+5.4731*0.000001*T[2]^2

C_P[2]=X_Protein[2]*C_P_Protein[2]+X_Ash[2]*C_P_Ash[2]+X_
Fat[2]*C_P_Fat[2]+X_Carbohydrate[2]*C_P_Carbohydrate[2]+X_
Water[2]*C_P_Water[2]
C_P_0[2]=X_Protein[2]*C_P_Protein_T_0+X_Ash[2]*C_P_Ash_T_0+X_
Fat[2]*C_P_Fat_T_0+X_Carbohydrate[2]*C_P_Carbohydrate_T_0+X_
Water[2]*C_P_Water_T_0
C_P_ref[2]=X_Protein[2]*C_P_Protein_T_ref+X_Ash[2]*C_P_Ash_T_
ref+X_Fat[2]*C_P_Fat_T_ref+X_Carbohydrate[2]*C_P_Carbohydrate_T_
ref+X_Water[2]*C_P_Water_T_ref
C_P_ave_en[2]=(C_P[2]+C_P_ref[2])/2
C_P_ave_ex[2]=(C_P[2]+C_P_0[2])/2

RHO_Protein[2]=1.3299*1000-0.5184*T[2]
RHO_Fat[2]=9.2559*100-0.41757*T[2]
RHO_Carbohydrate[2]=1.5991*1000-0.31046*T[2]
RHO_Ash[2]=2.4238*1000-0.28063*T[2]
RHO_Water[2]=9.9718*100+3.1439*0.001*T[2]-3.7574*0.001*T[2]^2
v[2]=X_Protein[2]/RHO_Protein[2]+X_Ash[2]/RHO_Ash[2]+X_Fat[2]/
RHO_Fat[2]+X_Carbohydrate[2]/RHO_Carbohydrate[2]+X_Water[2]/
RHO_Water[2]
h[2]=C_P_ave_en[2]*T[2]+v[2]*(P[2]-P[0])
ex[2]=C_P_ave_ex[2]*(T[2]-T[0]-T_K[0]*ln(T_K[2]/T_K[0]))+v[2]
*(P[2]-P[0])

{3}
P[3]=481.325
T[3]=105
T_K[3]=T[3]+273.15
h[3]=enthalpy(Water,T=T[3],P=P[3])
s[3]=entropy(Water,T=T[3],P=P[3])
ex[3]=h[3]-h[0]-T_K[0]*(s[3]-s[0])
```

```
{4}
P[4]=331.325
T[4]=85
T_K[4]=T[4]+273.15
h[4]=enthalpy(Water,T=T[4],P=P[4])
s[4]=entropy(Water,T=T[4],P=P[4])
ex[4]=h[4]-h[0]-T_K[0]*(s[4]-s[0])

{mass balance}
m_dot[1]=3.5
m_dot[2]=m_dot[1]
m_dot[3]=m_dot[4]

{energy balance}
m_dot[1]*h[1]+m_dot[3]*h[3]=m_dot[2]*h[2]+m_dot[4]*h[4]

{exergy balance}
Ex_dot[1]=m_dot[1]*ex[1]
Ex_dot[2]=m_dot[2]*ex[2]
Ex_dot[3]=m_dot[3]*ex[3]
Ex_dot[4]=m_dot[4]*ex[4]

Ex_dot_F=Ex_dot[3]-Ex_dot[4]
Ex_dot_P=Ex_dot[2]-Ex_dot[1]
Ex_dot_des=Ex_dot_F-Ex_dot_P
ETA_ex=(Ex_dot_P*100)/Ex_dot_F

{Cost balance}
PHI=1.05
y=6*30*24
CRF=(i*((1+i)^n))/(((1+i)^n)-1)
n=20
i=0.2
Z=2500
Z_dot=(Z*CRF*PHI)/(y*3600)

C_dot[1]=c[1]*Ex_dot[1]
C_dot[2]=c[2]*Ex_dot[2]
C_dot[3]=c[3]*Ex_dot[3]
C_dot[4]=c[4]*Ex_dot[4]

C_dot_F=C_dot[3]-C_dot[4]
C_dot[3]=0.1
c_F=C_dot_F/Ex_dot_F
C_dot_P=C_dot[2]-C_dot[1]
C_dot[1]=1
C_dot_F+Z_dot=C_dot_P
c[3]=c[4] {F rule}
C_dot_des=c_F*Ex_dot_des
C_dot_total=C_dot_des+Z_dot
```

2.6 Example 8.2

```
{0}
P[0]=101.325
T[0]=25
```

```
T_K[0]=T[0]+273.15
h[0]=enthalpy(Water,T=T[0],P=P[0])
s[0]=entropy(Water,T=T[0],P=P[0])

T_ref=0
C_P_Carbohydrate_T_ref=1.5488
C_P_Water_T_ref=4.1762

C_P_Carbohydrate_T_0=1.5488+1.9625*0.001*T[0]-5.9399*0.000001
*T[0]^2
C_P_Water_T_0=4.1762-9.0864*0.00001*T[0]+5.4731*0.000001*T[0]^2

{1}
P[1]=441.325
T[1]=118
T_K[1]=T[1]+273.15
X_Carbohydrate[1]=0.142
X_Water[1]=0.858

C_P_Carbohydrate[1]=1.5488+1.9625*0.001*T[1]-5.9399*0.000001
*T[1]^2
C_P_Water[1]=4.1762-9.0864*0.00001*T[1]+5.4731*0.000001*T[1]^2

C_P[1]=X_Carbohydrate[1]*C_P_Carbohydrate[1]+X_Water[1]*C_P_Water[1]
C_P_0[1]=X_Carbohydrate[1]*C_P_Carbohydrate_T_0+X_Water[1]*C_P_
Water_T_0
C_P_ref[1]=X_Carbohydrate[1]*C_P_Carbohydrate_T_ref+X_
Water[1]*C_P_Water_T_ref
C_P_ave_en[1]=(C_P[1]+C_P_ref[1])/2
C_P_ave_ex[1]=(C_P[1]+C_P_0[1])/2

RHO_Carbohydrate[1]=1.5991*1000-0.31046*T[1]
RHO_Water[1]=9.9718*100+3.1439*0.001*T[1]-3.7574*0.001*T[1]^2
v[1]=X_Carbohydrate[1]/RHO_Carbohydrate[1]+X_Water[1]/
RHO_Water[1]

h[1]=C_P_ave_en[1]*T[1]+v[1]*(P[1]-P[0])
ex[1]=C_P_ave_ex[1]*(T[1]-T[0]-T_K[0]*ln(T_K[1]/T_K[0]))+v[1]
*(P[1]-P[0])

{2}
P[2]=391.325
T[2]=124
T_K[2]=T[2]+273.15
X_Carbohydrate[2]=0.142
X_Water[2]=0.858

C_P_Carbohydrate[2]=1.5488+1.9625*0.00
1*T[2]-5.9399*0.000001*T[2]^2
C_P_Water[2]=4.1762-9.0864*0.00001*T[2]+5.4731*0.000001*T[2]^2

C_P[2]=X_Carbohydrate[2]*C_P_Carbohydrate[2]+X_Water[2]*C_P_
Water[2]
C_P_0[2]=X_Carbohydrate[2]*C_P_Carbohydrate_T_0+X_Water[2]*C_P_
Water_T_0
C_P_ref[2]=X_Carbohydrate[2]*C_P_Carbohydrate_T_ref+X_
Water[2]*C_P_Water_T_ref
```

```
C_P_ave_en[2]=(C_P[2]+C_P_ref[2])/2
C_P_ave_ex[2]=(C_P[2]+C_P_0[2])/2

RHO_Carbohydrate[2]=1.5991*1000-0.31046*T[2]
RHO_Water[2]=9.9718*100+3.1439*0.001*T[2]-3.7574*0.001*T[2]^2
v[2]=X_Carbohydrate[2]/RHO_Carbohydrate[2]+X_Water[2]/
RHO_Water[2]
h[2]=C_P_ave_en[2]*T[2]+v[2]*(P[2]-P[0])
ex[2]=C_P_ave_ex[2]*(T[2]-T[0]-T_K[0]*ln(T_K[2]/
T_K[0]))+v[2]*(P[2]-P[0])

{3}
T[3]=140.8
T_K[3]=T[3]+273.15
x[3]=1
h[3]=enthalpy(Water,T=T[3],x=x[3])
s[3]=entropy(Water,T=T[3],x=x[3])
ex[3]=h[3]-h[0]-T_K[0]*(s[3]-s[0])

{4}
T[4]=140.8
T_K[4]=T[4]+273.15
x[4]=0
h[4]=enthalpy(Water,T=T[4],x=x[4])
s[4]=entropy(Water,T=T[4],x=x[4])
ex[4]=h[4]-h[0]-T_K[0]*(s[4]-s[0])

{mass balance}
m_dot[1]=48.60
m_dot[2]=m_dot[1]
m_dot[3]=m_dot[4]

{energy balance}
m_dot[1]*h[1]+m_dot[3]*h[3]=m_dot[2]*h[2]+m_dot[4]*h[4]+Q_dot_L
Q_dot_L=4
T_b=56+273.15

{exergy balance}
Ex_dot[1]=m_dot[1]*ex[1]
Ex_dot[2]=m_dot[2]*ex[2]
Ex_dot[3]=m_dot[3]*ex[3]
Ex_dot[4]=m_dot[4]*ex[4]

Ex_dot_F=Ex_dot[3]-Ex_dot[4]
Ex_dot_P=Ex_dot[2]-Ex_dot[1]
Ex_dot_L=Q_dot_L*(1-(T_K[0]/T_b))
Ex_dot_des=Ex_dot_F-Ex_dot_P-Ex_dot_L
ETA_ex=(Ex_dot_P*100)/Ex_dot_F

PHI=1.05
y=3*30*24
CRF=(i*((1+i)^n))/(((1+i)^n)-1)
n=20
i=0.2
Z=8570
Z_dot=(Z*CRF*PHI)/(y*3600)
```

```
C_dot[1]=c[1]*Ex_dot[1]
C_dot[2]=c[2]*Ex_dot[2]
C_dot[3]=c[3]*Ex_dot[3]
C_dot[4]=c[4]*Ex_dot[4]

C_dot_F=C_dot[3]-C_dot[4]
C_dot[3]=4/3600
c[3]=c[4] {F rule}
c_F=C_dot_F/Ex_dot_F
C_dot_P=C_dot[2]-C_dot[1]
C_dot[1]=1
C_dot_L=c_F*Ex_dot_L
C_dot_F+Z_dot=C_dot_P+C_dot_L
C_dot_des=c_F*Ex_dot_des
C_dot_total=C_dot_des+C_dot_L+Z_dot
```

2.7 Example 8.3

```
{0}
P[0]=101.325
T[0]=25
T_K[0]=T[0]+273.15
h[0]=enthalpy(Water,T=T[0],P=P[0])
s[0]=entropy(Water,T=T[0],P=P[0])

C_P_Carbohydrate_T_0=1.5488+1.9625*0.001*T[0]-5.9399*0.000001
*T[0]^2
C_P_Water_T_0=4.1762-9.0864*0.00001*T[0]+5.4731*0.000001*T[0]^2

{1}
P[1]=101.325
T[1]=90
T_K[1]=T[1]+273.15
X_Carbohydrate[1]=0.142
X_Water[1]=0.858

C_P_Carbohydrate[1]=1.5488+1.9625*0.001*T[1]-5.9399*0.000001
*T[1]^2
C_P_Water[1]=4.1762-9.0864*0.00001*T[1]+5.4731*0.000001*T[1]^2

C_P[1]=X_Carbohydrate[1]*C_P_Carbohydrate[1]+X_Water[1]*C_P_
Water[1]
C_P_0[1]=X_Carbohydrate[1]*C_P_Carbohydrate_T_0+X_Water[1]*C_P_
Water_T_0
C_P_ave_ex[1]=(C_P[1]+C_P_0[1])/2

RHO_Carbohydrate[1]=1.5991*1000-0.31046*T[1]
RHO_Water[1]=9.9718*100+3.1439*0.001*T[1]-3.7574*0.001*T[1]^2
v[1]=X_Carbohydrate[1]/RHO_Carbohydrate[1]+X_Water[1]/
RHO_Water[1]
ex[1]=C_P_ave_ex[1]*(T[1]-T[0]-T_K[0]*ln(T_K[1]/
T_K[0]))+v[1]*(P[1]-P[0])

{2}
P[2]=551.325
```

```
T[2]=90.1
T_K[2]=T[2]+273.15
X_Carbohydrate[2]=0.142
X_Water[2]=0.858

C_P_Carbohydrate[2]=1.5488+1.9625*0.001*T[2]-5.9399*0.000001
*T[2]^2
C_P_Water[2]=4.1762-9.0864*0.00001*T[2]+5.4731*0.000001*T[2]^2

C_P[2]=X_Carbohydrate[2]*C_P_Carbohydrate[2]+X_Water[2]*C_P_
Water[2]
C_P_0[2]=X_Carbohydrate[2]*C_P_Carbohydrate_T_0+X_Water[2]*C_P_
Water_T_0
C_P_ave_ex[2]=(C_P[2]+C_P_0[2])/2

RHO_Carbohydrate[2]=1.5991*1000-0.31046*T[2]
RHO_Water[2]=9.9718*100+3.1439*0.001*T[2]-3.7574*0.001*T[2]^2
v[2]=X_Carbohydrate[2]/RHO_Carbohydrate[2]+X_Water[2]/
RHO_Water[2]
ex[2]=C_P_ave_ex[2]*(T[2]-T[0]-T_K[0]*ln(T_K[2]/
T_K[0]))+v[2]*(P[2]-P[0])

{mass balance}
m_dot[1]=48.60
m_dot[1]=m_dot[2]

{exergy balance}
Ex_dot[1]=m_dot[1]*ex[1]
Ex_dot[2]=m_dot[2]*ex[2]
W_dot=(m_dot[1]*v[1]*(P[2]-P[1]))/0.75

Ex_dot_F=W_dot
Ex_dot_P=Ex_dot[2]-Ex_dot[1]
Ex_dot_des=Ex_dot_F-Ex_dot_P
ETA_ex=(Ex_dot_P*100)/Ex_dot_F

PHI=1.05
y=3*30*24
CRF=(i*((1+i)^n))/(((1+i)^n)-1)
n=20
i=0.2
c_W=0.00000122
Z=1430
Z_dot=(Z*CRF*PHI)/(y*3600)

C_dot[1]=c[1]*Ex_dot[1]
C_dot[2]=c[2]*Ex_dot[2]

C_dot_F=c_W*W_dot
c_F=C_dot_F/Ex_dot_F
C_dot_P=C_dot[2]-C_dot[1]
C_dot[1]=1
C_dot_F+Z_dot=C_dot_P
C_dot_des=c_F*Ex_dot_des
C_dot_total=C_dot_des+Z_dot
```

2.8 Example 8.4

```
{0}
P[0]=101.325
T[0]=25
T_K[0]=T[0]+273.15
h[0]=enthalpy(Water,T=T[0],P=P[0])
s[0]=entropy(Water,T=T[0],P=P[0])

{1}
P[1]=81.325
x[1]=1
h[1]=enthalpy(Water,x=x[1],P=P[1])
s[1]=entropy(Water,x=x[1],P=P[1])
ex[1]=h[1]-h[0]-T_K[0]*(s[1]-s[0])

{2}
P[2]=601.325
T[2]=25
h[2]=enthalpy(Water,T=T[2],P=P[2])
s[2]=entropy(Water,T=T[2],P=P[2])
ex[2]=h[2]-h[0]-T_K[0]*(s[2]-s[0])

{3}
P[3]=121.325
T[3]=40
h[3]=enthalpy(Water,T=T[3],P=P[3])
s[3]=entropy(Water,T=T[3],P=P[3])
ex[3]=h[3]-h[0]-T_K[0]*(s[3]-s[0])

{mass balance}
m_dot[1]=0.82
m_dot[1]+m_dot[2]=m_dot[3]

{energy balance}
m_dot[1]*h[1]+m_dot[2]*h[2]=m_dot[3]*h[3]

{exergy balance}
Ex_dot[1]=m_dot[1]*ex[1]
Ex_dot[2]=m_dot[2]*ex[2]
Ex_dot[3]=m_dot[3]*ex[3]

Ex_dot_F=Ex_dot[1]+Ex_dot[2]
Ex_dot_P=Ex_dot[3]
Ex_dot_des=Ex_dot_F-Ex_dot_P
ETA_ex=(Ex_dot_P*100)/Ex_dot_F

PHI=1.05
y=6*30*24
CRF=(i*((1+i)^n))/(((1+i)^n)-1)
n=20
i=0.2
Z=8900
Z_dot=(Z*CRF*PHI)/(y*3600)
```

```
C_dot[1]=c[1]*Ex_dot[1]
C_dot[2]=c[2]*Ex_dot[2]
C_dot[3]=c[3]*Ex_dot[3]

C_dot_F=C_dot[1]+C_dot[2]
C_dot[1]=3
C_dot[2]=0.25
c_F=C_dot_F/Ex_dot_F
C_dot_P=C_dot[3]
C_dot_F+Z_dot=C_dot_P
C_dot_des=c_F*Ex_dot_des
C_dot_total=C_dot_des+Z_dot
```

2.9 Example 8.5

```
{0}
P[0]=101.325
T[0]=25
T_K[0]=T[0]+273.15
h[0]=enthalpy(Water,T=T[0],P=P[0])
s[0]=entropy(Water,T=T[0],P=P[0])

T_ref=0
C_P_Carbohydrate_T_ref=1.5488
C_P_Water_T_ref=4.1762

C_P_Carbohydrate_T_0=1.5488+1.9625*0.001*T[0]-5.9399*0.000001
*T[0]^2
C_P_Water_T_0=4.1762-9.0864*0.00001*T[0]+5.4731*0.000001*T[0]^2

{1}
P[1]=391.325
T[1]=124
T_K[1]=T[1]+273.15
X_Carbohydrate[1]=0.142
X_Water[1]=1-X_Carbohydrate[1]

C_P_Carbohydrate[1]=1.5488+1.9625*0.001*T[1]-5.9399*0.000001
*T[1]^2
C_P_Water[1]=4.1762-9.0864*0.00001*T[1]+5.4731*0.000001*T[1]^2

C_P[1]=X_Carbohydrate[1]*C_P_Carbohydrate[1]+X_Water[1]*C_P_
Water[1]
C_P_0[1]=X_Carbohydrate[1]*C_P_Carbohydrate_T_0+X_Water[1]*C_P_
Water_T_0
C_P_ref[1]=X_Carbohydrate[1]*C_P_Carbohydrate_T_ref+X_
Water[1]*C_P_Water_T_ref
C_P_ave_en[1]=(C_P[1]+C_P_ref[1])/2
C_P_ave_ex[1]=(C_P[1]+C_P_0[1])/2

RHO_Carbohydrate[1]=1.5991*1000-0.31046*T[1]
RHO_Water[1]=9.9718*100+3.1439*0.001*T[1]-3.7574*0.001*T[1]^2
v[1]=X_Carbohydrate[1]/RHO_Carbohydrate[1]+X_Water[1]/
RHO_Water[1]
```

```
h[1]=C_P_ave_en[1]*T[1]+v[1]*(P[1]-P[0])
ex[1]=C_P_ave_ex[1]*(T[1]-T[0]-T_K[0]*ln(T_K[1]/
T_K[0]))+v[1]*(P[1]-P[0])

{2}
P[2]=241.325
T[2]=126.6
T_K[2]=T[2]+273.15
X_Carbohydrate[2]=0.219
X_Water[2]=1-X_Carbohydrate[2]

C_P_Carbohydrate[2]=1.5488+1.9625*0.001*T[2]-5.9399*0.000001*T[2]^2
C_P_Water[2]=4.1762-9.0864*0.00001*T[2]+5.4731*0.000001*T[2]^2

C_P[2]=X_Carbohydrate[2]*C_P_Carbohydrate[2]+X_Water[2]*C_P_
Water[2]
C_P_0[2]=X_Carbohydrate[2]*C_P_Carbohydrate_T_0+X_Water[2]*C_P_
Water_T_0
C_P_ref[2]=X_Carbohydrate[2]*C_P_Carbohydrate_T_ref+X_
Water[2]*C_P_Water_T_ref
C_P_ave_en[2]=(C_P[2]+C_P_ref[2])/2
C_P_ave_ex[2]=(C_P[2]+C_P_0[2])/2

RHO_Carbohydrate[2]=1.5991*1000-0.31046*T[2]
RHO_Water[2]=9.9718*100+3.1439*0.001*T[2]-3.7574*0.001*T[2]^2
v[2]=X_Carbohydrate[2]/RHO_Carbohydrate[2]+X_Water[2]/
RHO_Water[2]
h[2]=C_P_ave_en[2]*T[2]+v[2]*(P[2]-P[0])
ex[2]=C_P_ave_ex[2]*(T[2]-T[0]-T_K[0]*ln(T_K[2]/
T_K[0]))+v[2]*(P[2]-P[0])

{3}
P[3]=241.325
x[3]=1
h[3]=enthalpy(Water,x=x[3],P=P[3])
s[3]=entropy(Water,x=x[3],P=P[3])
ex[3]=h[3]-h[0]-T_K[0]*(s[3]-s[0])

{4}
T[4]=140.8
x[4]=1
h[4]=enthalpy(Water,x=x[4],T=T[4])
s[4]=entropy(Water,x=x[4],T=T[4])
ex[4]=h[4]-h[0]-T_K[0]*(s[4]-s[0])

{5}
T[5]=140.8
x[5]=0
h[5]=enthalpy(Water,x=x[5],T=T[5])
s[5]=entropy(Water,x=x[5],T=T[5])
ex[5]=h[5]-h[0]-T_K[0]*(s[5]-s[0])

{mass balance}
m_dot[1]=36.63
m_dot[5]=m_dot[4]
m_dot[1]=m_dot[2]+m_dot[3]
X_Carbohydrate[1]*m_dot[1]=X_Carbohydrate[2]*m_dot[2]
```

```
{energy balance}
m_dot[1]*h[1]+m_dot[4]*h[4]=m_dot[2]*h[2]+m_dot[3]*h[3]+m_
dot[5]*h[5]+Q_dot_L
Q_dot_L=30

{exergy balance}
Ex_dot[1]=m_dot[1]*ex[1]
Ex_dot[2]=m_dot[2]*ex[2]
Ex_dot[3]=m_dot[3]*ex[3]
Ex_dot[4]=m_dot[4]*ex[4]
Ex_dot[5]=m_dot[5]*ex[5]

Ex_dot_F=Ex_dot[1]+(Ex_dot[4]-Ex_dot[5])
Ex_dot_P=Ex_dot[2]+Ex_dot[3]
Ex_dot_L=Q_dot_L*(1-((T_K[0])/(60+273.15)))
Ex_dot_des=Ex_dot_F-Ex_dot_P-Ex_dot_L
ETA_ex=(Ex_dot_P*100)/Ex_dot_F

PHI=1.05
y=3*30*24
CRF=(i*((1+i)^n))/(((1+i)^n)-1)
n=20
i=0.2
Z=86000
Z_dot=(Z*CRF*PHI)/(y*3600)

C_dot[1]=c[1]*Ex_dot[1]
C_dot[2]=c[2]*Ex_dot[2]
C_dot[3]=c[3]*Ex_dot[3]
C_dot[4]=c[4]*Ex_dot[4]
C_dot[5]=c[5]*Ex_dot[5]

C_dot[1]=0.85
C_dot[4]=0.03

C_dot_F=C_dot[1]+(C_dot[4]-C_dot[5])
c[4]=c[5] {F rule}
c_F=C_dot_F/Ex_dot_F
C_dot_P=C_dot[2]+C_dot[3]
c[2]=c[3] {P rule}
C_dot_F+Z_dot=C_dot_P+C_dot_L
C_dot_L=c_F*Ex_dot_L
C_dot_des=c_F*Ex_dot_des
C_dot_total=C_dot_des+Z_dot+C_dot_L
```

2.10 Example 8.6

```
{0}
P[0]=101.325
T[0]=25
T_K[0]=T[0]+273.15
h[0]=enthalpy(Water,T=T[0],P=P[0])
s[0]=entropy(Water,T=T[0],P=P[0])

{1}
P[1]=375
```

```
x[1]=0
h[1]=enthalpy(Water,x=x[1],P=P[1])
s[1]=entropy(Water,x=x[1],P=P[1])
ex[1]=h[1]-h[0]-T_K[0]*(s[1]-s[0])

{2}
P[2]=375
x[2]=0
h[2]=enthalpy(Water,x=x[2],P=P[2])
s[2]=entropy(Water,x=x[2],P=P[2])
ex[2]=h[2]-h[0]-T_K[0]*(s[2]-s[0])

{3}
P[3]=375
x[3]=0
h[3]=enthalpy(Water,x=x[3],P=P[3])
s[3]=entropy(Water,x=x[3],P=P[3])
ex[3]=h[3]-h[0]-T_K[0]*(s[3]-s[0])

{4}
P[4]=225
x[4]=0
h[4]=enthalpy(Water,x=x[4],P=P[4])
s[4]=entropy(Water,x=x[4],P=P[4])
ex[4]=h[4]-h[0]-T_K[0]*(s[4]-s[0])

{5}
P[5]=225
x[5]=1
h[5]=enthalpy(Water,x=x[5],P=P[5])
s[5]=entropy(Water,x=x[5],P=P[5])
ex[5]=h[5]-h[0]-T_K[0]*(s[5]-s[0])

{mass balance}
m_dot[1]+m_dot[2]+m_dot[3]=m_dot[4]+m_dot[5]
m_dot[1]=13.30
m_dot[2]=4.36
m_dot[3]=0.52

{energy balance}
m_dot[1]*h[1]+m_dot[2]*h[2]+m_dot[3]*h[3]=m_dot[4]*h[4]+m_
dot[5]*h[5]

{exergy balance}
Ex_dot[1]=m_dot[1]*ex[1]
Ex_dot[2]=m_dot[2]*ex[2]
Ex_dot[3]=m_dot[3]*ex[3]
Ex_dot[4]=m_dot[4]*ex[4]
Ex_dot[5]=m_dot[5]*ex[5]

Ex_dot_F=Ex_dot[1]+Ex_dot[2]+Ex_dot[3]
Ex_dot_P=Ex_dot[4]+Ex_dot[5]
Ex_dot_des=Ex_dot_F-Ex_dot_P
ETA_ex=(Ex_dot_P*100)/Ex_dot_F

PHI=1.05
y=3*30*24
```

```
CRF=(i*((1+i)^n))/(((1+i)^n)-1)
n=20
i=0.2
Z=3500
Z_dot=(Z*CRF*PHI)/(y*3600)

C_dot[1]=c[1]*Ex_dot[1]
C_dot[2]=c[2]*Ex_dot[2]
C_dot[3]=c[3]*Ex_dot[3]
C_dot[4]=c[4]*Ex_dot[4]
C_dot[5]=c[5]*Ex_dot[5]

C_dot[1]=0.003
C_dot[2]=0.001
C_dot[3]=0.0001

C_dot_F=C_dot[1]+C_dot[2]+C_dot[3]
c_F=C_dot_F/Ex_dot_F
C_dot_P=C_dot[4]+C_dot[5]
c[4]=c[5] {P rule}
C_dot_F+Z_dot=C_dot_P
C_dot_des=c_F*Ex_dot_des
C_dot_total=C_dot_des+Z_dot
```

2.11 Example 8.7

```
{0}
P[0]=101.325
T[0]=25
T_K[0]=T[0]+273.15
h[0]=enthalpy(Water,T=T[0],P=P[0])
s[0]=entropy(Water,T=T[0],P=P[0])

{1}
P[1]=800
T[1]=180
h[1]=enthalpy(Water,T=T[1],P=P[1])
s[1]=entropy(Water,T=T[1],P=P[1])
ex[1]=h[1]-h[0]-T_K[0]*(s[1]-s[0])

{2}
P[2]=300
h[2]=h[1]
s[2]=entropy(Water,P=P[2],h=h[2])
ex[2]=h[2]-h[0]-T_K[0]*(s[2]-s[0])

ex_des=ex[1]-ex[2]

Ex_dot[1]=m_dot*ex[1]
Ex_dot[2]=m_dot*ex[2]
Ex_dot_des=m_dot*ex_des
Ex_dot_F=Ex_dot[1]
Ex_dot_P=Ex_dot[2]

PHI=1.05
y=6*30*24
```

```
CRF=(i*((1+i)^n))/(((1+i)^n)-1)
n=20
i=0.2
Z=180
Z_dot=(Z*CRF*PHI)/(y*3600)

m_dot=1.42
c[1]=0.00005
C_dot[1]=c[1]*Ex_dot[1]
C_dot[2]=c[2]*Ex_dot[2]
C_dot_F=C_dot[1]
C_dot_P=C_dot[2]
C_dot_F+Z_dot=C_dot_P

c_F=C_dot_F/Ex_dot_F
C_dot_des=c_F*Ex_dot_des
C_dot_total=Z_dot+C_dot_des
```

Index

For Product Safety Concerns and Information please contact our EU representative GPSR@taylorandfrancis.com
Taylor & Francis Verlag GmbH, Kaufingerstraße 24, 80331 München, Germany

www.ingramcontent.com/pod-product-compliance
Lightning Source LLC
Chambersburg PA
CBHW060333220326
41598CB00023B/2694
* 9 7 8 1 0 3 2 5 4 3 9 7 0 *